Graduate Texts in Mathematics 16

David Winter

The Structure of Fields

Springer-Verlag New York · Heidelberg · Berlin

David Winter

Professor of Mathematics
University of Michigan
Ann Arbor, Michigan 48104

AMS Subject Classifications (1970)
12C05, 12E05, 12Fxx, 12H05

Library of Congress Cataloging in Publication Data

Winter, David J.
Structure of fields.

(Graduate texts in mathematics, v. 16)
1. Fields, Algebraic. 2. Galois theory.
I. Title. II. Series.
QA247.W55 512'.32 73-21824
ISBN 0-387-90074-4

Printed in the United States of America.

ISBN 0-387-90074-8 Springer-Verlag New York Heidelberg Berlin
ISBN 3-540-90074-8 Springer-Verlag Berlin Heidelberg New York

To Linda

Preface

The theory of fields is one of the oldest and most beautiful subjects in algebra. It is a natural starting point for those interested in learning algebra, since the algebra needed for the theory of fields arises naturally in the theory's development and a wide selection of important algebraic methods are used. At the same time, the theory of fields is an area in which intensive work on basic questions is still being done.

This book was written with the objective of exposing the reader to a thorough treatment of the classical theory of fields and classical Galois theory, to more modern approaches to the theory of fields and to one approach to a current problem in the theory of fields, the problem of determining the structure of radical field extensions.

I have written the book in the form of a text book, and assume that the reader is familar with the elementary properties of vector spaces and linear transformations. The other basic algebra needed for the book is developed in Chapter 0, although a reader with very little background in algebra should also consult other sources. Exercises varying from quite easy to very difficult are included at the end of each chapter. Some of these exercises supplement the text and are referred to at points where readers may want to see further discussion. Others are used to cover in outline form important material peripheral to the main themes in the book.

Chapters 1–4 give a comprehensive treatment of the more classical side of the theory of fields and Galois theory. Chapter 1 and 2 are concerned with the general structure of polynomials and extension fields. Galois theory is developed extensively in Chapter 3. Chapter 4 covers the fundamental theorems on algebraic function fields and relates algebraic function fields and affine algebraic varieties.

In Chapter 5, I discuss three modern versions of Galois theory, in which the Galois group of an extension is replaced by a ring, a Lie ring and a biring respectively. In Chapter 6, I describe the structure of radical extensions and their associated birings in terms of tori.

In Appendix S, I introduce the language of sets and describe the set theory needed for the book. Witt vectors are needed in 3.10, and their properties are developed in Appendix W. Tensor products are used quite often in Chapters 5 and 6, and are discussed in Appendix T.

In order to put the material of Chapters 5 and 6 in the proper formal framework, I have included a fairly thorough treatment of algebras, coalgebras and bialgebras in the appendices. In Appendix A, the structure of finite dimensional commutative algebras is determined. In Appendix C, I discuss coalgebras and develop the structure theory of cocommutative coalgebras.

In Appendix B, I develop a theory of K/k-bialgebras which generalizes the usual theory of k-bialgebras.

To those already familiar with the theory of fields, some further remarks may be of interest. In Chapter 2, the proof that the set k_{sep} of separable elements of a finite dimensional field extension of k is a simple field extension of k is simplified by a theorem on conjugates (see 2.2.10). At the beginning of Chapter 3, a generalization of the Dedekind Independence Theorem is proved (see 3.1.1). This is used to prove a theorem on Galois descent (see 3.2.5) which is then used to prove the Galois Correspondence Theorem (see 3.3.3). In 3.4, the proof of the Normal Basis Theorem is simplified by a theorem on conjugates (see 3.4.1). In Chapter 4, I prove that a p-basis of an arbitrary separable extension K/k is algebraically independent (see 4.3.17), which greatly simplifies the proofs of theorems on separating transcendency bases. In Chapter 5, I give a new treatment of the Jacobson-Bourbaki Correspondence Theorem (see 5.1.7) and an accompanying descent theorem (see 5.1.10), and of the Jacobson Differential Correspondence Theorem (see 5.2.6) and its accompanying descent theorem (see 5.2.9), inspired by work of Pierre Cartier and Gerhard Hochschild. In 5.3, I develop a Galois theory of normal extensions based on the biring $H(K/k)$ of an extension K/k. The structure of K/k is related to the structure of $H(K/k)$ (see 5.3.20), a Biring Correspondence Theorem is proved (see 5.3.12) and a radical splitting theorem for $H(K/k)$ is proved for finite dimensional normal extensions (see 5.3.21). This theory is parallel in some respects to a powerful Galois theory of normal extensions based on the universal cosplit measuring k-bialgebra of an extension K/k, developed by Moss Sweedler [20], but has the advantage that the biring $H(K/k)$ consists of linear transformations of K/k and is therefore more easily studied. In Chapter 6, I discuss in detail the structure of finite dimensional radical extensions K/k and their birings $H(K/k)$, in terms of tori. Tori are then used in proving a fairly deep generalization of a theorem of Jacobson on finite dimensional Lie rings of derivations of K (see 6.4.2). In Appendix B, I develop a formal theory of K/k-bialgebras, which reduces to the usual theory of k-bialgebras when $K = k$. I then define and discuss the K-measuring K/k-bialgebras and their k-forms, and determine the structure of the finite dimensional conormal K-measuring K/k-bialgebras and their cosplit k-forms. The theory thus developed places the material of Chapters 5 and 6 in a formal framework within which the structure of $H(K/k)$ can be more effectively studied.

Other approaches to the theory of radical extensions are outlined in E.5 and E.6 in the form of exercises. An outline of the proof of a theorem of Murray Gerstenhaber on subspaces of Der K closed under pth powers is given (see E.5.8). Higher derivations are introduced, and a sketch of the proof of Moss Sweedler's theorem characterizing in terms of higher derivations those finite dimensional radical extensions which are internal tensor products of simple extensions is given (see E.6.11, E.6.14). Moss Sweedler's universal cosplit measuring k-bialgebra is introduced and discussed in E.6.21 and E.6.22. The Pickert invariants of a radical extension are discussed in E.6.24 and E.6.25.

Reflected in this book are the ideas of many people who have influenced me directly and through their work in my thinking about fields. I would particularly like to mention George Seligman, with whom I first studied fields, Nathan Jacobson, whose work on fields is the basis for a large part of this book and Moss Sweedler, whose work on coalgebras, bialgebras and field theory is reflected in the last part of this book. Since a reflection is not real substitute for an original idea, readers are urged to explore the books and papers listed in the reference section, especially [2], [5], [9], [10], [11], [12], [18], [19], [20].

Much of this book is based on a course on bialgebras and courses on field theory given at the University of Michigan in 1969, 1971 and 1972. Most of the material of Chapters 5 and 6 and of Appendix B is the outgrowth of preliminary research described at the 1971 Conference on Lie Algebras and Related Topics at Ohio State University.

I would like to take this opportunity to express my thanks to my friend and former student, Pedro Sanchez, whose lecture notes to my courses made easier the writing of parts of this book, and to Hershey Kisilevsky, who showed me the irreducible polynomial used in proving 3.12.2. I also wish to thank the National Science Foundation for their support of research described here, and to express my appreciation to the California Institute of Technology, whose generous support during the academic year 1972–3 enabled the remaining research to be completed at this early date. Finally, I would like to express my thanks to Catherine Rader and Frances Williams, whose superb typing made as painless as possible the job of preparing the manuscript.

Ann Arbor, Michigan and
Pasadena, California, March 1, 1973 David J. Winter

TABLE OF CONTENTS

The Structure of Fields

0 Introduction

In this chapter, we give a brief but fairly self-contained introduction to abstract algebra, in order to develop the language, conventions and basic algebra used throughout the remainder of the book. Our notation for sets of objects and for operations on sets is given in Appendix S.

We begin with basic material on groups, rings and fields. We then briefly discuss transformation groups. Finally, we discuss the Krull Closure in a group in anticipation of its role in Chapter 3.

0.1 Basic algebra

A *product* on a set S is a mapping from $S \times S$ to S, which we may denote $(x, y) \mapsto x \circ y$. A subset T of a set S with product $x \circ y$ is *closed* (or *closed under* $x \circ y$) if $x \circ y \in T$ for all $x, y \in T$. A product $x \circ y$ on S is *associative* if $(x \circ y) \circ z = x \circ (y \circ z)$ for $x, y, z \in S$. An element e of a set S with product $x \circ y$ is an *identity* of S if $e \circ x = x \circ e = x$ for $x \in S$. One shows easily that S has at most one such e (see E.0.1). If such an e exists, S is said to *have an identity*.

A *monoid* (or *semigroup with identity*) is a set S with an associative product $x \circ y$ such that S has an identity. A *submonoid* of a monoid S is a closed subset T of S containing the identity of S. Such a T together with the product $x \circ y$ $(x, y \in T)$ is a monoid. For any element x of a monoid S, we let x^0 be the identity element of S and $x^n = x \circ x \circ \cdots \circ x$ (n times) for any positive integer n. In particular, $x^1 = x$ for $x \in S$. For $x \in S$, the set T consisting of $x^0, x^1, \ldots$ is a submonoid of S and $x^{m+n} = x^m \circ x^n$, $(x^m)^n = x^{mn}$ for all nonnegative integers m, n (see E.0.4). In a monoid S, an *inverse* of an element $x \in S$ is an element $y \in S$ such that $x \circ y = y \circ x = e$, e being the identity element of S. For each $x \in S$, x has at most one inverse y (see E.0.2). If $x \in S$ has an inverse, then we say that x is a *unit* or an *invertible element* of S, and we denote the inverse of x by x^-. The set S^* of units of S is a submonoid of S and $(x \circ y)^- = y^- \circ x^-$, $(x^-)^- = x$ for $x, y \in S^*$ (see E.0.3). For $x \in S^*$, we define $x^{(-n)} = (x^-)^n$ for any positive integer n. In particular, $x^{-1} = x^-$ for $x \in S^*$. We call x^n the nth *power* of x with respect to the product $\circ$. For $x \in S^*$, the set consisting of $x^0, x^{-1}, x^1, x^{-2}, x^2, \ldots$ is a submonoid of S and $x^{m+n} = x^m \circ x^n$, $(x^m)^n = x^{mn}$ for all integers m, n (see E.0.4). Elements x, y of monoid S *commute* if $x \circ y = y \circ x$. A monoid S is *Abelian* (or *commutative*) if $x \circ y = y \circ x$ for $x, y \in S$. If S is an Abelian monoid containing elements $x_1, \ldots, x_m$, we let $\prod_1^m x_i$ denote $x_1 \circ \cdots \circ x_m$ and then have $(\prod_1^m x_i)^n = \prod_1^m (x_i)^n$ for any nonnegative integer n (see E.0.5).

A *group* is a monoid S every element of which is a unit. Thus, a group is a monoid S such that $S = S^*$. For any monoid S, S^* is a group called the *group of units* of S. A *subgroup* of a group S is a submonoid T of S such that

$x^- \in T$ for all $x \in T$. A subgroup T of a group S with product $x \circ y$ $(x, y \in S)$ is a group with product $x \circ y$ $(x, y \in T)$. A group S is *Abelian* if it is Abelian as a monoid. If $x_1, \ldots, x_m$ are elements of an Abelian group S, then $(\prod_1^m x_i)^- = \prod_1^m (x_i)^-$ and $(\prod_1^m x_i)^n = \prod_1^m (x_i)^n$ for any integer n (see E.0.5).

A *ring* is a set A with two products $x + y$ and xy, called *addition* and *multiplication* respectively, such that A with addition is an Abelian group, A with multiplication is a monoid and $x(y + z) = xy + xz$, $(x + y)z = xz + yz$ for $x, y, z \in A$. A *subring* of a ring A is a subset B of A which is a subgroup of A with addition and a submonoid of A with multiplication. A subring B of a ring A together with the addition $x + y$ and multiplication xy $(x, y \in B)$ is a ring. The identities of a ring A with respect to addition and multiplication are denoted 0 and e respectively. The element e is the *identity* of the ring A. For $x \in A$, x^n is the nth power of x with respect to multiplication. We let $-x$ be the additive inverse of x in A, so that $x + (-x) = 0$, and we let $y - x = y + (-x)$ for $x, y \in A$. We then define $0 \cdot x = 0$, $n \cdot x = x + \cdots + x$ (n times) and $(-n) \cdot x = n \cdot (-x)$ for any positive integer n, so that $n \cdot x$ is the nth power of x with respect to addition. In particular, $1 \cdot x = x$ for $x \in A$. One easily proves the basic equations $x0 = 0x = 0$, $(-x)y = -(xy) = x(-y)$ for $x, y \in A$ and the basic equations $(m + n) \cdot x = m \cdot x + n \cdot x$, $m \cdot (n \cdot x) = (mn) \cdot x$, $m \cdot (x + y) = m \cdot x + m \cdot y$ for $x, y \in A$ and any integers m and n (see E.0.4).

A ring A is *commutative* if the monoid A with multiplication is commutative, that is, if $xy = yx$ for $x, y \in A$. An element x of A is a *unit* of the ring A if x is a unit in the monoid A with multiplication. The group of units of A is denoted A^*.

A ring A is an *integral domain* if A is commutative and $A - \{0\}$ is nonempty and closed under multiplication xy. In an integral domain, $e \neq 0$ (see E.0.7). A *field* is an integral domain K such that the group of units K^* is $K - \{0\}$, that is, such that each nonzero element is a unit. Every subring of a field is an integral domain. A *subfield* of a field K is a subring k of K such that $x \in k - \{0\} \Rightarrow x^{-1} \in k - \{0\}$. A subfield of a field K is a field. Every integral domain A is a subring of some field K such that $K = \{xy^{-1} | x \in A, y \in A - \{0\}\}$, and such a field K is a *field of quotients* of A (see E.0.10). Any two fields of quotients of A are essentially the same (see E.0.11).

A *homomorphism/isomorphism* from a monoid or group S with product $x \circ y$ and identity e to a monoid or group S' with product $x' \circ' y'$ and identity e' is a mapping/bijective mapping f from S to S' such that $f(x \circ y) = f(x) \circ f(y)$ and $f(e) = e'$. If an *isomorphism* from S to S' exists, S and S' are *isomorphic*. An *automorphism* of S is an isomorphism from S to S.

A *homomorphism/isomorphism* from a ring or field A to a ring or field A' is a mapping/bijective mapping f from A to A' such that f is a homomorphism/isomorphism of monoids from A with addition to A' with addition and from A with multiplication to A' with multiplication. If an isomorphism from A to A' exists, A and A' are *isomorphic*. An *automorphism* of A is an isomorphism from A to A.

An *ideal* of a ring A is a nonempty subset I of A closed under addition such that $xy \in I$ for $x \in A$, $y \in I$ and for $x \in I$, $y \in A$. The sets $\{0\}$ and A are ideals of A. In a commutative ring A, the set $xA = \{xa \mid a \in A\}$ $(x \in A)$ is an ideal of A called the *principle ideal* generated by x. If A is an integral domain and every ideal of A is principal, A is a *principle ideal domain.*

Suppose that S is an Abelian group with product $x + y$ and that T is a subgroup of S. We let $x + T = \{x + y \mid y \in T\}$ for $x \in S$. Then $x \in x + T$ and $x + T$ is the *coset* of T in S containing x. Two cosets $x + T$ and $x' + T$ are equal if and only if $x - x' \in T$. If $x - x' \notin T$, $x + T$ and $x' + T$ are disjoint (see E.0.17). Thus, an element x is contained in precisely one coset, namely $x + T$. We let S/T be the set $\{x + T \mid x \in S\}$ of cosets of T in S. We can define a product $(x + T) + (y + T) = (x + y) + T$ in T, and S/T with this product is an Abelian group (see E.0.17).

Next, let A be a ring and I an ideal of A. We can also define a product $(x + I)(y + I) = xy + I$, and A/I with the so defined additive and multiplicative products is a ring (see E.0.18). The mapping $f(x) = x + I$ $(x \in A)$ is a homomorphism from A to A/I. The ring A/I is the *quotient ring* of A by I, f the *quotient homomorphism.*

If $f: A \to B$ is a homomorphism from a ring A to a ring B, then the set $\text{Kernel}\, f = \{a \in A \mid f(a) = 0\}$ is an ideal of A called the *kernel* of f. The set $\text{Image}\, f = \{f(a) \mid a \in A\}$ is a subring of B called the *image* of f. There is an isomorphism from $A/\text{Kernel}\, f$ to $\text{Image}\, f$ which sends a $+\text{Kernel}\, f$ to $f(a)$ for $a \in A$ (see E.0.19). In particular, f is injective if and only if $\text{Kernel}\, f = \{0\}$.

Now suppose that A is a commutative ring and let I be an ideal of A. Then I is *maximal* if $I \neq A$ and the only ideals of A containing I are I and A. One shows easily that A is maximal if and only if A/I is a field (see E.0.23). If A/I is an integral domain, then I is a prime ideal. Equivalently, I is a *prime* ideal if $I \neq A$ and $xy \notin I$ for $x \notin I$ and $y \notin I$. The kernel of any homomorphism f from A into an integral domain is prime.

We now let K and L denote fields and let 1 denote the identity of K. Then K has no ideals other than $\{0\}$ and K, since $K/\{0\}$ is a field.

0.1.1 Proposition. Every homomorphism f from K to L is injective.

Proof. Kernel f is an ideal of K. Since $f(1) \neq 0$, *Kernel* $f \neq K$. Thus, $\text{Kernel}\, f = \{0\}$ and f is injective. □

For $a_0, \ldots, a_n \in K$, we let $\sum_0^n a_i X^i = a_0 X^0 + \ldots + a_n X^n$ denote the infinity-tple $(a_0, \ldots, a_n, 0, \ldots)$ (all entries are 0 after the $(n + 1)$-st). This is called the *polynomial with coefficients* $a_0, \ldots, a_n$. The polynomials $aX^0 (a \in K)$ are the *constant* polynomials, or the polynomials of *degree* 0. The *degree* of a nonconstant polynomial $\sum_0^n a_i X^i$, denoted $\text{Deg} \sum_0^n a_i X^i$, is the integer d such that $a_d \neq 0$ and $a_i = 0$ for $i > d$. The *leading coefficient* of $\sum_0^n a_i X^i$ is a_d where $d = \text{Deg} \sum_0^n a_i X^i$. If the leading coefficient of $\sum_0^n a_i X^i$ is 1, we say that $\sum_0^n a_i X^i$ is *monic*. One shows easily that two polynomials $\sum_0^n a_i X^i$ and $\sum_0^n b_i X^i$ are equal if and only if $a_i = b_i$ for $1 \leq i \leq n$. The set of polynomials with coefficients in K is denoted $K[X]$. We let

$$\sum_0^n a_i X^i + \sum_0^n b_i X^i = \sum_0^n (a_i + b_i) X^i$$

and

$$\left(\sum_0^m a_i X^i\right)\left(\sum_0^n b_j X_j\right) = \sum_0^{m+n} c_k X^k$$

where $c_k = \sum_{i+j=k} a_i b_j$, define addition and multiplication in $K[X]$. One easily shows that $K[X]$ with these products is a commutative ring. Note that $\mathrm{Deg}\,(f(X)g(X)) = \mathrm{Deg}\,f(X) + \mathrm{Deg}\,g(X)$ for nonzero $f(X)$, $g(X) \in K[X]$ (see E.0.24). It follows that $K[X]$ is an integral domain. It is convenient to "identify" a with aX^0 for $a \in K$ and 1 with X^0 (see E.0.9). Then K is the subset of constant polynomials and K is a subring of $K[X]$. The group of units of $K[X]$ is $K^* = K - \{0\}$ (see E.0.25).

0.1.2 Proposition. $K[X]$ is a principle ideal domain.

Proof. Let I be a nonzero ideal of $K[X]$. Take $f(X)$ to be a nonzero element of I of least degree, $g(X)$ a nonzero element of I. What we must show is that $g(X)$ is a multiple $f(X)h(X)$ of $f(X)$ (for some $h(X) \in K[X]$). Suppose not, and take the degree of $g(X)$ to be minimal such that $g(X) \in I - \{0\}$ and $g(X)$ is not a multiple of $f(X)$. Choose X^i such that $\mathrm{Deg}\,(f(X)X^i - (a_m/b_n)g(X)) < \mathrm{Deg}\,g(X)$ where a_m, b_n are the leading coefficients of $f(X)$, $g(X)$ respectively. By the minimality assumption, $f(X)X^i - (a_m/b_n)g(X)$ is a multiple of $f(X)$. But then $g(X)$ obviously is also a multiple of $f(X)$, a contradiction. Thus, every $g(X) \in I$ is a multiple of $f(X)$. □

The group of units of $K[X]$ is K^*. Elements $f(X)$, $g(X) \in K[X]$ are *associates* if $f(X) = cg(X)$ for some unit $c \in K^*$. Equivalently, $f(X)$ and $g(X)$ are associates if $f(X)$ divides $g(X)$ and $g(X)$ divides $f(X)$, where we say that $f(X)$ *divides* $g(X)$ if $g(X) = f(X)h(X)$ for some $h(X) \in K[X]$. If $f(X)$ is not a unit and if only units and associates of $f(X)$ divide $f(X)$, then $f(X)$ is *irreducible*.

0.1.3 Proposition. The following conditions are equivalent, for $f(X) \in K[X]$.

1. $f(X)$ is irreducible;
2. the ideal $f(X)K[X]$ is maximal;
3. the ideal $f(X)K[X]$ is prime.

Proof. Let $I = f(X)K[X]$. Suppose that $f(X)$ is irreducible and that J is an ideal of $K[X]$ containing I. Then the generator $g(X)$ of J divides $f(X)$ and is either a unit or an associate of $f(X)$, Thus, $J = A$ or $J = I$ is maximal. Suppose next that I is maximal. Then A/I is a field, so that I is prime. Finally, let I be prime and let $f(X) = g(X)h(X)$. Then $g(X) \in I$ or $h(X) \in I$. Thus, $f(X)$ divides $g(X)$ or $h(X)$. But $g(X)$ and $h(X)$ divide $f(X)$. Thus, $g(X)$ or $h(X)$ is an associate of $f(X)$ and $h(X)$ or $g(X)$ a unit. □

0.1.4 Proposition. Let $f(X)$ be irreducible and suppose that $f(X)$ divides $g(X)h(X)$. Then $f(X)$ divides $g(X)$ or $h(X)$.

Proof. Let $d(X)$ be the generator of the ideal $I = \{f(X)a(X) + g(X)b(X) \mid a(X), b(X) \in K[X]\}$ of $K[X]$. Then $d(X)$ divides each element of

I. Since $f(X), g(X) \in I$, $d(X)$ divides $f(X)$ and $g(X)$. Since $f(X)$ is irreducible, $d(X)$ is a unit c or $d(X)$ is an associate of $f(X)$. In the latter case, $f(X)$ divides $g(X)$ since $d(X)$ does. In the former case $c = f(X)a(X) + g(X)b(X)$ for some $a(X), b(X) \in K[X]$. Then $ch(X) = f(X)a(X)h(X) + g(X)h(X)b(X)$. Since $f(X)$ divides $g(X)h(X)$, $f(X)$ divides $ch(X)$, hence divides $h(X)$. ∎

0.1.5 Theorem. A nonconstant polynomial $f(X) \in K[X]$ can be factored into $f(X) = \prod_1^m g_i(X)$ where the $g_i(X)$ are monic irreducible elements of $K[X]$. Moreover, the factors $h_j(X)$ of any other such factorization $f(X) = \prod_1^n h_i(X)$ with $h_i(X) \in K[X]$ irreducible can be rearranged to $f(X) = \prod_1^n h_{i'}(X)$ so that $g_1(X) = h_{1'}(X) \ldots, g_m(X) = h_{n'}(X)$ (in particular, $m = n$).

Proof. The existence of the factorization is seen by a simple induction on $\operatorname{Deg} f(X)$. The uniqueness follows easily from 0.1.4 (see E.0.39). ∎

0.1.6 Proposition. Let R be a commutative ring containing x and containing the field k as subring. Then there is precisely one homomorphism $e: k[X] \to R$ such that $e(a) = a$ for $a \in k$ and $e(X) = x$.

Proof. Since each nonzero $f(X) \in k[X]$ has the form $\sum_0^n a_i X^i$ ($a_n \neq 0$) where n and the a_i are uniquely determined by $f(X)$, we may define e by $e(\sum_0^n a_i X^i) = \sum_0^n a_i x^i$. We leave the remaining details to the reader. ∎

The homomorphism e described in 0.1.6 is the *evaluation homomorphism* from $k[X]$ to R at x. It is convenient to denote $e(f(X))$ by $f(x)$ for $f(X) \in k[X]$.

Commutative rings isomorphic to $k[X]$ also have the properties described for $k[X]$ in the last few paragraphs. Such rings are used often in this book and are referred to as follows.

0.1.7 Definition. Let R be a commutative ring containing x and containing the field k as subring. Suppose that the evaluation homomorphism $f(X) \mapsto f(x)$ from $k[X]$ to R is an isomorphism. Then we say that x is an *indeterminant* over k and R is a polynomial ring over k in the indeterminant x, and we denote R by $k[x]$.

We now consider a polynomial ring $k[x]$ over k in an indeterminant x and its field of quotients $k(x)$. The elements of $k[x]$ are of the form $f(x) = \sum_0^n a_i x^i$ ($a_i \in k$ for all i) and the elements of $k(x)$ are of the form $u(x)/v(x)$ where $u(x) \in k[x]$ and $v(x) \in k[x] - \{0\}$. Let $k(x)[X]$ be the polynomial ring over the field $k(x)$ in an indeterminant X, and let $k[x][X]$ be the subring of $k(x)[X]$ consisting of the polynomials in X of the form $\sum_0^n a_i(x)X^i$ where the $a_i(x)$ are elements of $k[x]$ for $1 \leq i \leq n$.

0.1.8 Definition. An element $f(X) = \sum_0^n a_i(x)X^i$ of $k[x][X]$ is *primitive* if no irreducible element of $k[x]$ divides $a_i(x)$ for all i.

For any $f(X) \in k(x)[X]$, one can write $f(X) = a(x)f^*(X)$ where $f^*(X)$ is a primitive element of $k[x][X]$ and $a(x) \in k(x)$.

0.1.9 Proposition. Let $a(x)f^*(X) = b(x)g^*(X)$ where $f^*(X), g^*(X)$ are primitive elements of $k[x][X]$ and $a(x), b(x) \in k(x) - \{0\}$. Then $f^*(X) = dg^*(X)$ for some $d \in k$.

Proof. Let $a(x) = s(x)/t(x)$ and $b(x) = u(x)/v(x)$ where $s(x)$, $t(x)$, $u(x)$, $v(x) \in k[x]$. Then $s(x)v(x)f^*(X) = t(x)u(x)g^*(X)$. By 0.1.5, the coefficients in $k[x]$ of the left hand side $s(x)v(x)f^*(X)$ have a common divisor $m(x)$ of greatest degree, which is unique up to a constant multiple. Since $f^*(X)$ is primitive, $s(x)v(x)$ is such a common divisor, so that $s(x)v(x)$ is a constant multiple of $m(x)$. The same argument applies to the right hand side of the equation. Consequently, $s(x)v(x)d = t(x)u(x)$ for some $d \in k$. It follows that $s(x)v(x)f^*(X) = s(x)v(x)dg^*(X)$ and $f^*(X) = dg^*(X)$. □

0.1.10 Proposition. Let $f^*(X)$ and $g^*(X)$ be primitive elements of $k[x][X]$. Then $f^*(X)g^*(X)$ is a primitive element of $k[x][X]$.

Proof. Let $f^*(X) = \sum_0^m a_i(x)X^i$ and $g^*(X) = \sum_0^n b_j(x)X^j$. Let $c(x)$ be an irreducible element of $k[x]$, and let $a_i(x)$ and $b_j(x)$ be the first coefficients of $f^*(X)$ and $g^*(X)$ respectively which are not divisible by $c(x)$. Then the $(i + j)$th coefficient of $f^*(X)g^*(X)$ is $a_i(x)b_j(x) + \sum_{r=1}^{i} a_{i-r}(x)b_{j+r}(x) + \sum_{s=1}^{j} a_{i+s}(x)b_{j-s}(x)$, which is not divisible by $c(x)$ since the latter two sums are divisible by $c(x)$ and $a_i(x)b_j(x)$ is not divisible by $c(x)$ (see 0.1.4). □

0.1.11 Theorem. Let $f(X)$, $g(X)$, $h(X) \in k(x)[X]$ and let $f(X) = a(x)f^*(X)$, $g(X) = b(x)g^*(X)$, $h(X) = c(x)h^*(X)$ where $f^*(X)$, $g^*(X)$, $h^*(X)$ are primitive elements of $k[x][X]$. Then if $f(X)g(X) = h(X)$, we have $f^*(X)g^*(X) = dh^*(X)$, for some $d \in k$.

Proof. Let $f(X)g(X) = h(X)$. Then we have $a(x)b(x)f^*(X)g^*(X) = c(x)h^*(X)$. Since $f^*(X)g^*(X)$ and $h^*(X)$ are primitive, it follows that $f^*(X)g^*(X) = dh^*(X)$ for some $d \in k$, by 0.1.9. □

The observations that we have just made show that $k[x][X]$ has a unique factorization property analogous to the unique factorization property of $k(x)[X]$ described in 0.1.5. More generally, the integral domain $k[X_1, \ldots, X_n] = (\ldots((k[X_1])[X_2])\ldots[X_n])$ (constructed by iterating the construction of $k[x][X]$ and called the *polynomial ring* over k in the n indeterminants $X_1, \ldots, X_n$) has such a unique factorization property. (see E.0.49).

0.2 Groups

We now let G be a group with identity element e. It is often convenient to denote e by 1 and the subgroup $\{e\}$ by $\mathbf{1}$. *If* $\mathbf{S}$ is a collection of subgroups of G, then $\bigcap_{H \in \mathbf{S}} H$ is a subgroup of G. If $S \subset G$ and $\mathbf{S}$ is the collection of subgroups of G containing S, then $\langle S \rangle = \bigcap_{H \in \mathbf{S}} H$ is the subgroup of G *generated* by S. If $S = \{s_1, \ldots, s_n\}$, we denote $\langle S \rangle$ by $\langle s_1, \ldots, s_n \rangle$. In particular, $\langle g \rangle$ is the subgroup of G generated by g. If $G = \langle g \rangle$, then G is *cyclic* with *generator* g. The *order* of G is the cardinality (number of elements) of G and is denoted $|G|$. The *order* of an element of g of G is the order of $\langle g \rangle$ and is denoted $|g|$. The mapping $\alpha : \mathbb{Z} \to \langle g \rangle$ defined by $\alpha(m) = g^m$ for $m \in \mathbb{Z}$ is a homomorphism from $\mathbb{Z}$ as additive group onto $\langle g \rangle$. (See E.0.4). The kernel of α is an ideal I of $\mathbb{Z}$, so that $I = \{0\}$ or $I = \mathbb{Z}n$ (set of multiples of n) for some positive integer n (see E.0.30). Thus, $\langle g \rangle$ is isomorphic to $\mathbb{Z}$ or to the additive group

$\{\bar{0}, \bar{1}, \ldots, \overline{n-1}\}$ of integers modulo n. (See E.0.38). It follows that if $|g|$ is infinite, then $\langle g \rangle = \{g^m | m = 0, \pm 1, \pm 2, \ldots\}$ and the powers $g^m (m \in \mathbb{Z})$ are distinct. And if $|g|$ is finite, then $\langle g \rangle = \langle g^0, g^1, \ldots, g^{n-1} \rangle$ where $|g| = n$ and where n is the least positive integer such that $g^n = e$. Moreover, $g^m = e$ if and only if n divides m.

Let H be a subgroup of a group G and let $x \in G$. We let xH denote the set $\{xh | h \in H\}$ and call xH the *left coset* of H in G defined by x. The set of left cosets of H in G is denoted G/H. Left cosets xH and yH are equal if and only if $x^{-1}y \in H$. If $x^{-1}y \notin H$, the xH and yH are disjoint (see E.0.68). Thus, each element x of G is contained in precisely one left coset of H in G, namely xH. The *index* of H in G is the cardinality (number of elements) $|G/H|$ of G/H and is denoted $G{:}H$. The index $G{:}\mathbf{1}$ of $\mathbf{1}$ in G is the order of G. Since the cardinality of xH is $H{:}\mathbf{1}$ for all $x \in G$, we have the following theorem.

0.2.1 Theorem. Let G be a group, H a subgroup of G. Then $G{:}\mathbf{1} = (G{:}H)(H{:}\mathbf{1})$. In particular, the order $H{:}\mathbf{1}$ of any subgroup H and the order $|g|$ of any element g of a finite group G divide the order $G{:}\mathbf{1}$ of G. $\square$

If $G_1, \ldots, G_n$ are groups, the set $G_1 \times \ldots \times G_n$ together with the product $(g_1, \ldots, g_n)(h_1, \ldots, h_n) = (g_1h_1, \ldots, g_nh_n)$ is a group called the *outer direct product* of $G_1, \ldots, G_n$ and denoted $\prod_1^n G_i$ (outer direct product). If G is a group and if $G_1, \ldots, G_n$ are subgroups of G such that the mapping $f: \prod_1^n G_i$ (outer direct product) $\to G$ defined by $f(g_1, \ldots, g_n) = g_1 \ldots g_n$ is an isomorphism, then we say that G is the *inner direct product* of $G_1, \ldots, G_n$ and write $G = \prod_1^n G_i$ (inner direct product). Note that $|\prod_1^n G_i| = \prod_1^n |G_i|$ for any inner or outer direct product $\prod_1^n G_i$.

Suppose that G is a finite Abelian group. For any prime number p, the set $G_p = \{g \in G \mid g^{p^f} = e \text{ for some } f\}$ is a subgroup of G. The order of G_p is a power of p, as we now show by induction on the order of G_p. If $|G_p| = 1$, the assertion is trivial. Otherwise, let g be an element of $G_p - \{e\}$ and let $H = \langle g \rangle$. Since G is Abelian, we may pass from the group G_p to the group G_p/H. By induction, its order $G_p{:}H$ is a power of p. Since the order of $H = \langle g \rangle$ is a power of p, the order $G_p{:}\mathbf{1} = (G_p{:}H)(H{:}\mathbf{1})$ is a power of p.

We claim that $G = \prod_1^n G_{p_i}$ (internal direct product) where $|G| = \prod_1^n p_i^{f_i}$. To see this, consider the homomorphism $f: \prod_1^n G_{p_i}$ (outer direct product) $\to G$ defined by $f(g_1, \ldots, g_n) = g_1, \ldots, g_n$. We must show that Kernel $f = \mathbf{1}$ and Image $f = G$. Let p be a prime and let $(g_1, \ldots, g_n)$ be an element of $\prod_1^n G_{p_i}$ (outer direct product) of order p. Then $g_j{}^p = e$ for all j. Since $g_j \in G_{p_j}$, we have $g_j = e$ for $p \neq p_j$. But then $p = p_i$ and $f(g_1, \ldots, g_n) = g_i$ has order p for some i, so that $(g_1, \ldots, g_n) \notin$ Kernel f. If Kernel $f \neq \mathbf{1}$, then one sees easily that Kernel f would contain an element $(g_1, \ldots, g_n)$ of prime order, which we have just seen to be impossible. Thus, Kernel $f = \mathbf{1}$. Next, let $g \in G$ and note that the order of g is of the form $|g| = \prod_1^n p_i{}^{e_i}$, by 0.2.1. Since the integers $|g|/p_1{}^{e_1}, \ldots, |g|/p_n{}^{e_n}$ have greatest common divisor 1, we can express 1 as a linear combination $1 = m_1(|g|/p_1{}^{e_1}) + \cdots + m_n(|g|/p_n{}^{e_n})$ where $m_1, \ldots, m_n \in \mathbb{Z}$ (see E.0.41). Letting $g_i = g^{d_i}$ where $d_i = m_i(|g|/p_i{}^{e_i})$, we have $g = g^1 = g^{\sum_1^n d_i} = \prod_1^n g_j$ and $g_i{}^{p_i{}^{e_i}} = e$ for

$1 \leq i \leq n$. Thus, $g = f(g_1, \ldots, g_n)$ and $g \in \text{Image} f$. We have now shown that $G = \text{Image} f$ and $\mathbf{1} = \text{Kernel} f$, so that $G = \prod_1^n G_{p_i}$ (inner direct product).

The assumption in the preceding paragraph that G be a finite Abelian group can be replaced by the much weaker assumption that G be a finite *nilpotent* group, that is, that the subset $G_p = \{g \in G \mid g^{p^f} = e \text{ for some } f\}$ be a subgroup of G for every prime p. For then each G_p is a subgroup of G whose order is a power of p (see 0.3.2). And one sees easily that for any two distinct prime numbers p and q, the elements of G_p commute with the elements of G_q (see E.0.70), so that f is a homomorphism. The remainder of the discussion goes through as in the Abelian case, and again we have $G = \prod_1^n G_{p_i}$ (inner direct product). We state this for future reference.

0.2.2 Theorem. Let G be a finite nilpotent group. Then $G = \prod_1^n G_{p_i}$ (inner direct product) where $G:\mathbf{1} = \prod_1^n p_i^{e_i}$ is the prime decomposition of the order of G. ▯

A *basis* for a finite Abelian group G is a set of distinct nonidentity elements $g_1, \ldots, g_m$ of G such that $G = \langle g_1 \rangle \cdots \langle g_m \rangle$ (internal direct product). For distinct nonidentity elements $g_1, \ldots, g_m$ of G to be a basis for G, it is necessary and sufficient that $G = \langle g_1, \ldots, g_m \rangle$ and that $\prod_1^n g_i^{e_i} = e$ if and only if $g_i^{e_i} = e$ for $1 \leq i \leq m$, the e_i being integers for $1 \leq i \leq n$.

Every nontrivial finite Abelian group G has a basis. To prove this, we first note that since $G = \prod_1^n G_{p_i}$ (internal direct product) where $G:\mathbf{1} = \prod_1^n p_i^{e_i}$ is the prime decomposition of $G:\mathbf{1}$, it suffices to consider the case where $G = G_p$ and $G:\mathbf{1} = p^e$, p being a prime number. We now procede by induction on $G:\mathbf{1}$. If $G:\mathbf{1} = p$, then $G = \langle g \rangle$ for any $g \in G - \mathbf{1}$. Suppose that $G:\mathbf{1} = p^e > p$ and let $G^p = \{g^p \mid g \in G\}$. Then $G \supsetneqq G^p$, as one easily verifies, and we may assume that $G^p = \mathbf{1}$ or G^p has a basis $g_1, \ldots, g_r$. In the former case, the argument is as for vector spaces—in fact, Abelian groups G such that $G^p = \mathbf{1}$ may be regarded as vector spaces over the field $\{\bar{0}, \bar{1}, \ldots, \overline{p-1}\}$ of p elements (see E.0.38). In the latter case, let $h_1, \ldots, h_r$ be elements of G such that $h_i^p = g_i$ for $1 \leq i \leq r$, and let $H = \langle h_1, \ldots, h_r \rangle$. Then $h_1, \ldots, h_r$ is a basis for H. For suppose that $\prod_1^r h_i^{e_i} = e$. We must show that $h_i^{e_i} = e$ for $1 \leq i \leq r$. Taking pth powers, we have $\prod_1^r g_i^{e_i} = e$, so that $g_i^{e_i} = e$ and $p|e_i$ for $1 \leq i \leq r$. Letting $e_i = pf_i$, we have $e = \prod_1^r h_i^{e_i} = \prod_1^r g_i^{f_i}$. Thus, $e = g_i^{f_i}$ and $e = h_i^{e_i}$ for $1 \leq i \leq r$. Note that there is nothing more to prove if $G = H$, so that we may assume $G \supsetneqq H$. Letting $\bar{x}$ denote the coset xH for $x \in G$, we choose, by induction, a basis $\bar{x}_1, \ldots, \bar{x}_s$ for $\bar{G} = G/H$. Since $G^p = H^p$, there exist $u_1, \ldots, u_s \in H$ such that $x_j^p = u_j^p$ for $1 \leq j \leq s$. Letting $y_j = x_j u_j^{-1}$, we have $\bar{y}_j = \bar{x}_j$ and $y_j^p = e$ for $1 \leq j \leq s$. We claim that $h_1, \ldots, h_r, y_1, \ldots, y_s$ is a basis for G. It is clear that $G = \langle h_1, \ldots, h_r, y_1, \ldots, y_s \rangle$. Suppose that $e = \prod_1^r h_i^{e_i} \prod_1^s y_j^{f_j}$. Then $\bar{e} = \prod_1^s \bar{y}_j^{f_j}$, so that $\bar{e} = \bar{y}_j^{f_j}$ and $p|f_j$ for $1 \leq j \leq s$. But then $e = y_j^{f_j}$, since $e = y_j^p$, for $1 \leq j \leq s$. Thus, $e = \prod_1^s h_i^{e_i}$ and $e = h_i^{e_i}$ for $1 \leq i \leq r$. Thus, $h_1, \ldots, h_r, y_1, \ldots, y_s$ is a basis for G. We state this theorem for future reference.

0.2.3 Theorem. Every nontrivial Abelian group G has a basis. ▯

The *exponent* Exp G of a finite group G is the least integer m such that $g^m = e$ for all $g \in G$.

0.2.4 Theorem. Let G be a finite nilpotent group. Then G has an element of order Exp G.

Proof. We know that $G = \prod_1^n G_{p_i}$ (internal direct product) (see 0.2.2). Since the elements of G_{p_i} all have orders which are powers of p_i (see 0.3.2), G_{p_i} has an element g_i whose order is the exponent of G_{p_i}. Letting $|g_i| = p_i^{e_i}$, the element $g = \prod_1^n g_i$ has order $\prod_1^n p_i^{e_i}$, and one easily sees that $\prod_1^n p_i^{e_i}$ is the exponent of G. $\square$

We now turn to an arbitrary group G. For $x \in G$, we let Int $x(g) = {}^x g = xgx^{-1}$ for $g \in G$. Then Int x: $G \to G$ is an automorphism of G, called the *inner automorphism* of G determined by x. Note that Int e: $G \to G$ is id_G and $\text{Int}(xy) = \text{Int}\, x \circ \text{Int}\, y$. Thus, Int is a homomorphism from G to the group of bijections from G to G (see E.0.82). We let $\text{Int}\, G = \{\text{Int}\, g \mid g \in G\}$ and $C(G) = \{x \in G \mid \text{Int}\, x = \textit{int}\, e\} = \{x \in G \mid xg = gx \text{ for all } g \in G\}$. The subgroup $C(G)$ of G is called the *center* of G.

A subgroup H of a group G is *normal* in G if $\text{Int}\, x(H) = H$ for all $x \in G$. For a subgroup H of G to be normal, it is necessary and sufficient that $xH = Hx$ for all $x \in G$, where $Hx = \{hx \mid h \in H\}$. If H is a normal subgroup of G, then the product $(xH)(yH) = (xy)H$ $(x, y \in G)$ is well defined and $G/H = \{xH \mid x \in G\}$ together with this product is a group, called the *quotient group* of G by H (see E.0.69). For any normal subgroup H of a group G, the mapping f: $G \to G/H$ defined by $f(x) = xH$ $(x \in G)$ is a surjective homomorphism with Kernel H, and is called the *quotient homomorphism* from G to G/H.

If f is a homomorphism from a group G to a group G', then Kernel $f = \{x \in G \mid f(x) = e\}$ is a normal subgroup of G, Image $f = \{f(x) \mid x \in G\}$ is a subgroup of G' and there is an isomorphism from G/Kernel f to Image f mapping x Kernel f to $f(x)$ for all $x \in G$. In particular, f is injective if and only if Kernel $f = \mathbf{1}$.

If N and H are subgroups of a group G and if N is normal in G, then $NH = \{xy \mid x \in N, y \in H\}$ is a subgroup of G and N is a normal subgroup of NH. Furthermore, $N \cap H$ is a normal subgroup of H and there is an isomorphism from NH/N to $H/N \cap H$ mapping xN to $x(N \cap H)$ for all $x \in H$ (see E.0.71).

A *tower* in G is a chain $\mathbf{1} \subset G_1 \subset \cdots \subset G_n = G$ of subgroups of G. If G_i is normal in G_{i+1} and G_{i+1}/G_i is cyclic for $1 \leq i \leq n-1$, then this tower is *cyclic*. If G has a cyclic tower, G is *solvable*. If N is a normal subgroup of G, then G is solvable if and only if N and G/N are solvable (see E.0.76).

0.3 Transformation groups

Let G be a group, e the identity element of G. A *G-space* is a set X together with a product π: $G \times X \to X$, denoted $(g, x) \mapsto gx$ for $g \in G$, $x \in X$, such that $ex = x$ and $(gh)x = g(hx)$ for $g, h \in G$, $x \in X$. A G-space X determines a homomorphism from G into the group $F(X, X)^*$ of bijective functions from the set X to itself (see E.0.82). The kernel of this homomorphism is $N = \{g \in G \mid gx = x \text{ for } x \in X\}$, and is called the *kernel* of G on X. If $N = \mathbf{1}$, then X is *faithful*.

A *G-morphism* from a G-space X to a G-space Y is a mapping f from X to Y such that $f(gx) = gf(x)$ for $g \in G$, $x \in X$. A *G-isomorphism* from X to Y is a bijective G-morphism from X to Y. A *G-automorphism* of X is a G-isomorphism from X to X. The set of G-morphisms from X to Y/G-isomorphisms from X to Y/G-automorphisms of X is denoted $\text{Hom}_G(X, Y)/\text{Isom}_G(X, Y)/\text{Aut}_G X$.

A subset Y of a G-space X is *G-stable* (or *stable under* G) if $g(Y) = Y$ for $g \in G$. Such a Y together with $\pi|_{G \times Y}$ is a G-space called a *G-subspace* of X.

For $x \in X$, we let Gx denote $\{gx \mid g \in G\}$ and call Gx the *G-orbit* of x (or the *orbit* of x *under* G, or the *orbit* of G *containing* x). A subset Y of X is G-stable if and only if $Y = \bigcup_{y \in Y} Gy$. We let $X^G = \{x \in X \mid Gx = \{x\}\}$ and call X^G the *set of fixed points of* G in X.

We may regard G together with the group product $G \times G \to G$ as a G-space. More generally, G/H with the product $G \times G/H \to G/H$ given by $g(xH) = gxH$ $(g \in G, x \in G)$ is a G-space for any subgroup H of G.

We let $G_x = \{g \in G \mid gx = x\}$ and call G_x the *isotropy subgroup* of x. Then there is a G-isomorphism from G/G_x (as a G-space) to Gx (as G-space) mapping gG_x to gx for $g \in G$. In particular, $G:G_x = |Gx|$ (the cardinality of Gx) for $x \in G$. If $X = Gx$ for some (or every) $x \in X$, we say that G is *transitive* on X (or X is a *transitive* G-space). If $Gx = X$ and $G_x = \mathbf{1}$ for some (or every) $x \in X$, we say that G is *simply transitive* on X (or X is a *simply transitive* G-space). Thus, G is simply transitive on X if and only if the mapping $f_x: G \to X$ sending g to gx for $g \in C$ is a G-isomorphism for some (or every) $x \in X$. Also, G is simply transitive on X if and only if for any $x, y \in X$, there exists a unique $g \in G$ such that $gx = y$.

A *G-group* is a group H together with a product $G \times H \to H$ with respect to which H is a G-space such that $g(xy) = (gx)(gy)$ for $g \in G$, $x, y \in H$. For $g \in G$, the mapping $x \mapsto gx$ on a G-group H is an automorphism of H. Thus, products with respect to which a group H is a G-group correspond to homomorphisms from G to the group $\text{Aut}\, H$ of automorphisms of H. We carry over to G-groups the terminology *kernel*, *faithful*, *G-morphism*, *G-isomorphism*, etc. which we introduced for G-spaces. Note that if H is a G-group, the set H^G of fixed points of G in H is a subgroup of H.

A very important instance of a G-group is the group G itself, together with the product $G \times G \to G$ defined by $(g, x) \mapsto {}^g x = g \times g^{-1}$ $(g \in G, x \in G)$. In this case, the orbit of $x \in G$ is ${}^G x = \{gxg^{-1} | g \in G\}$, and is called the *conjugacy class* of x in G. The elements of ${}^G x$ are the *conjugates* of x in G. For ${}^G x$ to consist of the single point x, it is necessary and sufficient that x be an element of the center $C(G)$ of G. For a finite group, we therefore have the decomposition $G = C(G) \cup {}^G x_1 \cup \cdots \cup {}^G x_m$ (disjoint union) where ${}^G x_1, \ldots, {}^G x_m$ are those distinct orbits of G having two or more elements. Since $|{}^G x| = G:G_x$ for $x \in G$, this yields the *class equation* $G:\mathbf{1} = C(G):\mathbf{1} + G:G_{x_1} + \cdots + G:G_{x_m}$ of G. The subgroup G_x occurring in the class equation is the *centralizer* $\{g \in G \mid gx = xg\}$ of x in G. We can easily prove the following basic theorem, needed for the proof of 3.12.2.

0.3.1 Theorem. Let G be a finite group and let p be a prime number dividing $G{:}\mathbf{1}$. Then G has an element of order p.

Proof. We prove this by induction on $G{:}\mathbf{1}$. If $G{:}\mathbf{1} = 1$, the assertion is trivial. Suppose next that $G{:}\mathbf{1} > 1$ and consider the class equation $G{:}\mathbf{1} = C(G){:}\mathbf{1} + G{:}G_{x_1} + \cdots + G{:}G_{x_m}$ of G. If any of the proper subgroups G_{x_i} $(1 \leq i \leq m)$ has order divisible by p, then it has an element of order p, as desired. Otherwise, since p divides $G{:}\mathbf{1} = (G{:}G_{x_i})(G_{x_i}{:}\mathbf{1})$, p must divide $G{:}G_{x_i}$ for $1 \leq i \leq m$. From the class equation, it follows that p divides $C(G){:}\mathbf{1}$. Since $C(G)$ is Abelian, $C(G)$ therefore has an element of order p (see 0.2.2). □

A *p-group* is a group G such that the order of each element of G is a power of p.

0.3.2 Corollary. Let p be a prime number and let G be a finite p-group. Then the order $G{:}\mathbf{1}$ of G is a power of p.

Proof. Suppose not. Then there is a prime number q such that $q \neq p$ and q divides $G{:}\mathbf{1}$. But then G has an element of order q, a contradiction. □

Another important consequence of the class equation is the following theorem.

0.3.3 Theorem. The center $C(G)$ of a nontrivial finite p-group G is nontrivial.

Proof. Since $G{:}\mathbf{1}$ is a power of p, p divides $G{:}G_{x_i}$ for $1 \leq i \leq m$. Thus, p divides $C(G){:}\mathbf{1}$. (We refer, of course, to the class equation). □

0.3.4 Corollary. Every nilpotent group is solvable.

Proof. Suppose not and take a nonsolvable nilpotent group G of minimal order. We have seen that $G = \prod_1^n G_{p_i}$ (inner direct product), where we may suppose that G_{p_i} is nontrivial for $1 \leq i \leq n$. If $n \geq 2$, then G_{p_i} and $G/G_{p_1} \simeq \prod_2^n G_{p_i}$ are solvable, by the minimality assumption, so that G itself is solvable—a contradiction. Otherwise $n = 1$ and G is a p-group for $p = p_1$. But then $C(G) \neq \mathbf{1}$. Thus, $G/C(G)$ is solvable, by the minimality assumption. Since $C(G)$ is solvable (in fact Abelian), G is therefore solvable. □

We now let X be a set of n distinct elements. The *symmetric group* on X is the group $\mathbf{S}(X)$ of bijective mappings from X to X. Thus, for $\sigma, \tau \in \mathbf{S}(X)$, $\sigma\,\tau$ is the element of $\mathbf{S}(X)$ such that $(\sigma\tau)(x) = \sigma(\tau(x))$ for $x \in X$. We may regard X as $\mathbf{S}(x)$-space. If $x_1, \ldots, x_r$ are $r \geq 2$ distinct elements of X, then $[x_1, \ldots, x_r]$ denotes the element σ of $\mathbf{S}(X)$ such that $\sigma(x_1) = x_2$, $\sigma(x_2) = x_3$, $\ldots$, $\sigma(x_{r-1}) = x_r$, $\sigma(x_r) = x_1$ and $\sigma(x) = x$ for all $x \in X - \{x_1, \ldots, x_r\}$, and is called the *cycle* or *r-cycle* determined by $x_1, \ldots, x_r$. Two cycles $[x_1, \ldots, x_r]$

and $[y_1, \ldots, y_s]$ are *disjoint* if $x_i \neq y_j$ for $1 \leq i \leq r$ and $1 \leq j \leq s$. If $[x_1, \ldots, x_r] = \sigma$ and $[y_1, \ldots, y_s] = \tau$ are disjoint cycles, then $\sigma\tau = \tau\sigma$. Any nonidentity element $\sigma \in \mathbf{S}(X)$ can be expressed as a product $\sigma = [x_1, \ldots, x_r][y_1, \ldots, y_s] \cdots [z_1, \ldots. z_t]$ of pairwise disjoint cycles, and the pairwise disjoint cycles occurring in such a decomposition of σ are unique up to the order of their occurrence. When one regards X as a$\langle\sigma\rangle$-space, the orbits of two or more elements under the cycle group $\langle\sigma\rangle$ are then the sets $\{x_1, \ldots, x_r\}, \{y_1, \ldots, y_s\}, \ldots, \{z_1, \ldots, z_t\}$. To prove all this, take x_1 arbitrarily such that $\sigma(x_1) \neq x_1$ and let $x_1, x_2 = \sigma(x_1), x_3 = \sigma^2(x_1), \ldots, x_r = \sigma^{r-1}(x_1)$ be the orbit of x_1 under $\langle\sigma\rangle$. Then choose y_1 arbitrarily outside the orbit of x_1 such that $\sigma(y_1) \neq y_1$ and let $y_1, y_2 = \sigma(y_1), y_3 = \sigma^2(y_1), \ldots, y_s = \sigma^{s-1}(y_1)$ be the orbit of y_1 under $\langle\sigma\rangle$. Continuation of this process leads eventually to the last orbit $z_1, z_2 = \sigma(z_1), \ldots, z_t = \sigma^{t-1}(z_1)$ of two or more elements. Then $\sigma = [x_1, \ldots, x_r][y, \ldots, y_s] \ldots [z_1, \ldots, z_t]$, because the left hand side and right hand agree for all $x \in X$. It is clear that the disjoint cycles occurring in such a decomposition of σ are unique up to the order of their occurrence.

Note that for $r \geq 3$ and distinct elements $x_1, \ldots, x_r$ of X, $[x_2, x_1][x_1, \ldots, x_r] = [x_2, \ldots, x_r]$ and $[x_1, \ldots, x_r][x_r, x_{r-1}] = [x_1, \ldots, x_{r-1}]$. It follows that $[x_r, x_{r-1}] \cdots [x_2, x_1][x_1, \ldots, x_r] = id = [x_1, \ldots, x_r][x_r, x_{r-1}] \cdots [x_2, x_1]$, where *id* is the identity of $\mathbf{S}(X)$. Taking inverses, we have $[x_1, \ldots, x_r] = [x_1, x_2] \cdots [x_{r-1}, x_r]$.

For any two distinct elements x, y of X, the 2-cycle $[x, y]$ is called the *transposition* of x and y. It follows from the preceding paragraph that any r-cycle is the product of $r - 1$ transpositions. Consequently, any element $\sigma \in \mathbf{S}[X]$ can be written as a product of $\sum_1^m (r_i - 1)$ transpositions, where $r_1, \ldots, r_m$ are the number of elements in the m distinct orbits of $\langle\sigma\rangle$ in X.

Fix an ordering $x_1, \ldots, x_n$ of all n elements of X. For $i \neq j$, the *orientation* of the ordered pair (x_i, x_j) is $f(x_i, x_j) = +1$ if $i < j$ and $f(x_i, x_j) = -1$ if $j < i$. For $\sigma \in \mathbf{S}(X)$, we define $(-1)^\sigma = \prod_{i<j} f(\sigma(x_i), \sigma(x_j))$. Note that$(-1)^\sigma$ is -1 raised to the sth power where s is the number of pairs (x_i, x_j) $(i < j)$ whose orientation is changed by σ. If τ is a transposition, then one sees easily that $(-1)^{\tau\sigma} = -(-1)^\sigma = (-1)^\tau(-1)^\sigma$ for all $\sigma \in \mathbf{S}(X)$. If $\sigma = \tau_1, \tau_2, \ldots, \tau_r$ where $\tau_1, \tau_2, \ldots, \tau_r$ are transpositions, it follows that $(-1)^\sigma = (-1)^r$, so that $(-1)^\sigma$ is $+1$ if r is even and -1 if r is odd. It follows that if $\sigma = \tau_1, \ldots, \tau_r$ and $\sigma = \tau'_1, \ldots, \tau'_{r'}$ where the τ_i and $\tau'_{i'}$ are transpositions, then r is even/odd if and only if r' is even/odd. We say that σ is *even/odd* if σ has a decomposition $\tau = \tau_1, \ldots, \tau_r$ where r is even/odd. Note that every $\sigma \in \mathbf{S}(X)$ is either even or odd (but not both). Note also that $\sigma_1\sigma_2$ is even if and only if σ_1 and σ_2 are both even or both odd, for all $\sigma_1, \sigma_2 \in \mathbf{S}(X)$. Finally, note that $(-1)^\sigma = +1/(-1)^\sigma = -1$ if σ is even/odd, and that $(-1)^{\sigma_1\sigma_2} = (-1)^{\sigma_1}(-1)^{\sigma_2}$ for all $\sigma_1, \sigma_2 \in \mathbf{S}(X)$.

The subset $\mathbf{A}(X) = \{\sigma \in \mathbf{S}(X) \mid \sigma \text{ is even}\}$ is a subgroup of $\mathbf{S}(X)$, called the *alternating group* on X. If σ is any fixed transposition and $\tau \in \mathbf{S}(X) - \mathbf{A}(X)$, then $\sigma^{-1}\tau \in \mathbf{A}(X)$ (both σ^{-1} and τ are odd). Thus, $\mathbf{S}(X) = A(X) \cup \sigma\,\mathbf{A}(X)$. It follows that $\mathbf{S}(X){:}\mathbf{A}(X) = 2$ for $n \geq 2$. Furthermore, $\mathbf{A}(X)$ is a normal subgroup of $\mathbf{S}(X)$. For if $\tau \in \mathbf{A}(X)$ and $\sigma \in \mathbf{S}(X)$, then $(-1)^{\sigma\tau\sigma^{-1}} = (-1)^{\sigma}(-1)^{\tau}(-1)^{\sigma^{-1}} = (-1)^{\sigma}(-1)^{\sigma^{-1}} = 1$.

It is convenient at this point to take $X = \{1, \ldots, n\}$. The symmetric group $\mathbf{S}(\{1, \ldots, n\})$ on $\{1, \ldots, n\}$ is called the *symmetric group* on n letters and is denoted $\mathbf{S}(n)$. Similarly, the alternating group $\mathbf{A}(\{1, \ldots, n\})$ on $\{1, \ldots, n\}$ is called the *alternating group* on n letters and is denoted $\mathbf{A}(n)$.

For $n \geq 5$, one can prove that $\mathbf{A}(n)$ is *simple* in the sense that the only normal subgroups of $\mathbf{A}(n)$ are $\mathbf{A}(n)$ and $\mathbf{1}$. We do not prove this here, but we do note that $\mathbf{S}(n)$ is not solvable for $n \geq 5$. To see this, simply observe that $\mathbf{S}(n)$, for $n \geq 5$, contains a subgroup isomorphic to $\mathbf{A}(5)$. Since $\mathbf{A}(5)$ is simple (see E.0.84), $\mathbf{A}(5)$ is not solvable. Thus, $S(n)$ is not solvable.

We conclude this section with the following theorem, which we need in 3.12.

0.3.5 Theorem. Let p be a prime number. Then for any transposition σ and p-cycle τ in $\mathbf{S}(p)$, $\mathbf{S}(p)$ is generated by σ and τ.

Proof. For convenience, we reorder the elements so that $\sigma = [1, 2]$. Since some power of the p-cycle τ maps 1 to 2, we may replace τ by this power and reorder $3, \ldots, p$ so that $\tau = [1, 2, 3, \ldots, p]$. For any $\delta \in \mathbf{S}(p)$ and $i \neq j$, we have $\delta[i, j]\delta^{-1} = [\delta(i), \delta(j)]$. Upon using this formula repeatedly, we see that the subgroup H generated by σ and τ contains $\sigma = [1, 2]$, $\tau\sigma\tau^{-1} = [2, 3,]$ $\tau^2\sigma\tau^{-2} = [3, 4], \ldots, \tau^{p-2}\sigma\tau^{2-p} = [p-1, p]$. It follows that H contains $[1, 2]$, $[1, 3] = [1, 2][2, 3][1, 2]$, $[1, 4] = [1, 3][3, 4][1, 3], \ldots, [1, p] = [1, p-1][p-1, p][1, p-1]$. Finally, H contains $[i, j] = [1, i][1, j][1, i]$ for all $i \neq j$. It follows that $\mathbf{S}(p) = H$, since every element of $\mathbf{S}(p)$ is a product of transpositions. □

0.4 The Krull Closure in a group G

Let G be a group with identity element e, $\mathbf{N}$ the set of normal subgroups N of G of finite index. For any subset S of G, let $\bar{S} = \bigcap_{N \in \mathbf{N}} NS$ where NS denotes $\{xs \mid x \in N, s \in S\}$. The set $\bar{S}$ is the *Krull Closure* of S and S is *closed* if $S = \bar{S}$.

0.4.1 Proposition. Let H be a subgroup of G of finite index. Then H is closed.

Proof. G/H is finite. Regard G/H as G-space and let N be the kernel of G on G/H. Then G/N is isomorphic to a subgroup of the group $F(G/H, G/H)^*$ of transformations of the finite set G/H. Thus, $N \in \mathbf{N}$. But $NH = H$, by the definition of N, so that $\bar{H} = H$ and H is closed. □

For most of our purposes, the above discussion suffices. However, for the sake of completeness, we develop the above ideas further. The *Krull Topology* on G is the topology on G having the set $\{Ny \mid N \in \mathbf{N}, y \in G\}$ of cosets as base of open sets. The Krull Closure $\bar{S}$ of S is the closure of $\bar{S}$ of S in the Krull Topology (see E.0.85 and E.0.87).

0.4.2 Proposition. The open subgroups of G are the subgroups of G of finite index.

Proof. If H is an open subgroup of G, then $N \subset H$ for some $N \in \mathbf{N}$, so that H is of finite index. If H is a subgroup of G of finite index, then H is closed, by 0.4.1. But then the distinct cosets $H, Hy_2, \ldots, Hy_n$ of H in G are closed, and H is open as the complement of the closed subset $Hy_2 \cup \ldots \cup Hy_n$ of G. □

The following two corollaries are immediate consequences of the preceding proposition. They relate the Krull Topology to the discrete and product topologies (see E.0.85).

0.4.3 Corollary. A homomorphism f from G with the Krull Topology to a group with the discrete topology is continuous if and only if the kernel of f is a subgroup of G of finite index. □

0.4.4 Definition. Let X be a G-space and let $G/X/G \times X$ have the Krull/discrete/product topology. If the product mapping $G \times X \to X$ is continuous, we say that G acts *continuously* on X.

0.4.5 Corollary. Let X be a G-space. Then G acts continously on X if and only if the G-orbits in X are finite.

E.0 Exercises to Chapter 0

E.0.1 (Identity). Let S be a set with product $x \circ y$ $(x, y \in S)$. Show that if e and f are identities of S, then $e = f$.

E.0.2 (Inverses). Let S be a monoid and let $x \in S$. Show that if y and z are inverses of x, then $y = z$.

E.0.3 (Inverses). Let S be a monoid. Show that

(a) $(x \circ y)^- = y^- \circ x^-$ and $(x^-)^- = x$ for $x, y \in \mathbf{S}^*$;
(b) S^* is a submonoid of S.

E.0.4 (Powers). Let S be a monoid and let T be the set consisting of the powers $x^0, x^1, x^2, \ldots$ of a fixed element x of S. Show that

(a) $x^{m+n} = x^m \circ x^n$ and $(x^m)^n = x^{mn}$ for all nonnegative integers m, n;
(b) T is submonoid of S.

Suppose that $x \in S^*$ and let T' be the set of powers $x^0, x^{-1}, x^1, x^{-2}, x^2, \ldots$ of x. Show that

(c) $x^{m+n} = x^m \circ x^n$ and $(x^m)^n = x^{mn}$ for all integers m and n;
(d) $(x^n)^- = x^{-n}$ for all integers n;
(e) T' is a subgroup of S^*.

E.0.5. Let S be a monoid and let $x_1, \ldots, x_m \in S$. Suppose that the x_i commute pairwise. Show that $(x_1 \ldots x_m)^n = x_1{}^n \ldots x_m{}^n$ for any nonnegative integer n. (If the x_i are in S^*, this equation holds for any integer n).

E.0.6. Show that there is a ring A which possesses precisely one element. Such a ring is called a *null* ring. Show that any two null rings are isomorphic.

E.0.7. Let A be a ring and suppose the zero element 0 of A is also the identity element of A. Show that A is the null ring $A = \{0\}$.

E.0.8 (Imbeddings). Let A and B be rings and let $f: B \to A$ be an *imbedding* of B into A, that is let f be an injective homomorphism from B into A. Show that if A and B are disjoint and $A' = (A - f(B)) \cup B$, then A' can be regarded as a ring such that the mapping $f': A' \to A$ defined by $f'(a) = a$ for $a \in A' - B$ and $f'(b) = f(b)$ for $b \in B$ is an isomorphism.

E.0.9 (Imbeddings). For any two sets A and B, there exists a bijective function g from A to a set disjoint from B. Using this fact, show that if A and B are rings and $f: B \to A$ is an imbedding of B into A, then there exists a ring A' containing B as a subring and an isomorphism $f': A' \to A$ such that $f'(b) = f(b)$ for $b \in B$.

E.0.10 (Field of Quotients). Let A be an integral domain and let $x/y = \{(u, v) \mid u \in A,\ v \in A - \{0\}$ and $xv = uy\}$ for $x \in A$ and $y \in A - \{0\}$. Let $K = \{x/y \mid x \in A, y \in A - \{0\}\}$. Show that

(a) $(x, y) \in x/y\ (x \in A, y \in A - \{0\})$;

(b) if $x/y \neq x'/y'$, then x/y and x'/y' are disjoint $(x, x' \in A,\ y, y' \in A - \{0\})$;

(c) $x/y = x'/y'$ if and only if $xy' = xy' = x'y\ (x, x' \in A, y, y' \in A - \{0\})$;

(d) $x/y = xz/yz\ (x \in A, y, z \in A - \{0\})$;

(e) if $u/v = u'/v'$ and $x/y = x'/y'$, then $(uy + xv)/vy = (u'y' + x'v')/v'y'$ and $ux/vy = u'x'/v'y'\ (u, u', x, x' \in A, v, v', y, y' \in A - \{0\})$;

(f) K together with the products $u/v + x/y = (uy + xv)/vy$ and $(u/v)(x/y) = ux/vy$ is a field with zero and identity elements $0/1$ and $1/1$ and inverses $(x/y)^{-1} = y/x\ (x/y \in K - \{0/1\})$;

(g) the mapping $f: A \to K$ defined by $f(x) = x/1$ is an imbedding of A into K is a field of quotients of $f(A)$.

E.0.11 (Field of Quotients). Show that every integral domain A has a field of quotients. (Use E.0.9 and E.0.10). Show that if K and L are fields of quotients of an integral domain A, then there exists an isomorphism $f: K \to L$ such that $f(a) = a$ for $a \in A$.

E.0.12. Let $f: G \to H$ be a mapping from a group G to a group H such that $f(xy) = f(x)f(y)$ for $x, y \in G$. Show that f maps the identity element of G to the identity element of H and $f(x^-) = f(x)^-$ for $x \in G$.

E.0.13. Let A be a ring and let I be an ideal of A. Show that $0 \in I$ and that I is a subgroup of A as group with addition.

E.0.14. Let A be a commutative ring and let $x \in A$. Show that xA is an ideal of A.

E.0.15. Let A be a ring and let $I_1, \ldots, I_n$ be ideals of A. Let $I_1 + \cdots + I_n$ be the set $\{x_1 + \cdots + x_n \mid x_i \in I_i$ for $1 \leq i \leq n\}$ and let $I_1 \ldots I_n$ be the set of all finite sums of products $x_1 \ldots x_n$ $(x_i \in I_i$ for $1 \leq i \leq n)$. Show that

(a) $I_1 + \cdots + I_n$ is an ideal of A containing the ideals $I_1, \ldots, I_n$;

(b) $I_1 \cap \cdots \cap I_n$ is an ideal of A;

(c) $I_1, \ldots, I_n$ is an ideal of A contained in $I_1 \cap \cdots \cap I_n$.

E.0.16. Let R be a ring and let $\{R_\alpha \mid \alpha \in A\}$ be a collection of subrings of R. Show $\bigcap_{\alpha \in A} R_\alpha$ is a subring of R.

E.0.17. Let S be an Abelian group and let T be a subgroup of S. Show that

(a) $x + T$ and $x' + T$ are either equal or disjoint ($x, x' \in S$);
(b) $x + T = x' + T$ if and only if $x - x' \in T$ ($x, x' \in S$);
(c) if $x + T = x' + T$ and $y + T = y' + T$, then $(x + y) + T = (x' + y') + T$ ($x, y, x', y' \in S$);
(d) S/T together with the product $(x + T) + (y + T) = (x + y) + T$ is a group, the identity of S/T is $0 + T$ where 0 is the identity of S and the inverse of $x + T$ is $(-x) + T$ where $-x$ is the inverse of x ($x \in S$);
(e) the mapping $f: S \to S/T$ defined by $f(x) = x + T$ is a surjective homomorphism and Kernel $f = T$.

E.0.18. Let A be a ring, let I be an ideal of A and consider $A/I = \{x + I \mid x \in A\}$. Show that

(a) if $x + I = x' + I$ and $y + I = y' + I$, then $(xy) + I = (x'y') + I$ ($x, y, x', y' \in A$);
(b) A/I together with the products $(x + I) + (y + I) = (x + y) + I$ and $(x + I)(y + I) = (xy) + I$ is a ring, and the identity element of A/I is $e + I$ where e is the identity element of A;
(c) the mapping $f: A \to A/I$ defined by $f(x) = x + I$ is a surjective homomorphism and Kernel $f = I$.

E.0.19 (First Homomorphism Theorem). Let A and B be rings and $f: A \to B$ a homomorphism. Show that

(a) if a + Kernel $f = a'$ + Kernel f, then $f(a) = f(a')$ ($a, a' \in A$);
(b) the mapping $\bar{f}: A/\text{Kernel } f \to f(A)$ defined by $\bar{f}(a + \text{Kernel } f) = f(a)$ is an isomorphism;
(c) A and $A/\{0\}$ are isomorphic;
(d) f is injective if and only if Kernel $f = \{0\}$.

E.0.20 (Second Homomorphism Theorem). Let A be a ring, B a subring of A, I an ideal of A. Show that $B + I = \{b + x \mid b \in B, x \in I\}$ is a subring of A and $B \cap I$ is an ideal of B. Show that there is an isomorphism $f: B/B \cap I \to (B + I)/I$ such that $f(b + B \cap I) = b + I$ for all $b \in B$.

E.0.21. Describe and prove the analogues for groups of the First and Second Homomorphism Theorems for rings.

E.0.22. Let A be a ring with ideal A. Describe a natural bijection from the set of ideals of A containing I to the set of ideals of A/I.

E.0.23. Let A be a commutative ring with ideal I. Show that

(a) I is maximal if and only if A/I is a field;
(b) I is prime if and only if A/I is an integral domain;
(c) A is a field if and only if $A \neq \{0\}$ and the only ideals of A are $\{0\}$ and A.

E.0.24. Let K be a field and let $f(X), g(X) \in K[X] - \{0\}$. Show that

(a) $\text{Deg}(f(X)g(X)) = \text{Deg} f(X) + \text{Deg} g(X)$;
(b) $\text{Deg}(f(X) + g(X)) \leq \text{Max}(\text{Deg} f(X), \text{Deg} g(X))$ (the maximum of the two integers $\text{Deg} f(X)$ and $\text{Deg} g(X)$).

E.0.25. The group of units of $K[X]$, for K a field, is the set K^* of nonzero constant polynomials.

E.0.26. Let K be a field, $g(X)$ an irreducible element of $K[X]$ and $h_1(X), \ldots, h_n(X)$ elements of $K[X]$ such that $g(X)$ divides $\prod_1^n h_i(X)$. Show that $g(X)$ divides $h_i(X)$ for some i. Supposing, further, that $g(X)$ and the $h_i(X)$ $(1 \leq i \leq n)$ are monic and irreducible, show that $g(X) = h_i(X)$ for some i. Using these observations, give the details to the proof of 0.1.5.

E.0.27 (Ring of Integers). A *ring of integers* is a ring $\mathbb{Z}$ having a nonempty subset $\mathbb{N}$ closed under addition and multiplication and consisting of elements denoted $1, 2 = 1 + 1, 3 = 2 + 1, \ldots$ such that

1. (trichotomy) for $x \in \mathbb{Z}$, precisely one of the following possibilities occurs: $x = 0$ (x is *zero*), $x \in \mathbb{N}$ (x is *positive*), $-x \in \mathbb{N}$ (x is *negative*);
2. (induction) if S is a subset of $\mathbb{N}$ containing 1 such that $x \in S$ implies $x + 1 \in S$ for all $x \in \mathbb{Z}$, then $S = \mathbb{N}$.

The axioms of set theory imply that there exists a ring of integers. Show that

(a) $\mathbb{N}$ is infinite (e.g. show that $f(x) = x + 1$ is injective but not surjective from $\mathbb{N}$ to $\mathbb{N}$):

(b) (order properties) letting $x < y$ and $x \nless y$, mean that $y - x \in \mathbb{N}$ and $y - x \notin \mathbb{N}$ respectively, we have

  (i) (trichtomy) for $x, y \in \mathbb{N}$, precisely one of the following possibilities occurs: $x = y, x < y, y < x$;

  (ii) (antireflexitivity) $x \nless x$ $(x \in \mathbb{Z})$;

  (iii) (transitivity) $x < y$ and $y < z$ imply that $x < z$ $(x, y, z \in \mathbb{Z})$;

  (iv) (linearity) $u < v$ and $x < y$ imply that $u + x < v + y$; and $0 < w$ and $x < y$ imply that $wx < wy$ $(u, v, w, x, y \in \mathbb{Z})$;

(c) (least element property) for any subset T of $\mathbb{N}$, either T has a least element or T is empty (consider $S = \mathbb{N} - T$ and use the induction property of N);

(d) (universality property) for any ring A with identity element e, there is a unique homomorphism $f\colon \mathbb{Z} \to A$ (define f on $\mathbb{N}$ inductively by $f(1) = e$, $f(x + 1) = f(x) + e$, define $f(0) = 0$ and define $f(-x) = -f(x)$ for $x \in \mathbb{N}$).

E.0.28 (Ring of Integers). Show that if $\mathbb{Z}$ and $\mathbb{Z}'$ are both rings of integers, then there is a unique isomorphism from $\mathbb{Z}$ to $\mathbb{Z}'$. (Use the universality property).

E.0.29 (Euclidean Algorithm). Let $\mathbb{Z}$ be the ring of integers and let $a, b \in \mathbb{N}$. Show that there exist $m, r \in \mathbb{Z}$ such that $b = ma + r$ and $0 \leq r \leq a$. (Use the least element property to get a least remainder r).

E.0.30 (Ring of Integers). Show that the ring $\mathbb{Z}$ of integers is a principle ideal domain. (Use the Euclidean Algorithm and compare with 0.1.2).

E.0.31 (Ring of Integers). For integers m and n, we say that $a|b$ (a *divides* b) if $b = ma$ for some $m \in \mathbb{Z}$. A *prime* integer is an integer $p > 1$ such that $a|p$ if and only if $a = \pm 1$ or $a = \pm p$. Show that for any prime p, $p|xy$ implies

p divides x or p divides $y (x, y \in \mathbb{Z})$ (compare with 0.1.4). Show that a positive integer p is prime if and only if the ideal $\mathbb{Z}p$ is a prime ideal.

E.0.32 (Ring of Integers). Show that for any integer $b > 1$, b can be expressed uniquely as $b = \prod_1^m p_i^{e_i}$ where the p_i are primes, the e_i are positive integers and $p_1 < p_2 < \cdots < p_m$. (Compare with 0.1.5).

E.0.33 (Ring of Integers). Let $p_1, \ldots, p_n$ be distinct prime numbers. Show that $p_i \nmid (\prod_1^n p_j + 1)$ for $1 \leq i \leq n$. Use this to show that there are infinitely many prime numbers.

E.0.34 (Field of Rational Numbers). Let $\mathbb{Q}$ be the field of quotients of the ring $\mathbb{Z}$ of integers. We call $\mathbb{Q}$ the *field of rational numbers*. Let $\mathbb{P} = \{yx^{-1} \mid x, y \in \mathbb{N}\}$. Show that

(a) $\mathbb{P}$ is closed under addition and multiplication and $\mathbb{Q}$ satisfies the law of trichotomy with respect to $\mathbb{P}$ (see E.0.27);

(b) letting $x < y$ and $x \nless y$ mean that $y - x \in \mathbb{P}$ and $y - x \notin \mathbb{P}$ respectively, then $x < y$ $(x, y \in \mathbb{Q})$ satisfies the laws of trichotomy, antireflexitivity, transitivity and linearity, and the order $x < y (x, y \in \mathbb{Z})$ in $\mathbb{Z}$ is preserved by the imbedding $x \mapsto x/1$ of $\mathbb{Z}$ in $\mathbb{Q}$, that is, $x < y$ if and only if $x/1 < y/1$ for $x, y \in \mathbb{Z}$.

E.0.35 (Field of Real Numbers). An *ordered field* is a field K together with a subset P closed under addition and multiplication such that K satisfies the law of trichotomy with respect to P. Let $x < y$ and $x \nless y$ mean that $y - x \in P$ and $y - x \notin P$ respectively and note that $x < y$ $(x, y \in K)$ satisfies the laws of trichotomy, antireflexitivity, transitivity and linearity. A *sequence* in K is a function f from $\mathbb{N}$ to K and is denoted $f_1, f_2, \ldots$ where $f_i = f(i)$ $(i \in \mathbb{N})$. The set $R(K)$ of sequences in K is a ring with respect to the addition and multiplication defined by $(f + g)(i) = f(i) = g(i)$ and $(fg)(i) = f(i)g(i)$ $(i \in \mathbb{N}, f, g \in R(K))$. A sequence f_i in K is *convergent* if there exists $x \in K$ such that for each positive ε in K there exists $N \in \mathbb{N}$ such that $x - \varepsilon < f_i < x + \varepsilon$ for $i \geq N$. A sequence f_i in K is a *Cauchy sequence* if for each positive ε in K, there exists $N \in \mathbb{N}$ such that $-\varepsilon < f_j - f_i < \varepsilon$ for $i, j \geq N$. If every Cauchy sequence in K is convergent, then K is *complete*. Taking K to be the field $\mathbb{Q}$ of rational numbers, show that

(a) the set of R_C of Cauchy sequences in $\mathbb{Q}$ is a subring of $R(\mathbb{Q})$;

(b) the set $R_0 = \{f \in R(\mathbb{Q}) \mid f \text{ converges to } 0\}$ is an ideal in R_C;

(c) the ring $\mathbb{R} = R_C/R_0$ is a field, called the *field of real numbers*, and $\mathbb{R}$ together with $\mathbb{R}_+ = \{f + R_0 \mid f \notin R_0 \text{ and } 0 < f_i \text{ for all but finitely many } i\}$ is a complete ordered field;

(d) the function α from $\mathbb{Q}$ to $\mathbb{R}$ which maps each $x \in \mathbb{Q}$ to $f + R_0$ where f is the constant sequence $f_i = x$ for all i is an imbedding of $\mathbb{Q}$ in $\mathbb{R}$ mapping $\mathbb{P}$ to a subset of $\mathbb{R}_+$;

(f) $\alpha(\mathbb{Q})$ is dense in $\mathbb{R}$ in the sense that for every positive ε in $\mathbb{R}$ and every $y \in \mathbb{R}$, there exists $x \in \mathbb{Q}$ such that $y - \varepsilon < \alpha(x) < y + \varepsilon$;

(g) $\mathbb{R}$ is *Archimedian* in the sense that for x, y in $\mathbb{P}$, there exists n in $\mathbb{N}$ such that $y < nx$.

Can this discussion for $\mathbb{Q}$ be generalized to any Archimedian ordered field K?

E.0.36 (Field of Real Numbers). Show that any complete Archimedian ordered field is isomorphic to the field $\mathbb{R}$ of real numbers. Does this imply that every Archimedian ordered field is isomorphic to a subfield of the field $\mathbb{R}$ of real numbers?

E.0.37 (Field of Complex Numbers). Let $\mathbb{R}$ be the field of real numbers. Show that $\mathbb{C} = \mathbb{R}[X]/\mathbb{R}[X](X^2 + 1)$ is a field, called the *field of complex numbers*. Show that $\beta: \mathbb{R} \to \mathbb{C}$, defined by $\beta(x) = x + \mathbb{R}[X](X^2 + 1)$ for $x \in \mathbb{R}$, is an imbedding of $\mathbb{R}$ in $\mathbb{C}$. Show that $\mathbb{C}$ with addition and the scalar multiplication $x \cdot z$ ($x \in \mathbb{R}, z \in \mathbb{C}$) defined by $x \cdot z = \beta(x)z$ is a two dimensional real vector space with identity e and basis e, i where $i^2 = -e$. Thus, show that $\mathbb{C} = \{z | z = a \cdot e + b \cdot i \text{ with } a, b \in \mathbb{R}\}$, that $\beta: \mathbb{R} \to \mathbb{C}$ is given by $\beta(a) = a \cdot e$ ($a \in \mathbb{R}$) and that multiplication in $\mathbb{C}$ is given by $(a \cdot e + b \cdot i)(c \cdot e + d \cdot i) = (ac - bd) \cdot e + (ad + bc) \cdot i$ ($a, b, c, d \in \mathbb{R}$).

E.0.38 (Ring of Integers Modulo a). Let $a \in \mathbb{N}$. Then we let $\mathbb{Z}_n$ denote the ring $\mathbb{Z}_a = \mathbb{Z}/\mathbb{Z}a$ and call $\mathbb{Z}_a$ the *ring of integers modulo a*. Letting $\bar{b} = b + \mathbb{Z}a$ for $b \in \mathbb{Z}$, show that $\mathbb{Z}_a = \{\bar{0}, \bar{1}, \ldots, \overline{a - 1}\}$. Show that $\mathbb{Z}_a$ is an integral domain if and only if $\mathbb{Z}_a$ is a field if and only if a is a prime.

E.0.39 (Unique Factorization Domains). Let A be a commutative ring. We say that $a|b$ (read *a divides b*) if $b = m\,a$ for some $m \in A$. We write $a \nmid b$ if a does not divide b. If $a|b$ and $b|a$, then a and b are *associates*. If b is nonzero and b is not a unit, and if the only elements which divide b are units and associates of b, then b is *irreducible*. We say that A is a *unique factorization domain* if A is an integral domain and for each nonzero nonunit element b of A, $b = \prod_1^m b_i$ where the $b_1, \ldots, b_m$ are irreducible and are unique in the sense that if $b = \prod_1^n c_j$ where the $c_1, \ldots, c_n$ are irreducible, then $m = n$ and b_i is an associate of c_{j_i} for $1 \le i \le m$ for a suitable permutation $c_{j_1}, \ldots, c_{j_m}$ of the $c_1, \ldots, c_m$. Show that for an integral domain A to be a unique factorization domain, the following two conditions are necessary and sufficient:

1. each nonzero nonunit element b of A has a factorization $b = \prod_1^m b_i$ where the $b_1, \ldots, b_m$ are irreducible;
2. if b is irreducible and $b|xy$, the $b|x$ or $b|y$.

Show that in a unique factorization domain, a and b are associates if and only only if $a = bc$ where c is a unit.

E.0.40 (Unique Factorization Domains). Show that every principal ideal domain is a unique factorization domain. (Compare with 0.1.4).

E.0.41 (Unique Factorization Domains). Let A be a unique factorization domain. Then the *greatest common divisor* and *least common multiple* of $cb_1^{e_1} \ldots b_m^{e_m}$ and $b_1^{f_1} \ldots b_m^{f_m}$ ($e_i \ge 0, f_i \ge 0$ for all i; $b_1, \ldots, b_m$ irreducible and pairwise nonassociates, c a unit) are $b_1^{g_1} \ldots b_m^{g_m}$ and $b^{h_1} \ldots b_m^{h_m}$ respectively, where g_i is the lesser of e_i and f_i and h_i is the greater of e and f_i for all i. Elements $a_1, \ldots, a_n$ of A are *relatively prime* if the greatest common divisor of $a_1, \ldots, a_n$ is e (identity of A). Show that if A is a principle ideal domain, then $a_1, \ldots, a_n$ are relatively prime if and only if there exist $m_1, \ldots, m_n$ in A such that $e = m_1 a_1 + \cdots + m_n a_n$. (Note here that the greatest common divisor and least common multiple are only unique up to associates).

E.0.42 (Direct Product). Let $A_1, \ldots, A_n$ be rings. Then $A = A_1 \times \cdots \times A_n$ together with the addition and multiplication defined by $(a_1, \ldots, a_n) + (b_1, \ldots, b_n) = (a_1 + b_1, \ldots, a_n + b_n)$ and $(a_1, \ldots, a_n)(b_1, \ldots, b_n) = (a_1 b_1 \ldots, a_n b_n)$ is a ring (called the *direct product ring* of $A_1, \ldots, A_n$).

E.0.43 (Direct Limit). Let A be a totally ordered set. Let $\{R_a \mid a \in A\}$ be a collection of rings and let $\{\beta_{ba} \mid a, b \in A \text{ and } a \lneqq b\}$ be a collection of functions such that

1. $\beta_{ba}: R_a \to R_b$ is a homomorphism for $a \lneqq b$;
2. $\beta_{cb} \circ \beta_{ba}$ for $= \beta_{ca}$ for $a \lneqq b \lneqq c$.

Let R be the set of all functions f from A to $\bigcup_{a \in A} R_a$ such that $f(a) \in R_a$ for all $a \in A$. Define $f + g$ and fg for $f, g \in R$ by $(f + g)(a) = f(a) + g(a)$ and $(fg)(a) = f(a)g(a)$ $(a \in A)$. Show that

(a) R is a ring, called the *direct product* of the R_a. (Compare with E.0.42).

An element f of R is *almost coherent* if there exists $c \in A$ such that $\beta_{ba}(f(a)) = f(b)$ for $c \leq a \lneqq b$. Show that

(b) the set R_β of almost coherent elements of R is a subring of R.

A *null* element of R is an element f of R such that for some $c \in A$, $f(a) = 0$ for $c \leq a$. Show that

(c) the set R^0 of null elements of R is an ideal of R_β.

Let $\bar{R}$ be the ring R_β / R^0. Define $\beta_a: R_a \to \bar{R}$ by letting $\beta_a(x) = f + R^0$ where f is the element of R_β such that $f(c) = 0$ for $c \lneqq a$, $f(a) = x$ and $f(b) = \beta_{ba}(x)$ for $a \lneqq b$. Show that

(d) β_a is a homomorphism of rings and the diagram

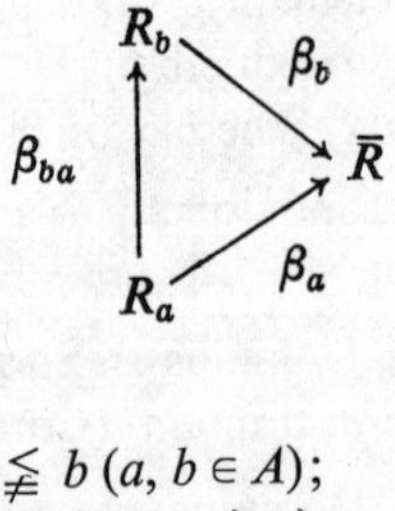

is commutative for $a \lneqq b$ $(a, b \in A)$;

(e) $\bar{R} = \bigcup_{a \in A} \beta_a(R_a)$ (ascending union);

(f) if the $R_a (a \in A)$ are fields, then the $\beta_a (a \in A)$ are injective and a field.

The ring $\bar{R}$ together with the functions $\beta_a (a \in A)$ is called the *direct limit* of the functions β_{ba} on the rings R_a and is denoted $\operatorname{Lim} \beta$.

E.0.44 (Chinese Remainder Theorem). Let A be a principal ideal domain with identity e, let a be a nonzero nonunit element of A and let $a = \prod_1^n x_i$ where x_i and x_j are relatively prime for $i \neq j$. Let $a_i = a/x_i$ and note that $a_1, \ldots, a_n$ are relatively prime, so that there exist $m_i \in A$ such that $e = m_1 a_1 +$

$\cdots + m_m a_n$ (see E.0.41). Let $e_i = m_i a_i$, so that $e = e_1 + \cdots + e_n$. Let $\bar{A} = A/Aa$ and $\bar{b} = b + Aa$ for $b \in A$. Show that

(a) $\bar{e} = \bar{e}_1 + \cdots + \bar{e}_n$, $\bar{e}_i\bar{e}_j = \bar{0}$ for $i \neq j$ and $\bar{e}_i^2 = \bar{e}_i$ for $1 \leq i \leq n$;
(b) $\bar{A} = \bar{A}\bar{e} + \cdots + \bar{A}\bar{e}_n$ (sum of ideals) and this sum is direct as a sum of additive groups;
(c) $\bar{A}\bar{e}_i$ is a ring with identity $\bar{e}_i$ and is isomorphic to A/Ax_i;
(d) A/Aa is isomorphic to the direct product ring $(A/Ax_1) \times \cdots \times (A/Ax_n)$.

E.0.45 (Rings of Integers Modulo a). Show that if $a = p_1^{e_1} \cdots p_n^{e_n}$ where $p_1, \ldots, p_n$ are distinct prime numbers and $e_1, \ldots, e_n$ are positive integers, then the ring $\mathbb{Z}_a$ of integers modulo a is isomorphic to $\mathbb{Z}_{p_1^{e_1}} \times \cdots \times \mathbb{Z}_{p_n^{e_n}}$ (direct product ring). (Use the Chinese Remainder Theorem).

E.0.46 (Simultaneous Congruences). Using the notation $u \equiv_a v$ (u is *congruent* to v modulo a) for $v - u \in \mathbb{Z}a$ ($a, u, v \in \mathbb{Z}$), show that if $a = x_1 \cdots x_n$ where x_i and x_j are positive and relatively prime for $i \neq j$ and if $u_1, \ldots, u_n \in \mathbb{Z}$, then

(a) there exists $v \in \mathbb{Z}$ such that $u_i \equiv_{x_i} v$ for $1 \leq i \leq n$;
(b) for $v' \in \mathbb{Z}$ to also satisfy $u_i \equiv_{x_i} v'$ for $1 \leq i \leq n$, it is necessary and sufficient that $v \equiv_a v'$.

(Use the Chinese Remainder Theorem).

E.0.47 (Euler Phi Function). For any integer $a > 1$, let $\varphi(a)$ be the number of integers b such that b is relatively prime to a and $1 \leq b < a$. Show that

(a) $\varphi(a)$ is the number of units in $\mathbb{Z}_a$;
(b) $\varphi(a\,a') = \varphi(a)\varphi(a')$ if a and a' are relatively prime;
(c) $\varphi(p^e) = p^{e-1}(p - 1)$ for p a prime.

(Use the Chinese Remainder Theorem for (b)).

E.0.48 (Ideal Structure of $\mathbb{Z}_a$). Determine the maximal ideals of $\mathbb{Z}_a$ for $a = p_1^{e_1} \cdots p_n^{e_n}$ (the p_1 being distinct primes and the e_i positive integers).

E.0.49 (Unique Factorization in $A[X]$). Let A be a unique factorization domain and let K be the field of quotients of A. Show that the polynomial ring $A[X] = \{f(X) \in K[X] \mid$ the coefficients of $f(X)$ are in $A\}$ is a unique factorization domain. (Compare with 0.1.11).

E.0.50 (Eisenstein's Criteria). Let $f(X) = \sum_0^d a_i X^i$ be an element of $\mathbb{Q}(X)$ and suppose that p is a prime number such that $p|a_1, \ldots, p|a_d$, but $p \nmid a_0$ and $p^2 \nmid a_d$ ($a \nmid b$ means that a does not divide b). Show that $f(X)$ is irreducible. (Consider a potential factorization $f(X) = (\sum_0^m b_r X^r)(\sum_0^n c_s X^s)$ where $p \mid b_m$ and $p \nmid c_n$, and consider a_{r+n} where r is minimal such that $p \nmid b_r$.)

E.0.51. Which of the following polynomials in $\mathbb{Q}[X]$ are irreducible?

(a) $X^4 - X^2 + 1$;
(b) $X^3 + X^2 + X + 1$;
(c) $X^{13} + 1$;
(d) $X^n - p$ $(n \geq 1, p$ prime).

E.0.52. Give necessary and sufficient conditions that $X^{p^m} - X$ divide $X^{p^n} - X$ in $\mathbb{Z}_p[X]$, p being a prime.

E.0.53. Let k be a field and consider the polynomial $X^3 + X + 1 \in k[X]$. Let I be the ideal $k[X]\,(X^3 + X + 1)$. Show that for every $f(X) \in k[X]$, $f(X) + I = (a\,X^2 + b\,X + c) + I$ for suitable $a, b, c \in k$ uniquely determined by $f(X)$. Express $(X^2 + 2X)^{12} + I$ in this form.

E.0.54. Let k be a field and let $f(X)$ be a polynomial of degree $n \geq 1$. Show that

(a) $k[X]$ together with the obvious scalar multiplication is a vector space over k containing $k[X]f(X)$ as subspace;
(b) the quotient vector space $k[X]/k[X]f(X)$ has basis $1 + I, X + I, \ldots, X^{n-1} + I$ where $I = k[X]f(X)$.

E.0.55 (Free Abelian Monoids). Let S be a set. Show that there is an Abelian monoid $\hat{S}$ containing S as subset such that

1. each element x of $\hat{S}$ can be written as $x = s_1{}^{e_1} \cdots s_m{}^{e_m}$ where the e_i are nonnegative integers and $s_1, \ldots, s_m$ are distinct elements of S;
2. if $s_1, \ldots, s_m$ are distinct elements of S and $s_1{}^{e_1} \cdots s_m{}^{e_m} = s_1{}^{f_1} \cdots s_m{}^{f_m}$ where the e_i and f_i are nonnegative integers, then $e_i = f_i$ for $1 \leq i \leq m$.

This monoid $\hat{S}$ is called the *free Abelian monoid* on S. (Consider the set T of functions from S to the set of nonnegative integers, make T into a monoid and imbed S in T).

E.0.56 (Polynomial Rings). Let K be a field. Define $K[X_1]$, $K[X_1, X_2] = (K[X_1])[X_2], \ldots, K[X_1, \ldots, X_n] = (K[X_1, \ldots, X_{n-1}])[X_n]$ successively according to 0.1 and E.0.49. Then $K[X_1, \ldots, X_n]$ is a ring consisting of the polynomials in the *indeterminants* $X_1, \ldots, X_n$ and is called the *polynomial ring* in n-variables over K. Show that

(a) $K[X_1, \ldots, X_n]$ is a unique factorization domain;
(b) the set $\hat{S}$ of *monomials* $X_1{}^{e_1} \cdots X_n{}^{e_n}$ (the e_i being nonnegative integers) is closed under multiplication and is a free Abelian monoid on the set $S = \{X_1, \ldots, X_n\}$;
(c) $\hat{S}$ is a basis for $K[X_1, \ldots, X_n]$.

E.0.57 (Polynomial Rings). Let K be a field. The polynomial rings $K[X_1, \ldots, X_n]$ are algebras over K. (See Appendix A). Show that the algebras $K[X_1, \ldots, X_n]$ and $K[X_1] \otimes \cdots \otimes K[X_n]$ are isomorphic. (See Appendices T and A).

E.0.58 (Group Rings). Let G be a group or monoid and let K be a field. Show that there exists a vector space $K[G]$ over K with basis G. Define $(\sum_{g \in G} a_g g)(\sum_{h \in G} b_h h) = \sum_{g,h \in G} a_g b_h (gh)$ for $a_g, b_h \in K (g, h \in G)$. Show that $K[G]$ with vector space addition and this multiplication is a ring (called the *group ring* or *monoid ring* of G over K). Show that $K[G]$ is an algebra over K in the sense of Appendix A.

E.0.59 (Polynomial Rings). Let K be a field. Show that the polynomial ring $K[X_1, \ldots, X_n]$ is isomorphic to the monoid ring $K[\hat{S}]$ of the free Abelian monoid on $S = \{X_1, \ldots, X_n\}$.

E.0.60 (Group Rings). Let G and H be groups and K a field. Let $K[G]$, $K[H]$, $K[G \times H]$ be the group rings of G, H, $G \times H$ (outer direct product). Show that $K[G \times H]$ is isomorphic to $K[G] \otimes K[H]$ (tensor product of K-algebras). (See Appendix A.)

E.0.61. Let G be a group whose order is a prime number p. Show that $G = \langle g \rangle$ for $g \in G - \mathbf{1}$.

E.0.62. Show that if G is a finite p-group, then the set $G^p = \{g^p \mid g \in G\}$ coincides with G only when $G = \mathbf{1}$.

E.0.63. Describe the noncyclic subgroups of $\mathbb{Z}_p \times \mathbb{Z}_p \times \mathbb{Z}_p$, p being a prime number.

E.0.64. Describe the decomposition $G = G_{p_1} \cdots G_{p_n}$ of 0.1 in each of the cases

(a) $G = \mathbb{Z}_{50}$ (as additive group);
(b) $G = \mathbb{Z}_{28} \times \mathbb{Z}_{12}$ (as additive group);
(c) $G = \mathbb{Z}_{24} \times \mathbb{Z}_{12} \times \mathbb{Z}_4$ (as additive group).

E.0.65. Let G be a commutative group, g, h elements of G of finite orders m, n. Show that

(a) $\langle g \rangle = \langle g^d \rangle$ if d and m are relatively prime;
(b) if m and n are relatively prime, then $\langle g, h \rangle = \langle gh \rangle = \langle g \rangle \langle h \rangle$ (internal direct product).

E.0.66. Let G be an Abelian group whose order is mn and n are relatively prime. Show that $G = G_m G_n$ (internal direct product) when G_m and G_n are subgroups of G of orders m and n respectively.

E.0.67. Let $G = G_m G_n$ (internal direct product) where G_m, G_n are subgroups of G of orders m, n respectively. Show that if m and n are relatively prime, then every subgroup H of G has the form $H = H_m H_n$ where $H_m \subset G_m$ and $H_n \subset G_n$.

E.0.68. Let G be a group with subgroup H, and let $x, y \in G$. Show that

(a) $xH = yH$ if and only if $x^{-1} y \in H$;
(b) xH and yH are disjoint if $x^{-1} y \notin H$;
(c) the function $L_x: G \to G$ defined by $L_x(y) = xy$ for $y \in G$ is a bijective function which maps H to xH.

E.0.69. Let G be a group, H a normal subgroup of G. Show that if $xH = x'H$ and $yH = y'H$, then $(xy)H = (x'y')H$. Show that G/H together with the product $(xH)(yH) = (xy)H$ is a group with identity H and inverses $(xH)^- = x^- H$ $(x \in G)$.

E.0.70. Let G be a group and let H and I be normal subgroups of G such that $H \cap I = \mathbf{1}$. Show that the elements of H commute with the elements of I. Show, in particular, that if p and q are two different prime numbers such that G_p and G_q are subgroups of G, then the elements of G_p commute with the elements of G_q.

E.0.71 (Second Homomorphism Theorem). Let G be a group and let N be a normal subgroup of G. Show that for any subgroup H of G, the mapping

$f: H/N \cap H \to NH/N$ mapping $x(N \cap H)$ to $xN (x \in H)$ is well defined and is an isomorphism from $H/N \cap H$ to NH/N.

E.0.72 (Semidirect Products). Let H be a group and let $\alpha: H \to \operatorname{Aut} N$ be a homomorphism from H to the group $\operatorname{Aut} N$ of automorphisms of a group N. Show that $N \times H$ together with the product $(u, x)(v, y) = (u\alpha(x)(v), xy)$ is a group (called the *semidirect product* of N and H with respect to α). Show that if G is a group, H is a subgroup of G and N is normal subgroup of G such that $N \cap H = \mathbf{1}$, then the mapping $f: N \times H \to NH$ defined by $f(u, x) = ux$ is an isomorphism from $N \times H$ (semidirect product) to NH where $\alpha: H \to \operatorname{Aut} N$ is defined by $\alpha(x) = \operatorname{Int} x|_N$ for $x \in H$. If $G = NH$ where H is a subgroup of G and N is a subgroup of G such that $N \cap H = \mathbf{1}$, we therefore write $G = NH$ (*internal semidirect product*).

E.0.73. Let $N = \mathbb{Z}_p \times \mathbb{Z}_p \times \cdots \times \mathbb{Z}_p$ (n times). Let $H = \mathbf{S}(n)$. For $\sigma \in H$ and $(a_1, \ldots, a_n) \in N$, let $\alpha(\sigma)(a_1, \ldots, a_n) = (a_{\sigma(1)}, \ldots, a_{\sigma(n)})$. Show that α is a homomorphism from H to $\operatorname{Aut} N$. Describe the center of the group $N \times H$ (semidirect product).

E.0.74 (Solvable Groups). Let G be a group. An *Abelian* tower in G is a tower $\mathbf{1} \subset G_1 \subset \cdots \subset G_n = G$ such that G_1 is normal in G_{i+1} and G_{i+1}/G_i is Abelian for $1 \leq i \leq n - 1$. Show that G is solvable if and only if G has an Abelian tower.

E.0.75 (Commutator Subgroups). Let G be a group. The *commutator subgroup* of G is the subgroup $G^{(1)}$ of G generated by the *commutators* $x \circ y = x^{-1}y^{-1}xy$ $(x, y \in G)$. Show that $G^{(1)}$ is a normal subgroup of G such that $G/G^{(1)}$ is Abelian, since $(yx)x \circ y = xy$ $(x, y \in G)$. Define $G^{(i+1)} = G^{(i)(1)}$ for $i = 1, 2, \ldots$. The series $G \supset G^{(1)} \supset G^{(2)} \supset \cdots$ is called the *commutator series*. Show that $G^{(i)}$ is a normal subgroup of G for all i.

E.0.76 (Solvable Groups). Show that

(a) G is solvable if and only if $G^{(n)} = \mathbf{1}$ for some n;
(b) for any normal subgroup N of G, $(G/N)^{(i)} = NG^{(i)}/N$ for all i;
(c) for any normal subgroup N of G, G is solvable if and only if N and G/N are solvable.

E.0.77. Show that $S(3)$ is not nilpotent.

E.0.78 (Nilpotent Groups). Show that a finite group G such that $G/C(G)$ is nilpotent is itself nilpotent. Show that if G is a finite nilpotent group and $G \neq \mathbf{1}$, then $C(G) \neq \mathbf{1}$.

E.0.79 (Nilpotent Groups). Show that every finite p-group is solvable. Use this to show that every finite nilpotent group is solvable.

E.0.80 (p-Groups). Let G be a group of order p^n where p is a prime and $n \geq 1$. Show that G has precisely one subgroup of order p^{n-1}.

E.0.81. Let G be a finite Abelian p-group and K the field $\mathbb{Z}_p$ of p elements. Show that $K[G]$ has only one maximal ideal.

E.0.82. Let X be a set and let $F(X, X)$ be the set of functions from X to X. Show that $F(X, X)$ together with the product $g \circ f$ (composition of functions) is a monoid and that the group $F(X, X)^*$ of units of $F(X, X)$ is the set of bijective functions from X to X. Show that if G is a group and X is a G-space, then the mapping $\rho: G \to F(X, X)$ defined by $\rho(g)(X) = gX$ for $g \in G$ $x \in X$ is a homomorphism from G into $F(X, X)^*$.

E.0.83. Show that $S(4)$ is solvable by showing that $S(4)$ has a cyclic tower $S(4) \supset A(4) \supset N \supset \mathbf{1}$ where N is an Abelian subgroup of $A(4)$ having 4 elements.

E.0.84. Show that $A(5)$ and $\mathbf{1}$ are the only normal subgroups of $A(5)$. (Letting $N \neq \mathbf{1}$ be a normal subgroup of $A(5)$, whow that N is transitive on $\{1, 2, 3, 4, 5\}$. Then show that the order of N is divisible by 5 and that N contains a 5-cycle. Use this to show that $N = A(5)$).

E.0.85 (Topologies). Let X be a set. A *topology* for X is a collection $\mathbf{U}$ of subsets of X such that

1. $X \in \mathbf{U}$ and $\varnothing \in \mathbf{U}$ ($\varnothing$ being the empty set);
2. if U_α $(\alpha \in A)$ are elements of $\mathbf{U}$, then $\bigcup_{\alpha \in A} U_\alpha$ is an element of $\mathbf{U}$;
3. if $U, V \in \mathbf{U}$, then $U \cap V \in \mathbf{U}$.

A *base* for X is a collection $\mathbf{U}_0$ of subsets of X such that if $U, V \in \mathbf{U}_0$, then $U \cap V \in \mathbf{U}_0$. Show that

(a) the set $\mathbf{U}$ of all subsets of X is a topology for X (called the *discrete topology* for X);
(b) for any base $\mathbf{U}_0$ for X, the set $\mathbf{U}$ consisting of X, $\varnothing$ and all sets of the form $\bigcup_{\alpha \in A} U_\alpha$ where the U_α $(\alpha \in A)$ are elements of $\mathbf{U}_0$ is a topology for X($\mathbf{U}_0$ is called a *base of open sets* for the topology $\mathbf{U}$);
(c) for $\mathbf{U}$ a topology for the set X and $\mathbf{V}$ a topology for the set Y, the collection $\mathbf{W}_0 = \{U \times V \mid U \in \mathbf{U}, V \in \mathbf{V}\}$ is a base for $X \times Y$ (and the topology $\mathbf{W}$ for $X \times Y$ having $\mathbf{W}_0$ as base of open sets is called the *product topology* for $X \times Y$).

E.0.86 (Topological Spaces). A *topological space* is a set X together with a topology $\mathbf{U}$. A subset U of a topological space X is *open* if $U \in \mathbf{U}$. A subset C of X is *closed* if $X - C$ is open. The *closure* of a subset S of X is the intersection $\bar{S}$ of all closed subsets of X containing S. Show that

(a) if C_α $(\alpha \in A)$ is a collection of closed subsets of X, then $\bigcap_{\alpha \in A} C_\alpha$ is a closed subset of X;
(b) if C and D are closed, then $C \cup D$ is closed;
(c) the closure $\bar{S}$ of any subset S of X is closed.

E.0.87 (Krull Topology). Let G be a group and let $\mathbf{N}$ be the set of normal subgroups of G of finite index. Show that $\mathbf{U}_0 = \{Ny \mid N \in \mathbf{N}, y \in G\}$ is a base for G. Show that the Krull topology $\mathbf{U}$ for G (the topology having $\mathbf{U}_0$ as base of open sets) has the property that the closure of a subset S of G is $\bigcap_{N \in \mathbf{N}} NS$.

E.0.88 (Cayley's Theorem). Every group G is isomorphic to a subgroup of the group $\mathbf{S}(G)$ of bijective linear transformations of G (Hint: For $g \in G$, consider the function $L_g\colon G \to G$ defined by $L_g(x) = gx$ for $x \in G$).

E.0.89. Every finite group G is isomorphic to a subgroup of $\mathbf{S}_n$, for any integer $n \geq G{:}\mathbf{1}$. (Hint: Use the preceding exercise, and embed $\mathbf{S}_m$ in $\mathbf{S}_n$ for $n \geq m$).

1 Some elementary field theory

In this chapter, we develop some of the basic theory of fields. We also describe the structure of an arbitrary field extension K/k (see 1.1.5) in terms of an algebraic field extension K/k' (see 1.2) and a purely transcendental field extension k'/k (see 1.6.4, 1.6.13). Throughout the chapter, K denotes a field.

1.1 Preliminaries

If S is a collection of subrings (respectively subfields) of K, then $\bigcap_{E \in \mathbf{S}} E$ is a subring (respectively subfield) of K (see E.0.15). In particular, the intersection $\pi(K)$ of all subfields of K is a subfield of K.

1.1.1 Definition. The subfield $\pi(K)$ of K is called the *prime field* of K. If $K = \pi(K)$, we say that K is a *prime field*.

If n is a positive integer and $x \in K$, we let $nx = x + \cdots + x$ (n times). We let $(-n)x = -(nx)$ and $0x = 0$. The mapping $\varphi : \mathbb{Z} \to \pi(K)$ from the ring $\mathbb{Z}$ of integers into the prime field $\pi(K)$ of K defined by $\varphi(n) = n1$, 1 being the identity element of K, is a homomorphism. The kernel I of φ is a prime ideal since the image of φ, being in the field $\pi(K)$, is an integral domain. Thus, either $I = \{0\}$ and φ is injective, or I is the set of multiples of a prime number p (see E.0.31). In the former case, one shows easily that

$$\{(m1)(n1)^{-1} \mid m, n \in \mathbb{Z}, n \neq 0\}$$

is a subfield of $\pi(K)$ isomorphic to the field $\mathbb{Z}_0 = \mathbb{Q} = \{mn^{-1} \mid m, n \in \mathbb{Z}, n \neq 0\}$ of rational numbers (see E.0.11, E.0.34), and this subfield must be $\pi(K)$ itself. In the latter case, the image subring $\{m1 \mid m \in \mathbb{Z}\}$ is isomorphic to the field $\mathbb{Z}_p = \mathbb{Z}/I$ of p elements, so that $\pi(K) = \{m1 \mid m = 0, 1, \ldots, p - 1\}$ and $\pi(K)$ is isomorphic to $\mathbb{Z}_p$. We state this for future reference.

1.1.2 Proposition. The prime field $\pi(K)$ of a field K is isomorphic to $\mathbb{Z}_0 = \mathbb{Q}$ or to $\mathbb{Z}_p$ for some prime number p.

The above proposition says that the prime fields, up to isomorphism, are the fields $\mathbb{Z}_0 = \mathbb{Q}$ and $\mathbb{Z}_p$, p a prime number.

1.1.3 Definition. The *characteristic* of K is 0 if $\pi(K)$ is isomorphic to $\mathbb{Z}_0$ and is p if $\pi(K)$ is isomorphic to $\mathbb{Z}_p$, p a prime number. The characteristic of K is denoted Char K.

In order to simplify later statements, we introduce an alternative to the notion of characteristic.

1.1.4 Definition. The *exponent characteristic* of K is 1 if $\pi(K)$ is isomorphic to $\mathbb{Z}_0$ and is p if $\pi(K)$ is isomorphic to $\mathbb{Z}_p$, p a prime number. The exponent characteristic of K is denoted Exp Char K.

If K is a field of exponent characteristic p, then the mapping $\pi: K \to K$, defined by $\pi(x) = x^p$, is a homomorphism, since the only nonzero terms in the binomial expansion of $(x + y)^p$ are x^p, y^p (see E.1.2). This homomorphism is called the *Frobenius homomorphism* of K.

Now let K be a field, k a subfield of K. Since K has the structure of additive group and, since products of elements of k with elements of K satisfy the conditions for a scalar product, we may regard K as a vector space over k.

1.1.5 Definition. An *extension field* (or *extension*) of k is a field K containing k as subfield together with the vector space structure over k described above. Such an extension field is denoted K/k or K. The *degree* of an extension K/k is the dimension of K over k. A *subextension* of an extension K of k is an extension k' of k such that k' is a subfield of K.

1.1.6 Definition. If K_1 and K_2 are extension fields of k, a *k-homomorphism/k-isomorphism* from K_1 to K_2 is a homomorphism/isomorphism from K_1 to K_2 which is k-linear (see 0.1.1, E.1.1).

If V is a vector space over K, then V may be regarded as a vector space over the subfield k of K. We then have the following transitivity property of dimension, $V:K$ being the notation for the dimension of V as vector space over K.

1.1.7 Proposition. Let V be a vector space over K, k a subfield of K. Then $(V:K)(K:k) = V:k$. In fact, if $\{e_a \mid a \in A\}$, $\{f_b \mid b \in B\}$ are bases for V over K and K over k respectively, then $\{e_a f_b \mid a \in A, b \in B\}$ is a basis for V over k.

Proof. In this proof, it is to be understood that only finitely many coefficients in a summation are nonzero. With this in mind, let $v = \sum_a v_a e_a$ be a typical element of V, the v_a being in K. Then let $v_a = \sum_b v_{ab} f_b$, the v_{ab} being in k. Then $v = \sum_{a,b} v_{ab} e_a f_b$. Thus, $\{e_a f_b \mid a \in A, b \in B\}$ spans V over k. Next, suppose that $\sum_{a,b} v_{ab} e_a f_b = 0$. Then we have $\sum_a (\sum_b v_{ab} f_b) e_a = 0$, so that $\sum_b v_{ab} f_b = 0$ for all $a \in A$, by the independence of the e_a's. Hence, $v_{ab} = 0$ for all $a \in A$, $b \in B$, by the independence of the f_b's, and the $e_a f_b$ $(a \in A, b \in B)$ are independent over k. This establishes the second assertion. The first follows from the second. □

We now let K/k be a field extension.

1.1.8 Definition. For $S \subset K$, we let $k\langle S\rangle$ be the k-span of S in K, $k[S]$ the intersection of all subrings of K containing k and S and $k(S)$ the intersection of all subfields of K containing k and S.

Obviously $k\langle S\rangle \subset k[S] \subset k(S)$. All three are k-subspaces of K, $k[S]$ is a subring and $K(S)$ a subfield of K.

1.1.9 Proposition. $k\langle S\rangle = \bigcup_{T\in\mathbf{S}} k\langle T\rangle$, $k[S] = \bigcup_{T\in\mathbf{S}} k[T]$ and $k(S) = \bigcup_{T\in\mathbf{S}} k(T)$ where $\mathbf{S}$ is the set of finite subsets of S.

Proof. The containment $\supset$ is clear. Since the right hand side contains S, it suffices for the containment $\subset$ to show that the right hand side is a k-subspace, subring containing k or subfield containing k respectively. But this is obvious, since the finitely many elements of S which one needs to consider form a finite set T. $\square$

It is convenient to introduce an alternate notation when S is finite.

1.1.10 Definition. Let $s_1, \ldots, s_n \in K$. Then $k\langle s_1, \ldots, s_n\rangle = k\langle S\rangle$, $k[s_1, \ldots, s_n] = k[S]$ and $k(s_1, \ldots, s_n) = k(S)$ where $S = \{s_1, \ldots, s_n\}$.

Any extension K of a field k can be built by constructing a finite or transfinite tower $k \subset k(s_1) \subset k(s_1)(s_2) \subset \cdots$ of "simple" extensions, which we now define and describe.

1.1.11 Definition. If $K = k(s)$, then K/k is a *simple extension* of k and s is a *primitive element* of the extension K/k.

To determine the simple extensions, let $K = k(s)$ and let $k[X]$ be the ring of polynomials $g(X)$ in the indeterminant X with coefficients in k. Consider the evaluation homomorphism $\varphi: k[X] \to k[s]$, that is, the mapping sending a polynomial $g(X) = \sum b_i X^i$ in $k[X]$ into the element $g(s) = \sum b_i s^i$ of $k[s]$ (see 0.1.6). The kernel I of φ is a prime ideal, since the image of φ is an integral domain (being contained in the field K). Thus, $I = \{0\}$, in which case no nonzero polynomial in $k[X]$ vanishes at s; or I is the set of multiples of an irreducible polynomial $f_s(X)$ vanishing at s, which we may take to be monic (see 0.1).

Note that every polynomial in $k[X]$ vanishing at s is divisible by $f_s(X)$ and that $f_s(X)$ is unique.

1.1.12 Definition. In the case $I = \{0\}$, s is *transcendental* over k. Otherwise s is *algebraic* over k and $f_s(X)$ is the *minimum polynomial* of s over k.

If s is transcendental over k, one shows easily that $\{g(s)h(s)^{-1} \mid g(X), h(X) \in k[X], h(X) \neq 0\}$ is a subfield of $k(s)$ k-isomorphic to the field $k[X]_0$ of quotients of $k[X]$ (see E.0.10) and this subfield must be $k(s)$ itself. If s is algebraic over k, then the image subring $\{g(s) \mid g(X) \in k[X]\}$ is isomorphic to $k[X]_{f_s} = k[X]/I$. And $k[X]_{f_s}$ is a field since I, being an ideal in a principle ideal domain generated by an irreducible element, is a maximal ideal (see 0.1.3). Thus, $k(s) = \{g(s) \mid g(X) \in k[X]\}$ and $k(s)$ is k-isomorphic to $k[X]_{f_s}$. (The field k is imbedded in $k[X]_{f_s}$ in a natural way, so that we may regard $k[X]_{f_s}$ as an extension of k.) It follows that $k(s) = k[s]$ and that $1, s, \ldots, s^{n-1}$ is a basis for $k(s)$ over k, where n is the degree $\operatorname{Deg} f_s(X)$ of $f_s(X)$. We now have proved the following proposition.

1.1.13 Proposition. Let k be a subfield of K, $s \in K$. If s is transcendental over k, then $k(s)$ is k-isomorphic to the field of quotients $k[X]_0$ of $k[X]$. If s is algebraic over k, then $k(s)$ is k-isomorphic to the quotient ring $k[X]_{f_s}$, $k(s) = k[s] = k\langle 1, s, \ldots, s^{n-1}\rangle$ and $k(s):k = n$ where $n = \operatorname{Deg} f_s(X)$. □

The above proposition shows that the simple extensions of k are, up to k-isomorphism, the extensions $k[X]_0$ and $k[X]_f$, $f(X)$ an irreducible element of $k[X]$.

We finally consider briefly the construction of roots of a polynomial $f(X)$ in $k[X]$. (See E.1.5).

1.1.14 Definition. Given a field k and a polynomial $f(X)$ in $k[X]$, a *root field* of $f(X)$ over k is a field extension K of k of the form $K = k(s)$ where $f(s) = 0$.

1.1.15 Proposition. Let k be a field, $f(X)$ a nonconstant polynomial in $k[X]$. Then $f(X)$ has a root field over k.

Proof. Let $g(X)$ be an irreducible factor of $f(X)$ in $k[X]$. Then $K = k[X]_g$ is a root field for $g(X)$ over k. But any root field for $g(X)$ is a root field for $f(X)$. Thus, $f(X)$ has a root field over k. □

If $f(X)$ is irreducible, root fields of $f(X)$ are unique up to isomorphism, which we now show.

1.1.16 Definition. Let φ be a homomorphism from a field k to a field k'. Then we let φ also denote the extension of φ to the homomorphism $\varphi\colon k[X] \to k'[X]$ such that $\varphi(\sum_0^n a_i X^i) = \sum_0^n \varphi(a_i) X^i$ for $\sum_0^n a_i X^i \in k[X]$ (see E.1.3).

1.1.17 Proposition. Let k be a field, $f(X)$ an irreducible polynomial in $k[X]$. Let $\varphi\colon k \to k'$ be an isomorphism, let K, K' be root fields for $f(X)$ over k and $\varphi(f(X))$ over k' respectively and let s, s' be roots of $f(X)$, $\varphi(f(X))$ in K, K' respectively. Then φ has a unique extension to an isomorphism φ from K to K' such that $\varphi(s) = s'$:

$$\begin{array}{ccc} K & \xrightarrow{\varphi} & K' \\ f(X)\Big| & & \Big|\varphi(f(X)) \\ k & \xrightarrow[\varphi]{} & k' \end{array}$$

Proof. Let s be a root of $f(X)$ in K, s' a root of $\varphi(f(X))$. Then $K = k[s]$, $K' = k'[s']$ and $\varphi(\sum a_i s^i) = \sum \varphi(a_i) s'^i$ defines an extension of φ to an isomorphism from K to K'. (This mapping is well defined since $\varphi\colon k[X] \to k'[X]$ maps I onto I' where I, I' are the kernels of the k-homomorphisms $k[X] \to k[s]$, $k'[X] \to k'[s']$ mapping X to s and X to s' respectively.) The unicity of the extension is obvious. □

1.2 Algebraic extensions

Throughout this section, k denotes a field, K an extension field of k. The following proposition is used later in describing subrings of K containing k.

1.2.1 Proposition. Let A be an integral domain containing k as subfield and suppose that $A:k < \infty$. Then A is a field.

Proof. Let a be a nonzero element of A. Then the left translation a_L of a in A, defined by $a_L(b) = ab$ for $b \in A$, is k-linear. It is injective, since A is an integral domain. Thus, it is surjective and $a_L(x) = 1$ has a solution x. Now $ax = 1$ and a is a unit in A. ▯

1.2.2 Proposition. Let $s \in K$. Then s is algebraic over k if and only if s is contained in some finite dimensional extension k' of k contained in K.

Proof. It suffices to show that s is algebraic over k if and only if $k(s):k$ is finite. But this is true, almost by definition (see 1.1.12). ▯

1.2.3 Proposition. Let S be a finite subset of K. Then the elements of S are all algebraic over k if and only if $k(S):k < \infty$.

Proof. One direction is clear. We prove the other by induction on the number n of elements of S. If $n = 1$, we apply the above proposition directly. If $n > 1$, let $s \in S$. Then $k(s)(S - \{s\}):k(s) < \infty$ and $k(s):k < \infty$, by induction. Since $k(S) = k(s)(S - \{s\})$, it follows from the transitivity of dimensions that $k(S):k < \infty$.

1.2.4 Definition. K is *algebraic* over k (or K/k is *algebraic*) if s is algebraic over k for every $s \in K$.

1.2.5 Proposition. Let S be a set of elements of K which are algebraic over k. Then $k(S) = k[S]$ and $k(S)/k$ is algebraic.

Proof. Let $\mathbf{S}$ be the set of finite subsets of S and let $T \in \mathbf{S}$. Then $k(T):k < \infty$ and $k(T)/k$ is algebraic, by 1.2.3. Now $k(T) = k[T]$, by 1.2.1. It follows that $k(S)/k$ is algebraic and $k(S) = k[S]$, since $k(S) = \bigcup_{T \in \mathbf{S}} k(T) = \bigcup_{T \in \mathbf{S}} k[T] = k[S]$, by 1.1.9. ▯

It follows from the above proposition that the set $k_{\text{alg}} = \{s \in K \mid s \text{ is algebraic over } k\}$ is a subfield of K containing k. Thus, k_{alg} is an algebraic extension of k.

1.2.6 Definition. k_{alg} is the *algebraic closure* of k in K.

We now prove a transitivity theorem for algebraic extensions.

1.2.7 Proposition. Let k' be a subfield of K containing k. Then K/k is algebraic if and only if K/k' and k'/k are algebraic.

Proof. One direction is trivial. Next, suppose that K/k' and k'/k are algebraic and let $s \in K$. Let $f(X) = \sum_0^n a_i X^i$ be the minimum polynomial of s over k' and let $k'' = k(a_0, \ldots, a_n)$. Since k'/k is algebraic, $k'':k < \infty$ by 1.2.3. But $k''(s):k'' < \infty$, since s is algebraic over k''. Thus, $k''(s):k < \infty$ and s is algebraic over k. Thus, K/k is algebraic. ▯

We conclude this section with the following proposition on groups G of automorphisms of K. In the proposition, K^G is the subfield $\{x \in K \mid \sigma(x) = x$

for $\sigma \in G\}$ of fixed points of G in K, Gx is the orbit $\{\sigma(x) \mid \sigma \in G\}$ of x under G for $x \in K$ and $\text{Roots}_K f(X)$ is the set of roots of $f(X)$ in K for $f(X) \in K[X]$.

1.2.8 Proposition. For $f(X) \in K^G[X]$, $\text{Roots}_K f(X)$ is G-stable. An element $x \in K$ is algebraic over K^G if and only if the G-orbit Gx of x in K is finite.

Proof. If $f(X) \in K^G[X]$ and $x \in \text{Roots}_K f(X)$, $\sigma \in G$, then $f(\sigma(x)) = \sigma(f(x)) = \sigma(0) = 0$ and $\sigma(x) \in \text{Roots}_K f(X)$. Thus, $\text{Roots}_K f(X)$ is G-stable. In particular, if x is algebraic over K^G and $f_x(X)$ is the minimum polynomial of x over K^G, then the G-orbit Gx of x is a subset of $\text{Roots}_K f_x(X)$ and is therefore finite. Suppose, conversely, that Gx is finite, and let $Gx = \{x_1, \ldots, x_n\}$ and $f(X) = \prod_1^n (X - x_i)$. For $\sigma \in G$,

$$\sigma(f(X)) = \prod_1^n \sigma(X - x_i) = \prod_1^n (X - \sigma(x_i)) = \prod_1^n (X - x_i) = f(X).$$

Thus, $f(X) \in K^G[X]$. Since x is a root of $f(X)$, x is therefore algebraic over K^G. □

1.3 Splitting fields

Let k be a field, $\mathbf{P} \subset k[X]$, $f(X) \in k[X]$.

1.3.1 Definition. $f(X)$ *splits* in the extension field K of k if $f(X) = c\prod_1^n (X - x_i)$ for suitable $c, x_i \in K$.

1.3.2 Definition. An extension field K of k is a *splitting field* for $\mathbf{P}$ over k if

1. $f(X) \in \mathbf{P} \Rightarrow f(X)$ splits in K;
2. $K = k(\text{Roots}_K \mathbf{P})$ where $\text{Roots}_K \mathbf{P}$ is the set $\{s \in K \mid f(s) = 0$ for some $f(X) \in \mathbf{P}\}$ of roots of $\mathbf{P}$ in K.

An extension field K of k is a *splitting field* for $f(X)$ over k if it is a splitting field for the one point set $\{f(X)\}$ over k.

Note that the second condition simply insures that K is as small an extension of k satisfying condition 1 as possible. It insures in particular that K/k is algebraic, by 1.2.5.

The following proposition follows from the unique factorization property of elements of $K[X]$.

1.3.3 Proposition. Let K be a splitting field for $\mathbf{P}$ over k and let K' be an extension field of K. Then $\text{Roots}_{K'} \mathbf{P} = \text{Roots}_K \mathbf{P}$. □

Splitting fields for $f(X)$ are constructed by induction from root fields for $f(X)$. The splitting fields for $\mathbf{P}$ are constructed by transfinite induction. Isomorphisms between any two such splitting fields are constructed similarly. We now give the details of these constructions.

1.3.4 Proposition. There exists a splitting field for $f(X)$ over k. If φ is an isomorphism from k to a field k' and if K, K' are splitting fields for $f(X)$

over k and $\varphi(f(X))$ over k' respectively, then φ has an extension to an isomorphism $\bar{\varphi}$ from K to K':

$$\begin{array}{ccc} K & \xrightarrow{\bar{\varphi}} & K' \\ f(X)\Big| & & \Big|\varphi(f(X)) \\ k & \xrightarrow[\varphi]{} & k' \end{array}$$

In particular, any two splitting fields for $f(X)$ over k are k-isomorphic.

Proof. The proof is by induction on $n = \operatorname{Deg} f(X)$. For $n = 1$, there is nothing to prove. Next, let $n > 1$ and let k_1 be a root field for $f(X)$ over k. Then $k_1 = k(s_1)$ where $f(s_1) = 0$. Now $f(X) = (X - s_1)f_1(X)$ where $f_1(X) \in k_1[X]$.

$$\begin{array}{ccc} & & K \\ & f_1(X) \diagup & \Big| \\ k_1 & & \Big| f(X) \\ & \diagdown & \Big| \\ & & k \end{array}$$

By induction, $f_1(X)$ has a splitting field K over k_1. Now

$$f_1(X) = c\prod_{i=2}^{n}(X - s_i)$$

for suitable $c, s_i \in K$ and $K = k_1(s_2, \ldots, s_n)$. Thus, $f(X) = c\prod_1^n (X - s_i)$ and $K = k(s_1, \ldots, s_n)$, so that K is a splitting field for $f(X)$ over k. Suppose next that $\varphi: k \to k'$ is an isomorphism and that K, K' are splitting fields for $f(X)$ over k and $\varphi(f(X))$ over k' respectively. Take $s_1, k_1, f_1(X)$ as before. Then $f(X) = g(X)h(X)$ where $g(X)$ is the minimum polynomial of s_1 over k and $h(X) \in k[X]$. Let s_1' be a root of $\varphi(g(X))$ in K'. Then φ has an extension to an isomorphism $\varphi: k(s_1) \to k(s_1')$ such that $\varphi(s_1) = s_1'$, by 1.1.17.

$$\begin{array}{ccc} K & \xrightarrow{\bar{\varphi}} & K' \\ f_1(X)\Big| & & \Big|\varphi(f_1(X)) \\ k_1 & \xrightarrow{\varphi} & \varphi(k_1) \\ \Big| & & \Big| \\ k & \xrightarrow[\varphi]{} & k' \end{array}$$

Now K, K' are splitting fields for $f_1(X)$ over k_1 and $\varphi(f_1(X))$ over $\varphi(k_1)$ respectively, so that φ has the desired extension $\bar{\varphi}$ by induction. $\square$

1.3.5 Theorem. There exists a splitting field K for $\mathbf{P}$ over k. If φ is an isomorphism from k to a field k' and if K, K' are splitting fields for $\mathbf{P}$ over k

and $\varphi(\mathbf{P})$ over k' respectively, then φ has an extension to an isomorphism $\bar{\varphi}$ from K to K'. In particular, any two splitting fields for $\mathbf{P}$ over k are k-isomorphic.

Proof. Suppose first that φ is an isomorphism from k to k' and that K, K' are splitting fields for $\mathbf{P}$ over k and $\varphi(\mathbf{P})$ over k'. For $\mathbf{S} \subset \mathbf{P}$, let $K_{\mathbf{S}} = k(\mathrm{Roots}_K \mathbf{S})$ and $K'_{\mathbf{S}} = k'(\mathrm{Roots}_{K'} \varphi(\mathbf{S}))$. Note that $K_{\mathbf{S}}, K'_{\mathbf{S}}$ are splitting fields for $\mathbf{S}$ over k and for $\varphi(\mathbf{S})$ over k'. Let $A = \{(\mathbf{S}, \alpha) \mid \mathbf{S} \subset \mathbf{P}$ and α is an isomorphism from $K_{\mathbf{S}}$ to $K'_{\mathbf{S}}$ extending $\varphi\}$ and let $\trianglelefteq$ be the partial order on A defined by letting $(\mathbf{S}, \alpha) \trianglelefteq (\mathbf{T}, \beta)$ if $\mathbf{S} \subset \mathbf{T}$ and $\beta|_{K_{\mathbf{S}}} = \alpha$ (see S.4).

It is easily seen that every chain in A has a least upper bound in A, so that A has a maximal element $(\mathbf{S}, \alpha)$, by Zorn's Lemma (see S.4). Suppose that $\mathbf{S} \subsetneqq \mathbf{P}$ and let $f \in \mathbf{P} - \mathbf{S}$. Let $\mathbf{T} = \mathbf{S} \cup \{f\}$ and note that $K_{\mathbf{T}}, K'_{\mathbf{T}}$ are splitting fields for f over $K_{\mathbf{S}}$ and $\varphi(f)$ over $K'_{\mathbf{S}}$ respectively. By 1.3.4, the isomorphism $\alpha: K_{\mathbf{S}} \to K'_{\mathbf{S}}$ can be extended to an isomorphism $\beta: K_{\mathbf{T}} \to K'_{\mathbf{T}}$. But then $(\mathbf{S}, \alpha) \ntrianglelefteq (\mathbf{T}, \beta)$, contradicting the assumption that $(\mathbf{S}, \alpha)$ be maximal. Thus, $\mathbf{S} = \mathbf{P}$. But then $K_{\mathbf{S}} = K$ and $K'_{\mathbf{S}} = K'$. Thus, α is an isomorphism from K to K' extending φ and we may let $\bar{\varphi} = \alpha$.

We next prove by transfinite induction that $\mathbf{P}$ has a splitting field K over k. For this, we take a well ordering $\preceq$ of $\mathbf{P}$ (see S.5). We may assume without loss of generality that $\mathbf{P}$ has a last element (otherwise, the original well ordering could be modified by placing the first element last, so that the second element becomes first, the third element second and so forth). Let $\mathbf{P}_f = \{g \in \mathbf{P} \mid g \preceq f\}$ for $f \in \mathbf{P}$ and note that $\mathbf{P} = \mathbf{P}_f$ when f is the last element of $\mathbf{P}$. It therefore suffices to show that for each $f \in \mathbf{P}$, $\mathbf{P}_f$ has a splitting field over k. If f is the first element of $\mathbf{P}$, then $\mathbf{P}_f = \{f\}$ and $\mathbf{P}_f$ has a splitting field over k, by 1.3.4. Assume next that f is not the first element of $\mathbf{P}$ and that for each $g \precneqq f$ in $\mathbf{P}$, there exists a splitting field K_g for $\mathbf{P}_g$ over k. We assert that $\mathbf{P}_f$ has a splitting field over k. For this, let A be the set of all pairs (g, β) where $g \preceq f$ and where β is a set $\{\beta_{ba} \mid a \precneqq b \precneqq g\}$ of k-isomorphisms $\beta_{ba}: K_a \to K_b$ such that $\beta_{cb} \circ \beta_{ba} = \beta_{ca}$ for all $a \precneqq b \precneqq c \precneqq g$. Let $\trianglelefteq$ be the partial order on A defined by $(g, \beta) \trianglelefteq (h, \gamma)$ if $g \preceq h$ and $\gamma_{ba} = \beta_{ba}$ for all $a \precneqq b \precneqq g$.

For each $(g, \beta) \in A$, the direct limit $\mathrm{Lim}\,\beta$ of the fields K_b $(b \precneqq g)$ (see E.0.43) is a field extension of k of the form $\mathrm{Lim}\,\beta = \bigcup_{b \precneqq g} \beta_b(K_b)$ where the β_b are k-homomorphisms from the K_b into $\mathrm{Lim}\,\beta$ such that the diagram

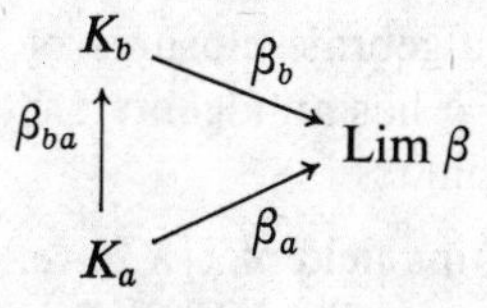

is commutative for $a \precneqq b \precneqq g$. Since K_b is a splitting field for

$$\mathbf{P}_b = \{a \in \mathbf{P} \mid a \preceq b\}$$

for $b \precneqq g$, $\mathrm{Lim}\,\beta = \bigcup_{b \precneqq g} \beta_b(K_b)$ is a splitting field for

$$\mathbf{P}_g - \{g\} = \{b \in \mathbf{P} \mid b \precneqq g\} = \textstyle\bigcup_{b \precneqq g} \mathbf{P}_b.$$

Letting $\bar{K}_g$ be a splitting field for g over $\operatorname{Lim}\beta$, we therefore obtain a splitting field $\bar{K}_g$ for $\mathbf{P}_g$ over k containing $\operatorname{Lim}\beta$. Suppose that $g \precneqq f$ and let g' be the element of $\mathbf{P}$ which immediately follows g. The field extensions K_g and $\bar{K}_g$ are k-isomorphic by the part of 1.3.5 which we have already proved. We may therefore fix a k-homomorphism ψ from $\operatorname{Lim}\beta$ into K_g and define β'_{ba} $(a \precneqq b \precneqq g')$ by $\beta'_{ba} = \psi \circ \beta_a$ $(a \precneqq g)$ and $\beta'_{ba} = \beta_{ba}$ $(a \precneqq b \precneqq g)$. Then $(g', \beta') \in A$, as one sees from the commutative diagram

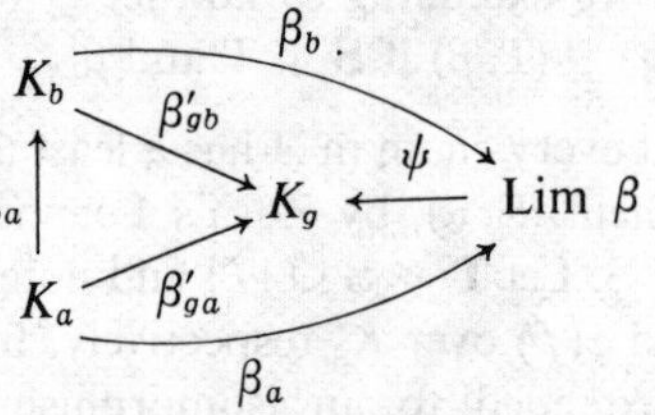

and $(g, \beta) \trianglelefteq (g', \beta')$, $(g,\beta) \neq (g',\beta')$. Each chain (g_i, β_i) $(i \in I)$ in A has a least upper bound $(\bar{g}, \bar{\beta})$ in A. In fact, $\bar{g}$ is the first element of $\mathbf{P}$ such that $g_i \preceq \bar{g}$ for all $i \in I$ and $\bar{\beta}$ is the set of k-homomorphisms $\bar{\beta}_{ba}$ $(a \precneqq b \precneqq \bar{g})$ such that $\bar{\beta}_{ba} = (\beta_i)_{ba}$ for all $a \precneqq b \precneqq g_i$ for all $i \in I$. By Zorn's Lemma, A therefore has a maximal element. We have seen that (g, β) is not maximal if $g \precneqq f$. Thus, A has a maximal element of the form (f, α). But then $\bar{K}_f$ (constructed as was $\bar{K}_g$ earlier) is a splitting field for $\mathbf{P}_f$ over k, which proves our assertion. We have now proved by transfinite induction that $\mathbf{P}_f$ has a splitting field over k for all $f \in \mathbf{P}$. Since $\mathbf{P} = \mathbf{P}_f$ when f is the last element of $\mathbf{P}$, $\mathbf{P}$ has a splitting field over k. ☐

1.4 Algebraic closure

Let k be a field.

1.4.1 Definition. An extension K of k is an *algebraic closure* of k if K/k is algebraic and the only extension K' of K such that K'/k is algebraic is $K' = K$. If k is an algebraic closure of itself, then we say that k is *algebraically closed.*

Note that the algebraically closed fields k are those fields k such that k is the only algebraic extension of k.

1.4.2 Theorem. The algebraic closures of k are the splitting fields of $k[X]$ over k. In particular, k has an algebraic closure and any two algebraic closures of k are k-isomorphic.

Proof. Let K be a splitting field of $k[X]$ over k and let K' be an extension of K such that K'/k is algebraic. Then $K' = \operatorname{Roots}_{K'} k[X] = \operatorname{Roots}_K k[X] = K$, by 1.3.3, and K is an algebraic closure of k. Conversely, let K be an algebraic closure of k. Let K' be a splitting field of $k[X]$ over K. Then K'/K is algebraic, so that K'/k is algebraic. Thus, $K' = K$ and K is a splitting field for $K[X]$ over K. Since K/k is algebraic so that $K = k(\operatorname{Roots}_K k[X])$, K is a splitting field of $k[X]$ over k. ☐

1.4.3 Corollary. k is algebraically closed if and only if each $f(X) \in k[X]$ has the form $f(X) = c \prod_i (X - s_i)$ for suitable $s_i \in k$.

1.4.4 Definition. k_{Alg} denotes an algebraic closure of k.

We now prove a number of simple properties of the algebraic closure.

1.4.5 Proposition. k_{Alg} is algebraically closed.

Proof. Let K' be an algebraic extension of k_{Alg}. Then K'/k is algebraic, by the transitivity of algebraic extensions. Thus, $K' = k_{\mathrm{Alg}}$. It follows that k_{Alg} is algebraically closed. □

1.4.6 Proposition. Let k' be an algebraic extension of k. Then k'_{Alg} is an algebraic closure of k.

Proof. k'_{Alg} is algebraically closed, and k'_{Alg}/k is algebraic by the transitivity of algebraic extensions. □

1.4.7 Corollary. If k' is an algebraic extension of k, then there exists a k-isomorphism from k' to a subfield of k_{Alg}.

Proof. Use the preceding proposition and the proposition that any two algebraic closures of k are k-isomorphic. □

1.4.8 Proposition. If k' is an extension of k contained in k_{Alg}, then k_{Alg} is an algebraic closure of k'.

Proof. k_{Alg} is algebraically closed and k_{Alg}/k' is algebraic. □

We conclude this section with a brief discussion of the group $G = \mathrm{Aut}_k k_{\mathrm{Alg}}$ of k-automorphisms of the algebraic closure k_{Alg} of k. Note that the k-linearity of $\sigma \in G$ is equivalent to the condition $\sigma(s) = s$ for $s \in k$. In fact the set $k_{\mathrm{Alg}}^G = \{s \in k_{\mathrm{Alg}} \mid \sigma(s) = s \text{ for all } \sigma \in G\}$ is an extension field of k. The extensions k_{Alg}^G/k and $k_{\mathrm{Alg}}/k_{\mathrm{Alg}}^G$ play an important role in Galois theory, and we describe them at the end of the section.

1.4.9 Theorem. The set $\mathrm{Roots}\, f(X)$ of roots of $f(X)$ in k_{Alg} is G-stable (see 0.3) for $f(X) \in k[X]$. If $f(X)$ is irreducible, G acts transitively on $\mathrm{Roots}\, f(X)$.

Proof. Let $\sigma \in G$ and $s \in \mathrm{Roots}\, f(X)$. Then we have $f(\sigma(s)) = \sigma(f(s)) = \sigma(0) = 0$, since $f(X) \in k[X]$ (see E.1.3). Thus, $\mathrm{Roots}\, f(X)$ is G-stable. Suppose next that $f(X)$ is irreducible and let $s, s' \in \mathrm{Roots}\, f(X)$. Then there exists a k-isomorphism σ_1 from $k(s)$ to $k(s')$ such that $\sigma_1(s) = s'$, by 1.1.17. But then σ_1 has an extension to an isomorphism σ from the splitting field k_{Alg} of $k[X]$ over $k(s)$ to the splitting field k_{Alg} of $k[X]$ over $k(s')$, by 1.3.5. Now $\sigma \in G$ and $\sigma(s) = s'$. □

1.4.10 Corollary. The orbits of G in k_{Alg} are finite. □

The following proposition is closely related to the above proposition and corollary.

1.4.11 Proposition. Let K be a finite dimensional extension of k. Then the set $\mathrm{Hom}_k(K, k_{\mathrm{Alg}})$ of k-homomorphisms from K into k_{Alg} is finite.

Proof. Let $s_1, \ldots, s_n$ be a basis for K/k. Then for $\sigma \in \mathrm{Hom}_k(K, k_{\mathrm{Alg}})$, we have $\sigma(s_i) \in \mathrm{Roots}_{k_{\mathrm{Alg}}} f_{s_i}(X)$ for $1 \leq i \leq n$, as in 1.4.9. But σ is completely determined by its values at $s_1, \ldots, s_n$ and, by what we have just seen, there are only finitely many possibilities for these values. Thus, $\mathrm{Hom}_k(K, k_{\mathrm{Alg}})$ is finite. □

1.4.12 Proposition. $k_{\mathrm{Alg}}^G = \{s \in k_{\mathrm{Alg}} \mid f_s(X) \text{ has only one root in } k_{\mathrm{Alg}}\}$, where $f_s(X)$ denotes the minimum polynomial of s over k for $s \in k_{\mathrm{Alg}}$.

Proof. This follows directly from 1.4.9. □

1.4.13 Proposition. For $s \in k_{\mathrm{Alg}}$, the minimum polynomial $f(X)$ of s over k_{Alg}^G has distinct roots and Roots $f(X)$ is the orbit of s under G (see 0.3).

Proof. Let $\{s_1, \ldots, s_n\}$ be the orbit of s under G, and let $g(X) = \prod_1^n (X - s_i)$. For $\sigma \in G$, we then have

$$\sigma(g(X)) = \prod_1^n \sigma(X - s_i) = \prod_1^n (X - \sigma(s_i)) = \prod_1^n (X - s_i) = g(X),$$

so that $g(X) \in k_{\mathrm{Alg}}^G[X]$ (see E.1.3). Since $g(s) = 0$, it follows that $f(X)$ divides $g(X)$. And, since Roots $f(X)$ is G-stable and contains s, each s_i is a root of $f(X)$. Thus, $f(X) = g(X)$. □

The above two propositions say, in the language of the next chapter, that k_{Alg}^G/k is a radical extension and that $k_{\mathrm{Alg}}/k_{\mathrm{Alg}}^G$ is a separable extension.

1.5 Finite fields

We now give a brief account of finite fields and their algebraic extensions, in terms of the preceding two sections. Throughout the section, p denotes a prime number and m, n denote positive integers.

Suppose that k is a field of dimension m over a prime field π of p elements. Then the number of elements of k is p^m. Thus, the multiplicative group $k^* = k - \{0\}$ of units of k has order $p^m - 1$ and $s^{p^m - 1} = 1$ for all $s \in k^*$. Thus, $s^{p^m} = s$ for all $s \in k$. Since $X^{p^m} - X$ has at most p^m roots in k_{Alg}, it follows that $k = \mathrm{Roots}_{k_{\mathrm{Alg}}}(X^{p^m} - X)$. Consequently, k is a splitting field for $X^{p^m} - X$ over π.

Suppose conversely that k is a splitting field for $X^{p^m} - X$ over the field $\mathbb{Z}_p$ of p elements. Since $s \mapsto s^{p^m}$ is a homomorphism on k, $\mathrm{Roots}_k(X^{p^m} - X)$ is a subfield of k. Thus, $k = \mathrm{Roots}_k(X^{p^m} - X)$. Now $X^{p^m} - X$ has no multiple roots, since $X^{p^m} - X$ and its derivative $p^m X^{p^m - 1} - X = -X$ are relatively prime (see E.1.8). Thus, $X^{p^m} - X = \prod_1^{p^m} (X - s_i)$ where the $s_1, \ldots, s_{p^m}$ are distinct elements of the splitting field k. Thus, $k = \{s_1, \ldots, s_{p^m}\}$ and k has p^m elements. From the preceding paragraph, it follows that $k : \mathbb{Z}_p = m$.

We have now established the following characterization of finite fields.

1.5.1 Proposition. If k is a field extension of dimension m of a prime field π of p elements, then k has p^m elements, k is the splitting field of $X^{p^m} - X$ over π and $k = \text{Roots}_{k_{\text{Alg}}}(X^{p^m} - X)$. Conversely, if p is a prime and m a positive integer, the splitting field of $X^{p^m} - X$ over $\mathbb{Z}_p$ is a field extension of dimension m of $\mathbb{Z}_p$.

1.5.2 Corollary. The number of elements of a finite field of characteristic p is p^m for some m. For every m, there exists a field of characteristic p having p^m elements, and any two fields of p^m elements are isomorphic.

Proof. If k is a finite field of characteristic p, then k has p^m elements where $m = k:\pi(k)$. Thus, a field has p^m elements if and only if it is the splitting field of $X^{p^m} - X$ over a prime field π of p elements. In particular, any two fields of p^m elements are isomorphic, by the theorem on isomorphisms of splitting fields. □

1.5.3 Corollary. Let k be a finite field and let $s \in k_{\text{Alg}}$. Then the roots in k_{Alg} of the minimum polynomial $f_x(X)$ of s over k are distinct.

Proof. Let the number of elements of $k(s)$ be p^m. Then s is a root of $X^{p^m} - X$, so that $f_s(X)$ divides $X^{p^m} - X$. But $X^{p^m} - X$ has distinct roots in k_{Alg}, as was noted in the proof of 1.5.1. Thus, $f_s(X)$ has distinct roots in k_{Alg}. □

Next, let k be a finite field of $q = p^m$ elements. If K is a field extension of k of dimension n, then K has q^n elements and $K = \text{Roots}_{K_{\text{Alg}}}(X^{q^n} - X)$, by 1.5.1, so that K is the splitting field of $X^{q^n} - X$ over k. Conversely, any splitting field of $X^{q^n} - X$ over k is a field extension of k of dimension n, by 1.5.1. We now have the following generalization of 1.5.1, in view of the theorem of isomorphism of splitting fields over k.

1.5.4 Proposition. Let k be a field of $q = p^m$ elements. Then there exists a field extension K of k of dimension n, namely any splitting field of $X^{q^n} - X$ over k. Moreover, any two extension fields of k of dimension n are k-isomorphic.

Since any algebraic extension of k is a union of finite dimensional extensions of k, the above proposition yields a description of all algebraic extensions of a finite field k. In particular, every nonzero element $s \in k_{\text{Alg}}$ is a root of unity, that is, satisfies an equation $s^d = 1$ for some positive integer d, since this is true of the non-zero elements of any finite dimensional extension K of k.

We conclude this section with a brief account of the group $\text{Aut}_k K$ of k-automorphisms of a finite dimensional extension field K of a finite field k. For this, let the number of elements of k be $q = p^m$ and let the number of elements of K be q^n. The pth power homomorphism $\sigma: K \to K$, defined by $\sigma(s) = s^p$ for $s \in K$, is injective (as a homomorphism), hence surjective since K is finite. Thus, $\sigma \in \text{Aut}\, K$. By 1.5.1, k is the set of fixed points of $\tau = \sigma^m$ and the order of τ is n. We claim that $\text{Aut}_k K$ is the cyclic group generated by τ. For this, we need the corollary to the following proposition.

1.5.5 Proposition. Let A be a finite subgroup of the group L^* of units of a field L. Then A is cyclic.

Proof. Let m be the exponent of A (see 0.2.4). Then $A \subset \text{Roots}_L(X^m - 1)$ so that A has at most m elements. But A has a cyclic subgroup of order m, by 0.2.4. Thus, A is cyclic of order m. □

1.5.6 Corollary. K^* is cyclic and K/k is a simple extension.

Proof. Since K^* is finite, it is cyclic by 1.5.5. Let K^* be generated by s, we have $K = k(s)$. □

1.5.7 Proposition. $\text{Aut}_k K$ is the cyclic group generated by τ and $k = K^{\text{Aut}_k K}$.

Proof. Let $K = k(s)$. Then the minimum polynomial $f_s(X)$ of s over k has the form $f_s(X) = \prod_1^n (X - s_i)$. Now each $\sigma \in \text{Aut}_k K$ maps s into $\sigma(s) = s_i$ for some i and σ is completely determined by $\sigma(s)$. Thus, $\text{Aut}_k K$ has at most n elements. But $1, \tau, \ldots, \tau^{n-1}$ are n distinct elements of $\text{Aut}_k K$. Thus, $\text{Aut}_k K$ is the cyclic group generated by τ. In particular,

$$K^{\text{Aut}_k K} = \{s \in K \mid \tau(s) = s\} = \text{Roots}_K(X^q - X) = k. \qquad \square$$

1.6 Transcendency basis of a field extension

Let K be an extension field of k. A subset S of K is algebraically independent over k if no element s of S is algebraic over $K(S - \{s\})$ (see 1.6.3). An extension field k' of k is purely transcendental if $k' = k(S)$ for some algebraically independent subset S of k'. In this section, we show that any field extension K/k can be broken up into an algebraic extension K/k' and a purely transcendental extension k'/k by suitably choosing a subfield k' of K containing k (see 1.6.13). We also give a simple description of all purely transcendental extensions of k (see 1.6.6).

1.6.1 Definition. An element x of K is *algebraically dependent* on a subset S of K over k (written $x \prec S$) if x is algebraic over $k(S)$. If x is not algebraically dependent on S over k, we write $x \not\prec S$. A subset S of K is algebraically dependent on a subset T of K over k (written $S \prec T$) if $s \prec T$ for all $s \in S$.

1.6.2 Theorem. Let $x \in K$, $S \subset K$ and $T \subset K$. Then:

1. $s \prec S$ for all $s \in S$;
2. if $x \prec S$ and $S \prec T$, then $x \prec T$;
3. if $x \prec S$, then $x \prec S^0$ for some finite subset S^0 of S;
4. if $x \prec S$ and $x \not\prec S - \{s\}$, then $s \prec (S - \{s\}) \cup \{x\}$.

Proof. (1) is clear since $s \in k(S)$ for all $s \in S$. For (2), simply note that if x is algebraic over $k(S)$ and s is algebraic over $k(T)$ for all $s \in S$, then x is algebraic over $k(T)$ by 1.2.7. For (3), let x be algebraic over $k(S)$ and let $f(X) = \sum_{i=0}^n a_i X^i$ be a monic polynomial in X with coefficients in $k(S)$ such that $f(x) = 0$. Since $k(S) = \bigcup_{S^0 \in \mathbf{S}} k(S^0)$ where $\mathbf{S}$ is the set of finite subsets of S (see 1.1.9), we can find $S^0 \in \mathbf{S}$ such that $a_0, \ldots, a_n \in k(S^0)$. Then $x \prec S^0$. For (4), suppose that x is algebraic over $k(S)$ but x is not algebraic over k' where $k' = k(S - \{s\})$. Then there exists a nonzero polynomial $f(X, s) = \sum_{i=0}^{m} a_i(s)X^i$ in X with coefficients $a_i(s)$ in $k'(s) = k(S)$ such that $f(x, s) = 0$, and we may take the $a_i(s)$ to be polynomials in s over k' (by clearing denominators). Then $f(X, s) = \sum_0^n b_j(X)s^j$ for suitable polynomials $b_j(X)$ in X with coefficients in k' where $b_n(X) \neq 0$. Since x is not algebraic over k', we have $b_n(x) \neq 0$. But $0 = f(x, s) = \sum_0^n b_j(x)s^j$. Thus, s is algebraic over $k'(x)$ and $s \prec (S - \{s\}) \cup \{x\}$. □

1.6.3 Definition. A subset S of K is *algebraically independent* over k if $s \not\prec S - \{s\}$ for all $s \in S$.

1.6.4 Definition. K/k is *purely transcendental* if $K = k(S)$ for some algebraically independent subset S of K over k.

1.6.5 Theorem. A subset S of K is algebraically independent over k if and only if for any distinct elements $s_1, \ldots, s_n$ of S, the monomials $s_1^{e_1} \cdots s_n^{e_n}$ ($e_1, \ldots, e_n$ nonnegative integers) are distinct and linearly independent over k.

Proof. Suppose that S is not algebraically independent over k and choose $s \in S$ such that $s \prec S - \{s\}$. Then there exist $s_1, \ldots, s_n$ in $S - \{s\}$ such that s is algebraic over $k(s_1, \ldots, s_n)$ (see 1.6.2). But then one sees easily that the monomials $s_1^{e_1} \ldots s_n^{e_n} s^e$ are not distinct and linearly independent over k. Suppose conversely that S has distinct elements $s_1, \ldots, s_n$ such that $\sum a_{e_1 \cdots e_n} s_1^{e_1} \cdots s_n^{e_n} = 0$ is a nontrivial linear combination of the monomials $s_1^{e_1} \cdots s_n^{e_n}$. Then it follows that s_i is algebraic over $k(s_1, \ldots, s_{i-1}, s_{i+1}, \ldots, s_n)$ and $s_i \prec S - \{s_i\}$ for some i. □

1.6.6 Corollary. For any set S and field k, there is a purely transcendental extension K of k containing S such that S is algebraically independent over k and $K = k(S)$. The monomials $s_1^{e_1} \cdots s_n^{e_n}$ ($s_1, \ldots, s_n$ being distinct elements of S and $e_1, \ldots, e_n$ being nonegative integers) are distinct and form a k-basis for $k[S]$, and $k(S)$ is isomorphic to the field of quotients of $k[S]$ (see 0.1 and E.0.10). If $K_1 = k(S_1)$ and $K_2 = k(S_2)$ are purely transcendental extensions of k and if S_1 and S_2 have the same number of elements (cardinality) and are algebraically independent over k, then there is a k-isomorphism f from $K_1 = k(S_1)$ to $K_2 = k(S_2)$ which maps S_1 bijectively to S_2 and $k[S_1]$ isomorphically to $k[S_2]$.

Proof. Let $\hat{S}$ be the free Abelian monoid on S (see E.0.55), so that $\hat{S}$ consists of the monomials $s_1^{e_1} \cdots s_n^{e_n}$ ($s_1, \ldots, s_n$ being distinct elements of S and $e_1, \ldots, e_n$ being nonnegative integers), two such monomials $s_1^{e_1} \cdots s_n^{e_n}$ and $s_1^{f_1} \cdots s_n^{f_n}$ are equal if and only if $e_i = f_i$ for $1 \leq i \leq n$ and $s_1^{e_1} \cdots s_n^{e_n} s_1^{f_1} \cdots s_n^{f_n} = s_1^{e_1+f_1} \cdots s_n^{e_n+f_n}$ for any two such monomials. Let $k[S]$ be a vector space over k with basis $\hat{S}$, so that an element f of $k[S]$ is a linear combination $f = \sum a_{e_1 \cdots e_n} s_1^{e_1} \cdots s_n^{e_n}$ (all but finitely many coefficients being 0) and the coefficients $a_{e_1 \cdots e_n}$ of f are uniquely determined by f. The multiplication in $\hat{S}$ can be extended uniquely to a commutative multiplication in

$k[S]$ such that $f(ag + bh) = afg + bfh$ for $a, b \in k$ and $f, g \in \hat{S}$. Together with this multiplication and vector space addition, $k[S]$ is an integral domain (see 0.1) ($k[S]$ is also the monoid k-algebra of the monoid $\hat{S}$ in the sense of E.0.58). Letting K be the field of quotients of $k[S]$ (see 0.1 and E.0.10) and identifying each $f \in k[S]$ with $f/1 \in K$ and each $a \in k$ with $a1_K \in K$ where 1_K is the identity of K, so that $S \subset K$ and K is a field extension of k, one shows easily that $K = k(S)$ and S is algebraically independent over k. The remaining assertions about the $s_1^{e_1} \cdots s_n^{e_n}$ and about $k(S)$ and $k[S]$ are clear from the construction. Now suppose that $K_1 = k(S_1)$ and $K_2 = k(S_2)$ are purely transcendental extensions of k and that S_1 and S_2 are algebraically independent subsets of K_1 and K_2 respectively over k having the same number of elements (cardinality). Let f be a bijective function from S_1 to S_2. By 1.6.5, we can define $f(s_1^{e_1} \cdots s_n^{e_n}) = f(s_1)^{e_1} \cdots f(s_n)^{e_n}$ for $s_1, \ldots, s_n$ in S. And by 1.6.5, we can further extend f to a bijective k-linear mapping from $k[S_1]$ (k-span of the $s_1^{e_1} \cdots s_n^{e_n}$ with $s_1, \ldots, s_n \in S_1$) to $k[S_2]$ (k-span of the $t_1^{e_1} \cdots t_n^{e_n}$ with $t_1, \ldots, t_n \in S_2$). Since $K_1 = k(S_1)$ (a field of quotients of $k[S_1]$) and $K_2 = k(S_2)$ (a field of quotients of $k[S_2]$), f can further be extended to a k-isomorphism from K_1 to K_2 (see 0.1 and E.0.10). □

1.6.7 Definition. If $K = k(S)$ and $R = k[S]$ where S is algebraically independent over k, we say that K is a *field of rational expressions* and R a *ring of polynomials* in the algebraically independent elements s of S.

1.6.8 Proposition. Let $S \subset T \subset K$ and suppose that T is algebraically independent over k and $T \prec S$. Then $S = T$.

Proof. Let $t \in T$. Since $S \subset T$ and $t \not\prec T - \{t\}$, we have $t \not\prec S - \{t\}$. Since $t \prec S$, it follows that $t \in S$. Thus, $T \subset S$ and $S = T$. □

1.6.9 Proposition. Let $x \in K$ and $S \subset K$. Suppose that S is algebraically independent over k and $x \not\prec S$. Then $S \cup \{x\}$ is algebraically independent.

Proof. Suppose that $y \in S \cup \{x\}$ and $y \prec (S \cup \{x\}) - \{y\}$. Since $x \not\prec S$, we have $y \neq x$. Thus, $y \in S$ and $y \prec (S - \{y\}) \cup \{x\}$. Since $x \not\prec S - \{y\}$, it follows from 1.6.2 that $x \prec S$, a contradiction. □

1.6.10 Proposition. Let S_a ($a \in A$) be a collection of subsets of K such that $S_a \subset S_b$ or $S_b \subset S_a$ for every pair of elements $a, b \in A$. Suppose that S_a is algebraically independent over k for all $a \in A$. Then $S = \bigcup_{a \in A} S_a$ is algebraically independent over k.

Proof. Suppose that $s \in S$ and $s \prec S - \{s\}$. Then $s \prec S^0$ for some finite subset S^0 of $S - \{s\}$. We can choose $a \in A$ such that $S^0 \subset S_a$ and $s \in S_a$, by the inclusion condition on the S_a, S_b with $a, b \in A$. Then $s \prec S_a - \{s\}$ and S_a is not algebraically independent over k, a contradiction. Thus, $s \not\prec S - \{s\}$ for $s \in S$ and S is algebraically independent over k. □

1.6.11 Theorem. For $S \subset K$, S has a subset S^0 such that S^0 is algebraically independent over k and $S \prec S^0$.

Proof. By 1.6.10 and a straightforward application of Zorn's Lemma (see S.4), S has a maximal subset S^0 algebraically independent over k. Suppose that $s \in S$ and $s \not\prec S^0$. Then $S^0 \cup \{s\}$ is algebraically independent over k by 1.6.9, contradicting the maximality of S^0. Thus, $S \prec S^0$. ☐

1.6.12 Definition. A *transcendency basis* for K/k is a subset S of K algebraically independent over k such that $K/k(S)$ is algebraic.

1.6.13 Theorem. *K/k has a transcendency basis S. If S and T are transcendency bases for K/k, then S and T have the same number of elements (cardinality).*

Proof. By 1.6.11, K has a subset S algebraically independent over k such that $K \prec S$. Then $K/k(S)$ is algebraic and S is a transcendency basis for K/k. Next, let S and T be transcendency bases for K/k. To show that S and T have the same number of elements (cardinality), we follow the proof of the invariance of dimension of a vector space. Suppose first that S and T are infinite. Then for each $s \in S$, there exists a finite subset T_s of T such that $s \prec T_s$. Since $K \prec S$ and $S \prec \bigcup_{s \in S} T_s$, we have $K \prec \bigcup_{s \in S} T_s$ and $T \prec \bigcup_{s \in S} T_s$, so that $T = \bigcup_{s \in S} T_s$ by 1.6.8. Similarly, $S = \bigcup_{t \in T} S_t$ where the S_t are finite subsets of S. Since S and T are infinite, it follows that S and T have the same cardinality. Suppose next that, say, S is finite. We prove by induction on the number i of elements of $S - T$ that S and T have the same number of elements. If $S \subset T$, then $S = T$ since $T \prec S$ (see 1.6.8). Thus, the assertion is true if $i = 0$. Next, let $i > 0$ so that $S \not\subset T$. Take $s \in S - T$. Then $t \not\prec S - \{s\}$ for some $t \in T$ (otherwise $T \prec S - \{s\}$ and $S \prec S - \{s\}$ since $S \prec T$, contradicting 1.6.8). For such a t, $S' = (S - \{s\}) \cup \{t\}$ is independent (see 1.6.9) and $s \prec S'$ (see 1.6.2). Thus, $S \prec S'$ and S' is a transcendency basis for K/k. Since the number of elements of $S' - T$ is $i - 1$, S' and T have the same number of elements by induction. Thus, S and T have the same number of elements. ☐

1.6.14 Definition. The *transcendency degree* of K/k is the number of elements of a transcendency basis for K/k.

E.1 Exercises to Chapter 1

E.1.1. Let K and L be field extensions of a field k and let $f: K \to L$ be an isomorphism of fields. Show that f is linear over k if and only if $f(a) = a$ for all $a \in k$.

E.1.2 (Frobenius Homomorphism). Let K be a field of exponent characteristic p. Show that the mapping $\pi: K \to K$ defined by $\pi(x) = x^p$ is a homomorphism. (Consider the binomial expansion of $(x + y)^p$ and show that p divides the binomial coefficients of the unwanted terms).

E.1.3. Let φ be a homomorphism from a field k to a field k'. Define $\varphi: k[X] \to k'[X]$ by $\varphi(\sum_0^n a_i X^i) = \sum_0^n \varphi(a_i) X^i$ for $\sum_0^n a_i X^i \in k[X]$. Show that $\varphi: k[X] \to k'[X]$ is a homomorphism.

E.1.4 (Euclidean Algorithm). Let $f(X), g(X)$ be elements of $k[X]$, k being a field, and let $\operatorname{Deg} f(X) \geq 1$. Show that $g(X) = m(X)f(X) + r(X)$ for suitable $m(X), r(X) \in k[X]$ such that $\operatorname{Deg} r(X) < \operatorname{Deg} f(X)$.

E.1.5 (Roots). Let $g(X)$ be a polynomial with coefficients in a field k and let K be an extension field of k. Then an element $a \in K$ is a *root* of $f(X)$ if $g(a) = 0$. Show that if a is a root of $g(X)$ in K, then $g(X) = m(X)(X - a)$ for some $m(X) \in k[X]$. (Use the Euclidean Algorithm for $k[X]$).

E.1.6 (Partial Fraction Decomposition). Let k be a field and let $f(X), g(X)$ be nonzero elements of $k[X]$ such that $\operatorname{Deg} f(X) < \operatorname{Deg} g(X)$. Show that

(a) if $g(X) = a(X)b(X)$ where $a(X), b(X)$ are relatively prime elements of $k[X]$, then $f(X)/g(X) = m(X)/a(X) + n(X)/b(X)$ where $\operatorname{Deg} m(X) < \operatorname{Deg} a(X)$ and $\operatorname{Deg} n(X) < \operatorname{Deg} b(X)$ (consider the equation $f(X) = n(X)a(X) + m(X)b(X)$);

(b) if $g(X) = c \prod_1^n g_i(X)^{e_i}$ where the $g_i(X)$ are distinct monic irreducible elements of $k(X)$ and $c \in k$, then

$$\frac{f(X)}{g(X)} = \sum_0^n \frac{f_i(X)}{g_i(X)^{e_i}}$$

for suitable $f_i(X)$ such that $\operatorname{Deg} f_i(X) < \operatorname{Deg} g_i(X)^{e_i}$ for $1 \leq i \leq n$ (use part (a) inductively).

E.1.7 (Differentiation). Let k be a field and let $f(X) = \sum_0^n a_i X^i$ be an element of $k[X]$. Define $f(X)' = \sum_0^n i a_i X^{i-1}$ (the formal *derivative* of $f(X)$ with respect to X). Show that $(cf(X))' = cf(X)'$, $(f(X) + g(X))' = f(X)' + g(X)'$ and $(f(X)g(X))' = f(X)'g(X) + f(X)g(X)'$ for all $f(X), g(X) \in k[X]$ and $c \in k$.

E.1.8 (Multiple Roots). Show that if an element $f(X)$ in the polynomial ring $k[X]$ over the field k is relatively prime to $f(X)'$, then $f(X)$ has no multiple roots in a splitting field for $f(X)$ over k.

E.1.9 (Automorphisms of $\mathbb{R}$). Let σ be an automorphism of the field $\mathbb{R}$ of real numbers. Show that $\sigma = id_{\mathbb{R}}$ by showing that

(a) $\sigma(a) = a$ for $a \in \mathbb{Q}$;

(b) a is positive if and only if $\sigma(a)$ is positive for $a \in \mathbb{R}$;

(c) $a < b$ if and only if $\sigma(a) < \sigma(b)$ for $a, b \in \mathbb{R}$;

(d) if $a_1, a_2, \ldots$ is a sequence in $\mathbb{Q}$ converging to a, then $\sigma(a) = a$.

E.1.10 (Quadratic Extensions). A *quadratic extension* is a field extension of degree 2. Show that if K/k is a quadratic extension of characteristic not 2, then

(a) K/k has a basis $1, y$ such that $y^2 \in k$;

(b) $\operatorname{Aut}_k K$ consists of id_K and σ where σ is the *conjugation* $\sigma(a + by) = a - by$ $(a, b \in k)$.

Show that if K/k is a quadratic extension of characteristic 2, then $\operatorname{Aut}_k K$ has order 1 or 2 and

(c) if $\operatorname{Aut}_k K$ has order 1, K has a basis $1, y$ such that $y^2 \in k$;

(d) if $\operatorname{Aut}_k K$ has order 2, then K has a basis $1, y$ such that $y^2 - y \in k$.

E.1.11. Show that the splitting field of an irreducible polynomial $f(X)$ in $k[X]$ of degree 2 is a quadratic extension.

E.1.12. Construct a splitting field over $\mathbb{Q}$ for

(a) $X^2 - 5$;
(b) $X^2 - p$, p being any prime;
(c) $X^4 - 8X^2 + 15$;
(d) $X^4 + X + 1$.

E.1.13. Let p be a prime number and let $a \neq 1$ be a pth root of unity in $\mathbb{C}$, that is, a root in $\mathbb{C}$ of $X^p - 1$. Show that

(a) the minimum polynomial of a over $\mathbb{Q}$ is $X^{p-1} + X^{p-2} + \cdots + X + 1$;
(b) $\mathbb{Q}(a)$ is a splitting field for $X^p - 1$ over $\mathbb{Q}$.

E.1.14. Let K/k be a field extension and let $K = k(a)$. Suppose that K contains n distinct roots of the minimum polynomial $f(X)$ of a over k, where n is the degree of K/k. Show that

(a) $G = \operatorname{Aut}_k K$ acts transitively on the set $\operatorname{Roots}_K f(X)$ of roots of $f(X)$ in K;
(b) $K^G = k$ where $K^G = \{x \in K \mid \sigma(x) = x \text{ for } \sigma \in G\}$.

E.1.15. Let K/k be a field extension of characteristic $p > 0$ and let $K = k(a)$ where $a^p \in k$. Show that

(a) $\operatorname{Aut}_k K = \mathbf{1}$ (has only one element);
(b) the minimum polynomial of a is $X^p - a^p$ if $a \notin k$ and $X - a$ if $a \in k$.

E.1.16. Let k be a field of characteristic $p > 0$ and let $b \in k$. Show that either $X^p - b$ is irreducible in $k[X]$ or $b = a^p$ and $X^p - b = (X - a)^p$ for some $a \in k$.

E.1.17. Prove that the algebraic closure of a countable field is countable.

E.1.18. Describe a field extension K/k which is not purely transcendental but is transcendental in the sense that each element $x \in K - k$ is transcendental.

E.1.19 (Transcendency Degree). Let L/K and K/k be field extensions with transcendency bases T and S respectively. Show that

(a) $S \cup T$ is a transcendency basis for L/k;
(b) the transcendency degree of L/k is the sum of the transcendency degrees of L/K and K/k.

E.1.20 (Abstract Dependence Relations). Let K be any set. Suppose that for any element x of K and any subset S of K, it is specified whether $x \prec S$ is true or false, and suppose that the following conditions hold for $x \in K$, $S \subset K$ and $T \subset K$:

1. $s \prec S$ for all $s \in S$;
2. if $x \prec S$ and $S \prec T$, then $x \prec T$ (where $S \prec T$ means that $s \prec T$ for all $s \in S$);
3. if $x \prec S$, then $x \prec S^0$ for some finite subset S^0 of S;
4. if $x \prec S$ and $x \not\prec S - \{s\}$, then $s \prec (S - \{s\}) \cup \{x\}$.

Then $\prec$ is called a *dependence relation* between the elements of K and the subsets of K. A subset S of K is *independent* if $s \nprec S - \{s\}$ for all $s \in S$. A *basis* for K is an independent subset S of K such that $K \prec S$. Following the format of 1.6, prove that K has a basis and that any two bases of K have the same cardinality. Use this to prove the Basis Theorem for finite and infinite dimensional vector spaces.

E.1.21 (Lüroth Polynomial). Let K/k be a field extension, x a transcendental element of K over k. Let $k(x)[X]$ be the polynomial ring in an indeterminant X over the field $k(x)$. Let $y \in k(x) - k$ and $y = u(x)/v(x)$ where $u(x), v(x)$ are relatively prime elements of $k[x]$. Let $P(X) = yv(X) - u(X)$. Show that

(a) $P(X) \neq 0$ and $P(x) = 0$, so that x is algebraic over $k(y)$;
(b) y is transcendental over k;
(c) if $yv(X) - u(X) = Q(X)R(X)$ where $Q(X) \in k[X]$ and $R(X) \in k[y][X]$, then $Q(X) \in k$ (otherwise, try replacing X by a root of $Q(X)$ in the algebraic closure of k);
(d) $P'(X) = u(x)v(X) - v(x)u(X)$ is primitive in $k[x][X]$ (consider the equation $u(x) - v(x)z = d(x)(R(X)/v(X))$ where $z = u(X)/v(X)$ and compare with (c)).

E.1.22 (Resultant). Let k be a field and let $f(X) = \sum_0^m a_i X^i$, $g(X) = \sum_0^n b_j X^j$ where either $a_m \neq 0$ or $b_n \neq 0$. Show that $f(X)$ and $g(X)$ have a common factor of positive degree if and only if there exist $n(X), m(X) \in k[X]$ (not both 0) such that $n(X)f(X) = m(X)g(X)$, $\operatorname{Deg} m(X) < n$ and $\operatorname{Deg} m(X) < m$. Letting $n(X) = \sum_0^{n-1} n_j X^j$ and $m(X) = \sum_0^{n-1} m_i X^i$ where the n_j, m_i are elements of k (not all zero), describe the $n + m$ linear equations in the n_j, m_i which are satisfied if and only if $n(X)f(X) = m(X)g(X)$. Conclude that $f(X)$ and $g(X)$ have a common factor of positive degree if and only if the *resultant* determinant

$$R(f,g) = \begin{vmatrix} a_m & a_{m-1} & \cdots & a_0 & & & & \\ & a_m & a_{m-1} & \cdots & a_0 & & & \\ & & \ddots & \ddots & & \ddots & & \\ & & & a_m & a_{m-1} & \cdots & & a_0 \\ b_n & b_{n-1} & \cdots & b_1 & b_0 & & & \\ & b_n & b_{n-1} & \cdots & b_1 & b_0 & & \\ & & \ddots & \ddots & & \ddots & \ddots & \\ & & & b_n & b_{n-1} & \cdots & b_1 & b_0 \end{vmatrix}$$

of $f(X)$ and $g(X)$ vanishes.

E.1.23. Interpret the vanishing of the determinant

$$\begin{vmatrix} 1 & 0 & -1 & 0 & 0 \\ 0 & 1 & 0 & -1 & 0 \\ 0 & 0 & 1 & 0 & -1 \\ 1 & 3 & 3 & 1 & 0 \\ 0 & 1 & 3 & 3 & 1 \end{vmatrix}$$

in the context of the preceding problem.

E.1.24 (Symmetric Polynomials). Let k be a field and let $R = k[X_1, .,, X_n]$ be the polynomial ring in n variables over k. Let G be the symmetric group on the set $\{X_1, \ldots, X_n\}$ and define $\sigma(f(X))$ for $\sigma \in G$ and $f(X) = \sum a_{i_1 \cdots i_n} X_1^{i_1} \cdots X_n^{i_n}$ by $\sigma(f(X)) = \sum a_{i_1 \cdots i_n} \sigma(X_1)^{i_1} \cdots \sigma(X_n)^{i_n}$.
Show that

(a) $\sigma\colon R \to R$ as defined above is an automorphism of R;
(b) G acts as a transformation group on R;
(c) The set $R^G = \{f(X) \mid \sigma(f(X)) = f(X) \text{ for all } \sigma \in G\}$ is a subring of R and subspace of R over k;
(d) R^G contains the *elementary symmetric polynomials*

$$\begin{aligned} f_1 &= X_1 + \cdots + X_n \\ f_2 &= X_1X_2 + X_1X_3 + \cdots + X_2X_3 + \cdots + X_{n-1}X_n \\ &\vdots \\ f_n &= X_1X_2 \cdots X_n \end{aligned}$$

(e) every *symmetric polynomial* $f(X)$ (element $f(X)$ of R^G) can be expressed as a polynomial in the elementary symmetric polynomials $f_1, \ldots, f_n$, that is, R^G is *generated* by the n elements $f_1, \ldots, f_n$.

To prove (e), order the given $f(X) \in R^G$ *lexographically*, placing a monomial term $aX_1^{i_1} \cdots X_n^{i_n}$ occurring in $f(X)$ before another $bX_1^{j_1} \cdots X_n^{j_n}$ if the first nonvanishing difference $i_m - j_m$ is positive. Each occurring monomial term $aX_1^{i_1} \cdots X_n^{i_n}$ is accompanied by all expressions obtained from this one by permuting the exponents i_m. Write only the lexographically first such term, so that $i_1 \geq i_2 \geq \cdots \geq i_n$. Letting $aX_1^{i_1} \cdots X_n^{i_n}$ be the first term occurring in $f(X)$, show that $g = af_1^{i_1 - i_2} f_2^{i_2 - i_3} \cdots f_n^{i_n}$ has the same first occurring term $aX_1^{i_1} \cdots X_n^{i_n}$. Examine the difference polynomial $f(X) - g$ and conclude that one can prove by induction that $f(X)$ can be expressed as a polynomial in the $f_1, \ldots, f_n$. Show, finally, that

(*f*) the expression of a nonzero $f(X) \in R^G$ as polynomial in $f_1, \ldots, f_n$ over k is unique (that is, the coefficients of the monomials in $f_1, \ldots, f_n$ are uniquely determined).

E.1.25. Express the following symmetric polynomials in terms of elementary symmetric polynomials in $k[X, Y, Z]$;

(a) $X^2 + Y^2 + Z^2$
(b) $X^2YZ + XY^2Z + XYZ^2 + XYZ$

E.1.26 (Generic Discriminant). Let $k[X_1, \ldots, X_n, X]$ be the polynomial ring over k in $n + 1$ variables and let $f = \prod_0^n (X - X_i)$. Show that $f = X^n - f_1X^{n-1} + f_2X^{n-2} - \cdots + (-1)^n f_n$ where the $f_1, \ldots, f_n$ are the symmetric polynomials in $X_1, \ldots, X_n$. The symmetric polynomial

$$D = \prod_{j<i} (X_j - X_i)^2$$

can be expressed as a polynomial $D = D(g_0, \ldots, g_{n-1})$ in the coefficients $g_i = (-1)^{n-i}f_n$ of $f = X^n + g_{n-1}X^{n-1} + g_{n-2}X^{n-2} + \cdots + g_0$, called the *generic discriminant* or the *discriminant* of f. Show that

(a) the discriminant of $f = X^2 + bX + c$ is $D = b^2 - 4c$;
(b) the discriminant of $f = X^3 + aX^2 + bX + c$ is $D = a^2b^2 - 4b^3 - 4a^3c - 27c^3 + 18abc$.

(Here, it is understood that $f = (X - X_1)(X - X_2)$ in (a) and $f = (X - X_1) \times (X - X_2)(X - X_3)$ in (b)).

E.1.27 (Discriminant). Let $f(X) = a_0 + a_1X + \cdots + a_nX^n$ $(a_n \neq 0)$ where the a_i are in a field k. The *discriminant* of $f(X)$ is

$$D(f(X)) = a_n^{2n-2} D\left(\frac{a_0}{a_n}, \frac{a_1}{a_n}, \ldots, \frac{a_{n-1}}{a_n}\right)$$

where $D(g_0, \ldots, g_{n-1})$ is as defined in the preceding exercise. Show that

(a) $D(f(X))$ is a polynomial in $a_0, \ldots, a_n$;
(b) if $f(X) = a_n \prod_0^n (X - x_i)$ with $x_1, \ldots, x_n$ in an extension field K of k, then $D(f(X)) = a_n^{2n-2} \prod_{j<i} (x_j - x_i)^2$;
(c) $D(f(X)) = 0$ if and only if $f(X)$ has multiple roots in a splitting field.

Describe the discriminant of $aX^2 + bX + c$ for $a, b, c \in k$ and $a \neq 0$.

E.1.28 (Generic Resultant). Let $f(X) = a_m \prod_1^m (X - X_i)$ and $g(X) = b_n \prod_1^n (X - Y_j)$ in the polynomial ring $k[X_1, \ldots, X_m, Y_1, \ldots, Y_n, X]$ in $m + n + 1$ variables. Then the *generic resultant* of f and g is

$$S(f, g) = a_m^m b_n^n \prod_{i,j} (X_i - Y_j)$$

Show that $R(f, g) = S(f, g)$ by showing that

(a) $R(f, g)$ is $a_m^m b_n^n$ times a symmetric polynomial in the X_i, Y_j with coefficients in the prime field;
(b) $S(f, g)$ divides $R(f, g)$ (show that $X_i - Y_j$ divides $R(f, g)$ for all i, j);
(c) $S(f, g) = a_m^m \prod_i g(X_i) = (-1)^{mn} b_n^n \prod_j f(Y_j)$ and $S(f, g)$ is therefore $a_m^m b_n^n$ times a symmetric polynomial in the X_i, Y_j with coefficients in the prime field;
(d) $R(f, g) = S(f, g)$ (compare degrees of homogeneity and constants).

E.1.29 (Resultant and Discriminant). Let $f(X) = \sum_0^n a_iX^i$ $(a_n \neq 0)$ be a polynomial with coefficients in a field k. Show that

$$R(f, f') = \pm a_n D(f, f')$$

where $f = f(X)$, $f' = f(X)'$. (Show that

$$R(f, f') = a_n^{n-1} \prod_i f'(x_i) = a_n^{2n-1} \prod_{i \neq j} (x_j - x_i) = a_n a_n^{2n-2} \prod_{j<i} (x_j - x_i)^2$$

where $x_1, \ldots, x_n$ are the roots of $f(X)$ in some extension field K). Use this to show again that f has multiple roots in a splitting field if and only if f and f' have a common factor of positive degree.

E.1.30 (Interpolation). Let $x_0, \ldots, x_n$ be distinct elements of a field k and let $y_0, \ldots, y_n \in k$. Show that there is precisely one element $f(X) \in k[X]$ of degree at most n such that $f(x_i) = y_i$ for $0 \leq i \leq n$, namely

$$f(X) = \sum_0^n \frac{f(x_i)}{\prod_{j \neq i} (x_i - x_j)} \prod_{j \neq i} (X - x_j)$$

E.1.31. Regard $\mathbb{Q}$ as the prime subfield of the field $\mathbb{C}$ of complex numbers. What is $K:\mathbb{Q}$ where K is the subfield of $\mathbb{C}$ generated by $\sqrt{2}$ and $i = \sqrt{-1}$?

E.1.32. Describe a splitting field K for $X^4 + 1$ over $\mathbb{Q}$.

E.1.33. Construct a splitting field K for $X^3 + X^2 + 1$ over $\mathbb{Z}_3$.

E.1.34. Construct a splitting field over $\mathbb{Z}_5$ for $X^4 + X^3 + X^2 + X + 1$.

E.1.35. Let K/k be a finite dimensional field extension and let k' be a subfield of K containing k. Show that $k':k$ divides $K:k$.

E.1.36. Let K/k be a field extension and let $a \in K$. Suppose that $k(a):k$ is odd. Show that $k(a) = k(a^2)$.

E.1.37 (Kaplanski). Let k be a field, $a \in k$, m and n relatively prime positive integers. Show that

(a) x is a root of $X^{mn} - a$ only if x^m is a root of $X^n - a$;
(b) if $X^m - a$ and $X^n - a$ are irreducible, then the degrees of $k(x^n)$, $k(x^m)$ and $k(x)$ over k are m, n and mn respectively;
(c) $X^{mn} - a$ is irreducible if and only if $X^m - a$ and $X^n - a$ are irreducible.

E.1.38. Let k be an infinite field and let $K = k(X)$ be the field of rational expression in an indeterminant X over k. Let $G = \mathrm{Aut}_k K$. Show that

(a) $K^G = k$ where K^G is the fixed field $K^G = \{a \in K \mid \sigma(a) = a$ for $\sigma \in G\}$;
(b) the only finite dimensional subextension k'/k of K/k is the extension k/k.

E.1.39. Show that for any prime number p, $\mathbb{Z}_p[X]$ has irreducible elements of degree p^n for every positive integer n.

E.1.40. Let K be the algebraic closure of a finite field $\mathbb{Z}_p$ (p prime) and let GL_nK be the group of nonsingular $n \times n$ matrices with coefficients in K. Prove that

(a) every element of GL_nK has finite order;
(b) for $n \geq 2$, GL_nK has an infinite p-subgroup;

(c) for each $g \in GL_nK$, the cyclic subgroup $\langle g \rangle$ contains elements t, u such that $g = tu$ and such that t and u are conjugate to elements of GL_nK of the form

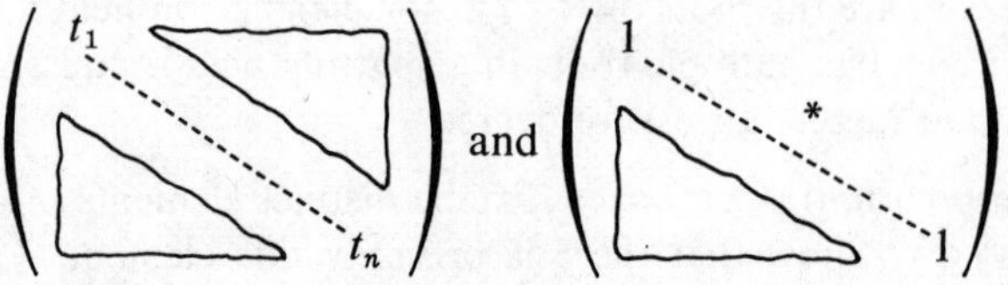

respectively.

E.1.41. Show that if k is an infinite field and $f(X)$ a nonzero element of $k[X]$, then $f(t) \neq 0$ for some $t \in k$. Describe for any finite field k a nonzero polynomial $f(X)$ such that $f(t) = 0$ for all $t \in k$.

E.1.42. Let K/k be a field extension and let $\sigma \in \mathrm{Aut}_k K$. Show that for any subfield k' of K containing k, $\sigma(\mathrm{Aut}_{k'} K)\sigma^{-1} = \mathrm{Aut}_{\sigma(k')} K$.

E.1.43. Prove that the discriminant of $X^5 + cX + d$ is $4^4c^5 + 5^5d^4$.

E.1.44. Determine all irreducible polynomials of degree at most 3 over $\mathbb{Z}_3$.

E.1.45. Let K/k be a field extension of prime degree q. Show that K and k are the only subfields of K containing k.

E.1.46. Let $f(X)$ be an element of the polynomial ring $k[X]$ over the field k. Let K be a splitting field for $f(X)$ over k. Then $K{:}k \leq (\mathrm{Deg}\, f(X))!$

2 The Structure of algebraic extensions

In this chapter, we give an account of the structure of algebraic extensions. The first three sections are concerned with describing the structure of an algebraic extension K/k in terms of the structure of the irreducible polynomials in $k[X]$ having roots in K. In the remaining section, properties of subfields A, B of K containing k are compared with properties of their composite AB.

Throughout the chapter, k is a field, K an algebraic extension of k, k_{Alg} an algebraic closure of k containing K. (Note here that any two algebraic closures of k containing K are isomorphic, by 1.4.2, 1.4.8.) We let p denote the exponent characteristic of k (see 1.1.4), in order that $s^{p^e} = s$ for $s \in k$ when k has characteristic 0.

2.1 The structure of an irreducible polynomial

Let $f(X)$ be an irreducible polynomial in $k[X]$ and let $f(X) = c\prod_1^n (X - s_i)$ where the s_i are the roots of $f(X)$ in k_{Alg}.

2.1.1 Definition. We say that $f(X)$ is *separable* over k if the s_i are pairwise distinct, *radical* (or *purely inseparable*) over k if the s_i are all equal.

Note that $f(X)$ is separable and radical over k if and only if $f(X)$ is linear, that is, has the form $f(X) = c(X - s)$.

2.1.2 Proposition. $f(X)$ is separable over k if and only if $f'(X) \neq 0$ (see E.1.7).

Proof. $f(X)$ has a multiple root if and only if $f(X)$ and $f'(X)$ have a common root (see E.1.8), that is, if and only if $f(X)$ and $f'(X)$ are not relatively prime. Since $f(X)$ is irreducible, this happens if and only if $f'(X) = 0$. □

2.1.3 Corollary. Let $f(X) = g(X^{p^e})$ with $g(Y) \in k[Y]$ and e maximal. Then $g(Y)$ is irreducible and separable over k.

Proof. The irreducibility is clear. The separability follows from the proposition, for if $g'(Y) = 0$, then $g(Y) = \sum_0^n c_i Y^i$ where $c_i = 0$ or p divides i for all i. But then e would not be maximal. □

2.1.4 Definition. The above e is called the *radical exponent* of $f(X)$.

2.1.5 Proposition. Let the radical exponent of $f(X)$ be e. Then $f(X) = c\prod_1^m (X - s_j)^{p^e}$, where the $s_1, \ldots, s_m$ are the distinct roots of $f(X)$ in k_{Alg}.

Proof. Let $f(X) = g(X^{p^e})$ as in 2.1.3. Let $g(Y) = c\prod_1^m (Y - t_j)$ where the t_j are distinct elements of k_{Alg}. Letting s_j be a solution in k_{Alg} to $s_j^{p^e} = t_j$ $(1 \le j \le m)$, we then have

$$f(X) = g(X^{p^e}) = c\prod_1^m (X^{p^e} - s_j^{p^e}) = c\prod_1^m (X - s_j)^{p^e}$$

(see E.1.2). ▯

2.1.6 ***Corollary.*** Let e be the radical exponent of $f(X)$. Then $f(X)$ is separable over k if and only if $e = 0$, and $f(X)$ is radical over k if and only if $f(X) = c(X - s)^{p^e}$ for some $s \in k_{\text{Alg}}$.

2.2 Separable and radical extensions

The structure of algebraic extensions K of k is closely related to the structure of irreducible polynomials in $k[X]$, since we can pass from elements $s \in K$ to their minimum polynomials $f_s(X)$ over k. We exploit this relationship here and begin by introducing terminology similar to the terminology introduced in 2.1.

2.2.1 ***Definition.*** An element $s \in K$ is *separable* (respectively *radical*) over k if its minimum polynomials $f_s(X)$ over k is separable (respectively radical) over k.

Note that s is separable and radical over k if and only if $s \in k$.

The next proposition is a restatement of part of 1.4.12.

2.2.2 ***Proposition.*** An element $s \in k$ is radical over k if and only if $s \in k_{\text{Alg}}^G$, where $G = \text{Aut}_k\, k_{\text{Alg}}$.

2.2.3 ***Proposition.*** If an element $s \in K$ is separable (respectively radical) over k, then s is separable (respectively radical) over k' for any subfield k' of K containing k.

Proof. This is clear since the minimum polynomial of s over k' divides the minimum polynomial of s over k. ▯

2.2.4 ***Proposition.*** Let $s \in K$ and let e be the radical exponent of $f_s(X)$. Then s^{p^e} is separable over k and s is radical over $k(s^{p^e})$.

Proof. $f_s(X) = g(X^{p^e})$ where $g(Y)$ is an irreducible and separable polynomial in $k[Y]$. Since $g(Y)$ is the minimum polynomial for s^{p^e}, s^{p^e} is separable over k. And s is radical over $k(s^{p^e})$ since $s^{p^e} \in k(s^{p^e})$ (see 2.1.6). ▯

2.2.5 ***Proposition.*** Let $s \in K$. Then s is separable over k if and only if $k(s) = k(s^p)$, and s is radical over k if and only if $s^{p^e} \in k$ for some e.

Proof. Suppose first that s is separable over k. Then s is separable and radical over $k(s^p)$, so that $s \in k(s^p)$ and $k(s) = k(s^p)$. Suppose next that s is not separable over k. Then $f_s(X)$ has radical exponent $e \ge 1$ and $f_s(X) =$

$g(X^p)$ where $g(Y)$ is the minimum polynomial of s^p over k. Thus, $k(s){:}k = \operatorname{Deg} f_s(X) \neq \operatorname{Deg} g(Y) = k(s^p){:}k$ and $k(s) \neq k(s^p)$. The remaining assertion is clear from 2.1.6. □

We now turn to the study of separable and radical extensions K of k. We let k_{Alg} be an algebraic closure of k containing K.

2.2.6 ***Definition.*** The extension K/k is *separable* (respectively *radical*) if each $s \in K$ is separable (respectively radical) over k.

Note that the extension K/k is separable and radical if and only if $K = k$. The following two propositions are restatements of 2.2.2 and 2.2.3.

2.2.7 ***Proposition.*** The extension K/k is radical if and only if K is a subfield of k_{Alg}^G, where $G = \operatorname{Aut}_k k_{\mathrm{Alg}}$.

2.2.8 ***Proposition.*** If the extension K/k is separable/radical, then K/k' is separable/radical for any subfield k' of K containing k.

2.2.9 ***Theorem.*** Let $s \in K$. Then s is separable/radical over k if and only if $k(s)/k$ is separable/radical.

Proof. One direction is clear. We now prove the other. Suppose first that s is radical over k. Then $s^{p^e} \in k$ for some e. Let $t \in k(s)$ and express t as $t = \sum_0^{n-1} a_i s^i$ for suitable $a_i \in k$. Then $t^{p^e} = \sum_0^{n-1} a_i^{p^e} s^{p^e i} \in k$ and t is radical over k. Thus, $k(s)/k$ is radical. Suppose next that s is separable over k and let $t \in k(s)$. We show that t is separable over k by showing that $k(t) = k(t^p)$ (see 2.2.5). But $k(s) = k(s^p)$. Since $1, s, \ldots, s^{n-1}$ is a basis for $k(s)/k$ and $1, s^p, \ldots, (s^p)^{n-1}$ is a basis for $k(s^p)/k$ ($n = k(s){:}k = k(s^p){:}k$), this implies that if $r_0, \ldots, r_{n-1}$ span $k(s)$ over k, so do $r_0{}^p, \ldots, r_{n-1}^p$. Since $n = k(s){:}k$, the latter must also be linearly independent over k. Taking $r_i = t^i$ for $0 \leq i \leq m - 1$ ($m = k(t){:}k$), it follows that the elements $1, t^p, \ldots, (t^p)^{m-1}$ of $k(t)$ are linearly independent over k, hence form a basis for $k(t)/k$. Thus, $k(t) = k(t^p)$ and t is separable over k. Thus, $k(s)/k$ is separable. □

2.2.10 ***Theorem.*** Let k be infinite, let K_1, K_2 be two extensions of k and let $\sigma_1, \ldots, \sigma_n$ be distinct elements of the set $\operatorname{Hom}_k(K_1, K_2)$ of k-homomorphisms from K_1 to K_2. Then there exists $s \in K_1$ such that $\sigma_1(s), \ldots, \sigma_n(s)$ are distinct. If K_1/k is algebraic, s can be taken to be separable over k.

Proof. Let $s_0, \ldots, s_m$ be elements of K such that σ_i and σ_j take on the same values at $s_0, \ldots, s_m$ if and only if $i = j$. Let $h(X) = \sum_{r=0}^m s_r X^r$ and note that $\sigma_i(h(X)) = \sigma_j(h(X))$ if and only if $i = j$. Thus, $\prod_{i \neq j} (\sigma_i(h(X) - \sigma_j h(X)))$ is a nonzero polynomial in $K_2[X]$. Since k is infinite, there exists $t \in k$ such that $\prod_{i \neq j} (\sigma_i(h(t)) - \sigma_j(h(t))) \neq 0$. Letting $s = h(t)$, $\sigma_1(s), \ldots, \sigma_n(s)$ are distinct. If K_1/k is algebraic, s^{p^e} is separable over k for some e and $\sigma_1(s^{p^e}), \ldots, \sigma_n(s^{p^e})$ are distinct (see E.1.2). Thus, we may replace s by s^{p^e} and s can be taken to be separable over k. □

2.2.11 *Definition.* We let $k_{\text{sep}} = \{s \in K \mid s \text{ is separable over } k\}$ and $k_{\text{rad}} = \{s \in K \mid s \text{ is radical over } k\}$. The sets k_{sep} and k_{rad} are called the *separable* and *radical closures* of k in K respectively.

2.2.12 *Definition.* We let $k_{\text{Sep}} = \{s \in K_{\text{Alg}} \mid s \text{ is separable over } k\}$ and $k_{\text{Rad}} = \{s \in k_{\text{Alg}} \mid s \text{ is radical over } k\}$. These sets are called the *separable closure* and *radical closure* of k respectively.

It is easily seen that k_{rad} and k_{Rad} are field extensions of k (see E.1.2). That k_{sep} is a subfield of K is a consequence of the following important theorem.

2.2.13 *Theorem.* Let $K{:}k < \infty$. Then $k_{\text{sep}} = k(s)$ for some $s \in k$, the extension K/k_{sep} is radical and the mapping $\sigma \mapsto \sigma|_{k_{sep}}$ from $\text{Hom}_k(K, k_{\text{Alg}})$ to $\text{Hom}_k(k_{\text{sep}}, k_{\text{Alg}})$ is a bijection.

Proof. If k is finite, then $K = k(s)$ for some $s \in K$ and K/k is separable, by 1.5.6. Thus, $k_{\text{sep}} = K = k(s)$ and there is nothing more to prove. Suppose next that k is infinite. Since $K{:}k < \infty$, $\text{Hom}_k(K, k_{\text{Alg}})$ has only finitely many elements $\sigma_1, \ldots, \sigma_n$, by 1.4.11, and we may take $\sigma_1 = id_K$, the inclusion mapping from K into k_{Alg}. Now there exists $s \in k_{\text{sep}}$ such that $\sigma_1(s), \ldots, \sigma_n(s)$ are distinct, by 2.2.10. Now $k(s) \subset k_{\text{sep}}$, by 2.2.9, and we claim that $k(s) = k_{\text{sep}}$. For this, it suffices to show that $K/k(s)$ is radical, for then if $t \in k_{\text{sep}}$, t is separable and radical over $k(s)$ so that $t \in k(s)$. Let us now regard k_{Alg} as the algebraic closure $k(s)_{\text{Alg}}$ of $k(s)$ (see 1.4.8). Letting $\sigma \in \text{Aut}_{k(s)}\, k(s)_{\text{Alg}} = \text{Aut}_{k(s)} k_{\text{Alg}}$, we have $\sigma(s) = s = id_K(s) = \sigma_1(s)$, so that $\sigma|_K = \sigma_1 = id_K$ by the distinctness of the $\sigma_j(s)$ for $\sigma_j \in \text{Hom}_k(K, k_{\text{Alg}})$. It follows that $K \subset k(s)_{\text{Alg}}^G$ where $G = \text{Aut}_{k(s)}\, k(s)_{\text{Alg}}$. Thus, $K/k(s)$ is radical and $k(s) = k_{\text{sep}}$.

We have shown in particular that K/k_{sep} is radical. It is worth noting that this also is an immediate consequence of the fact that for $t \in K$, $t^{p^e} \in k_{\text{sep}}$ for some e (see 2.2.4).

It remains to show that $\sigma \mapsto \sigma|_{k_{sep}}$ is a bijection from $\text{Hom}_k(K, k_{\text{Alg}})$ to $\text{Hom}_k(k_{\text{sep}}, k_{\text{Alg}})$. Suppose that $\sigma, \tau \in \text{Hom}_k(K, k_{\text{Alg}})$ and $\sigma|_{k_{\text{sep}}} = \tau|_{k_{\text{sep}}}$. Let $t \in K$ and choose e such that $t^{p^e} \in k_{\text{sep}}$. Then $\sigma(t)^{p^e} = \sigma(t^{p^e}) = \tau(t^{p^e}) = \tau(t)^{p^e}$. Since the p^eth power mapping is injective (as a homomorphism of fields), $\sigma(t) = \tau(t)$. Thus, $\sigma = \tau$ and $\sigma \mapsto \sigma|_{k_{\text{sep}}}$ is injective on $\text{Hom}_k(K, k_{\text{Alg}})$. We finally show that $\sigma \mapsto \sigma|_{k_{\text{sep}}}$ is surjective from $\text{Hom}_k(K, k_{\text{Alg}})$ to $\text{Hom}_k(k_{\text{sep}}, k_{\text{Alg}})$. Thus, let $\varphi \in \text{Hom}_k(k_{\text{sep}}, k_{\text{Alg}})$. Since k_{Alg} is an algebraic closure of k_{sep} and of $\varphi(k_{\text{sep}})$, φ has an extension to an automorphism $\bar{\varphi}$ of k_{Alg}. Letting $\sigma = \bar{\varphi}|_K$, we have $\sigma|_{k_{\text{sep}}} = \varphi$. (It is perhaps worth noting here that we could, instead, define σ directly in terms of φ. For $t \in K$, take e such that $t^{p^e} \in k_{\text{sep}}$ and let $\sigma(t)$ be the p^e-th root of $\varphi(t^{p^e})$.) ☐

2.2.14 *Corollary.* Every finite dimensional separable extension is simple.

2.2.15 *Proposition.* Let S be a subset of $k_{\text{sep}}/k_{\text{rad}}$. Then $k(S)/k$ is separable/radical. In particular, k_{sep} and k_{rad} are subfields of K.

Proof. Since $k(S) = \bigcup_{T \in \mathbf{S}} k(T)$ where $\mathbf{S}$ is the set of finite subsets of S, we may assume without loss of generality that S is finite. Then $k(S)/k$ is finite dimensional, so that $k(S)_{\text{sep}} = k(s)$ for some $s \in k_{\text{sep}}$. If $S \subset k_{\text{sep}}$, then $S \subset k(S)_{\text{sep}} = k(s)$ so that $k(S) = k(S)_{\text{sep}}$ and $k(S)/k$ is separable. If $S \subset k_{\text{rad}}$, then obviously $k(S)/k$ is radical. □

2.2.16 *Definition.* We let $(K:k)_{\text{sep}} = k_{\text{sep}}:k$ and $(K:k)_{\text{rad}} = K:k_{\text{sep}}$. The integer $(K:k_{\text{sep}}/(K:k)_{\text{rad}}$ is called the *separability degree/radical degree* of K over k.

Note that $K:k = (K:k)_{\text{sep}}(K:k)_{\text{rad}}$, by the transitivity of degree. We prove later on that separability degree and radical degree are transitive (see 2.2.21).

2.2.17 *Theorem.* Let $K:k < \infty$. Then the number of elements of $\text{Hom}_k(K, k_{\text{Alg}})$ is $(K:k)_{\text{sep}}$. If G is a subgroup of $\text{Aut}_k K$, then $G:\mathbf{1} = K:K^G$. Two subgroups G and H of $\text{Aut}_k K$ are equal if and only if their fixed fields K^G and K^H are equal.

Proof. Since the expression $\sigma \mapsto \sigma|_{k_{\text{sep}}}$ maps $\text{Hom}_k(K, k_{\text{Alg}})$ bijectively into $\text{Hom}_k(k_{\text{sep}}, k_{\text{Alg}})$, by 2.2.13, it suffices to show that $\text{Hom}_k(k_{\text{sep}}, k_{\text{Alg}})$ has $k_{\text{sep}}:k$ elements. Take $s \in k_{\text{sep}}$ such that $k_{\text{sep}} = k(s)$ and let $f_s(X)$ be the minimum polynomial of s over k. Then $f_s(X) = \prod_1^n (X - s_i)$ where $n = k_{\text{sep}}:k$ and $s_1, \ldots, s_n$ are distinct elements of k_{Alg}. For each i, there exists a unique k-isomorphism $\sigma_i: k(s) \to k(s_i)$ such that $\sigma_i(s) = s_i$, by 1.1.17. And each $\sigma \in \text{Hom}_k(k(s), k_{\text{Alg}})$ maps s to $\sigma(s) = s_i$ for some i. Thus, $\text{Hom}_k(k(s), k_{\text{Alg}}) = \{\sigma_1, \ldots, \sigma_n\}$ and $\text{Hom}_k(k_{\text{sep}}, k_{\text{Alg}})$ has $k_{\text{sep}}:k$ elements.

Suppose next that G is a subgroup of $\text{Aut}_k K$. Let $s \in K$ and $g(X) = \prod_1^n (X - s_i)$ where $s_1, \ldots, s_n$ is the orbit of s under G. Then $g(s) = 0$, and $g(X) \in K^G[X]$ since $\sigma(g(X)) = g(X)$ for all $\sigma \in G$. (Compare with the proof of 1.2.8). Thus, $g(X)$ is the minimum polynomial of s over K^G, so that s is separable over K^G. Thus, K/K^G is separable. It follows that s may be chosen such that $K = K^G(s)$. Now $K:K^G = \text{Deg}\, g(X) = n$. Since $s_1, \ldots, s_n$ is the orbit of s under G and $K = K^G(s)$, there exists for each i precisely one element $\sigma \in G$ such that $\sigma(s) = s_i$ $(1 \leq i \leq n)$. That is, G acts simply transitively on the roots of the minimum polynomial over K^G of the generator s of K over K^G. It follows that $G:\mathbf{1} = K:K^G$. If H is also a subgroup of $\text{Aut}_k K$, and if $K^G = K^H$, it follows that $K^G = K^H = K^I$ and $G:\mathbf{1} = H:\mathbf{1} = I:\mathbf{1} = K:K^G$ where I is the subgroup of $\text{Aut}_k K$ generated by $G \cup H$. But then $G = I = H$, since $G \subset I \supset H$. □

The above theorem establishes a *Galois Correspondence* $G \leftrightarrow K^G$ and is called the *Galois Correspondence Theorem* for finite dimensional extensions. A thorough discussion of this correspondence is given from another point of view in 3.3.

2.2.18 *Proposition.* Let $K:k < \infty$. Then $(K:k)_{\text{rad}} = p^e$ for some e and $x^{p^e} \in k_{\text{sep}}$ for $x \in K$.

Proof. Since $(K:k)_{\text{rad}} = K:k_{\text{sep}}$, we may replace k by k_{sep}. That is, it suffices to show that if K/k is a finite dimensional radical extension, then $K:k = p^e$ for some e and $x^{p^e} \in K$ for all $x \in k$. Now for each $x \in K$, $x^{p^f} \in K$ for some f (see 2.2.4). Thus, there exists a tower $k = K_0 \subset K_1 \subset \cdots \subset K_e$ of distinct subfields $K_0, K_1, \ldots, K_e$ and elements $x_1, \ldots, x_e$ such that $K_i = K_{i-1}(x_i)$ and $x_i^p \in K_{i-1}$ for $1 \le i \le e$. But then $K_i:K_{i-1} = p$ for $1 \le i \le e$ and $K:k = \prod_1^e (K_i:K_{i-1}) = p^e$. For the minimal polynomial of x_i over K_{i-1} is $X^p - x_i^p$ (see 2.1.6) and $K_i:K_{i-1} = \operatorname{Deg}(X^p - x_i^p) = p$ $(1 \le i \le e)$. Note also that the mapping $\pi(x) = x^p$ maps K_i into K_{i-1}, so that $\pi^e(K) \subset k$ and $x^{p^e} = \pi^e(x) \in k$ for $x \in K$. □

We conclude this section in giving some simple structural properties of K in terms of k_{sep} and k_{rad}.

2.2.19 ***Proposition.*** $K = k_{\text{sep}}k_{\text{rad}}$ if and only if K/k_{rad} is separable. A basis for k_{sep} over k is a basis for $k_{\text{sep}}k_{\text{rad}}$ over k_{rad}.

Proof. Since $K = \bigcup_{S \in \mathbf{S}} k_{\text{rad}}(S)$ where $\mathbf{S}$ is the set of finite subsets of K, it suffices to prove the proposition in the case where $K:k_{\text{rad}} < \infty$. Thus, let us assume that $K:k_{\text{rad}} < \infty$. If $K = k_{\text{sep}}k_{\text{rad}}$, then $K = k_{\text{rad}}(k_{\text{sep}})$ and K/k_{rad} is separable by 2.2.15. Suppose conversely that K/k_{rad} is separable and choose $x \in K$ such that $K = k_{\text{rad}}(x)$. Taking x^{p^e} to be separable over k, by 2.2.18, we have $K = k_{\text{rad}}(x) = k_{\text{rad}}(x^{p^e}) \subset k_{\text{sep}}k_{\text{rad}}$, by 2.2.5, and $K = k_{\text{sep}}k_{\text{rad}}$. It remains to show that if $x_1, \ldots, x_n$ is a basis for k_{sep} over k, then $x_1, \ldots, x_n$ is a basis for $k_{\text{sep}}k_{\text{rad}}$ over k_{rad}. Suppose not. Then there exist $y_i \in k_{\text{rad}}$, not all 0, such that $\sum_1^n x_i y_i = 0$. Choose $e \ge 0$ such that $y_i^{p^e} \in k$ for $1 \le i \le n$. Then $\sum_1^n x_i^{p^e} y_i^{p^e} = 0$ and $x_1^{p^e}, \ldots, x_n^{p^e}$ are linearly dependent over k. Thus, the k-span k' of $\{z^{p^e} \mid z \in k_{\text{sep}}\}$ is a proper subfield of k_{sep}. But k_{sep}/k' is radical and separable (see 2.2.8) and $k_{\text{sep}} = k'$, a contradiction. Thus, $x_1, \ldots, x_n$ is a basis for $k_{\text{sep}}k_{\text{rad}}$ over k_{rad}. □

The above proposition says that the k-homomorphism from $k_{\text{sep}} \otimes_k k_{\text{rad}}$ to K which maps $x \otimes y$ to xy $(x \in k_{\text{sep}}, y \in k_{\text{rad}})$ is injective and is a k-isomorphism if and only if K/k_{rad} is separable. Here, we are regarding $k_{\text{sep}} \otimes_k k_{\text{rad}}$ as vector space over k and as ring such that $(x \otimes y)(x' \otimes y') = xx' \otimes yy'$ for $x, x' \in k_{\text{sep}}$, $y, y' \in k_{\text{rad}}$ (see A.2).

2.2.20 ***Corollary.*** $k_{\text{Alg}} = k_{\text{Sep}}k_{\text{Rad}}$ and k_{Alg} is k-isomorphic to $k_{\text{Sep}} \otimes_k k_{\text{Rad}}$.

Proof. $k_{\text{Alg}}/k_{\text{Rad}}$ is separable, by 2.2.2 and 1.4.13. □

We now prove the transitivity of separability degree and radical degree.

2.2.21 ***Theorem.*** Let $K:k < \infty$ and let k' be a subfield of K containing k. Then $(K:k)_{\text{sep}} = (K:k')_{\text{sep}}(k':k)_{\text{sep}}$ and $(K:k)_{\text{rad}} = (K:k')_{\text{rad}}(k':k)_{\text{rad}}$.

Proof. Since $(K:k) = (K:k)_{\text{sep}}(K:k)_{\text{rad}}$ relates separability degree and radical degree (see 2.2.16), and since degree is transitive, it suffices to prove that separability degree is transitive. For this, consider the diagram

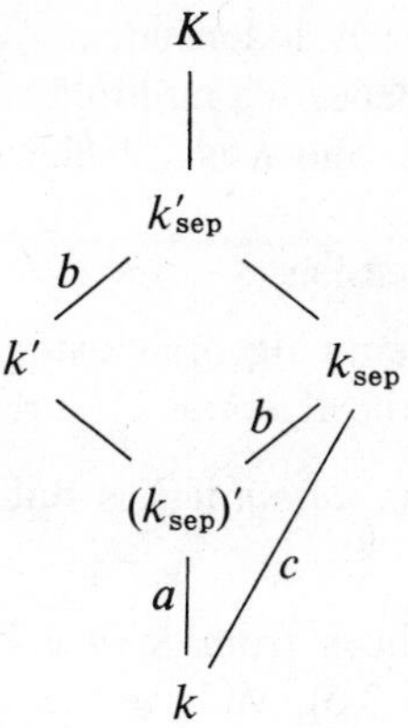

where $(k_{\text{sep}})'$ is the separable closure of k in k' and k'_{sep} is the separable closure of k' in K. Note that $k'_{\text{sep}} = k'k_{\text{sep}}$, since $x \in k'_{\text{sep}}$ implies that $x^{p^e} \in k_{\text{sep}}$ (for suitable e), which in turn implies that $x \in k'(x^{p^e}) \subset k'k_{\text{sep}}$.

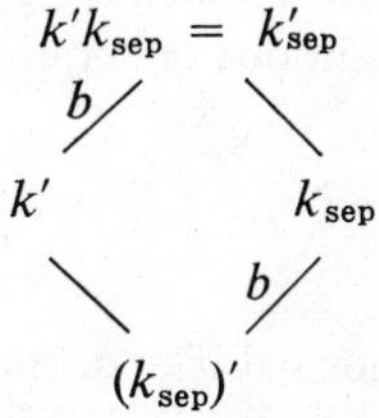

Then $k'/(k_{\text{sep}})'$ is radical (see 2.2.13) and $k_{\text{sep}}/(k_{\text{sep}})'$ is separable (see 2.2.8), so that $k'_{\text{sep}}:k' = k_{\text{sep}}:(k_{\text{sep}})'$, by 2.2.19. Now $(K:k')_{\text{sep}}(k':k)_{\text{sep}} = (k'_{\text{sep}}:k') \times ((k_{\text{sep}})':k) = (k_{\text{sep}}:(k_{\text{sep}})')((k_{\text{sep}})':k) = k_{\text{sep}}:k = (K:k)_{\text{sep}}$ as asserted. A sketch of this proof is that $ab = c$ where a, b, c are as indicated by the diagrams. □

We close with the following theorem, which underlies part of the comparison of normal extensions, and Galois and radical extensions, discussed at the end of 2.3. It is convenient here to use the language of k-algebras given in Appendix A.

2.2.22 ***Theorem.*** If k_s/k and k_r/k are finite dimensional separable and radical extensions of k respectively, then $\bar{K} = k_s \otimes_k k_r$ (tensor product of k-algebras) is a field extension of $k = 1 \otimes k$ such that $\bar{K}_{\text{sep}} = k_s \otimes 1$ and $\bar{K}_{\text{rad}} = 1 \otimes k_r$.

Proof. We identify k with the subring $1 \otimes k$ of $\bar{K}$ via the mapping $y \mapsto 1 \otimes y$. We wish to show that the k-algebra $\bar{K}$ is a field. Thus, let $\sum_1^n x_i \otimes y_i \in \bar{K} - \{0\}$. We may assume that $x_1, \ldots, x_n$ are linearly independent over k. Choose e such that $y_i^{p^e} \in k$ for $1 \leq i \leq n$. Then

$$\left(\sum_1^n x_i \otimes y_i\right)^{p^e} = \sum_1^n x_i^{p^e} \otimes y_i^{p^e} = \sum_1^n x_i^{p^e} y_i^{p^e} \otimes 1.$$

Since $x_1^{p^e}, \ldots, x_n^{p^e}$ are linearly independent over k (by the separability of k_s/k), the latter is nonzero, hence is a multiplicatively invertible element of $\bar{K}$. Thus, $\sum_1^n x_i \otimes y_i$ is invertible and $\bar{K}$ is a field. □

2.3 Normal and Galois extensions

We continue letting K be an algebraic extension of a field k of exponent characteristic p and k_{Alg} be the algebraic closure of k containing K.

2.3.1 Definition. K/k is *normal* if K is stable under the group $\text{Aut}_k\, k_{\text{Alg}}$ of k-automorphisms of k_{Alg}.

Since any k-homomorphism from K into k_{Alg} can be extended to a k-automorphism of k_{Alg} (see 1.3.5), we have the following alternate conditions for K/k to be normal.

2.3.2 Proposition. K/k is normal if and only if $\text{Aut}_k K$ is the set $\text{Aut}_k\, k_{\text{Alg}}|_K$ of restrictions of elements of $\text{Aut}_k\, k_{\text{Alg}}$ to K.

2.3.3 Proposition. Let K be a normal algebraic extension of k containing k'. Then the following conditions are equivalent:

1. k'/k is normal;
2. k' is stable under $\text{Aut}_k K$;
3. $\text{Aut}_k k' = \text{Aut}_k K|_{k'}$.

2.3.4 Theorem. K/k is normal if and only if K is the splitting field of some set $\mathbf{P}$ of irreducible polynomials in $k[X]$.

Proof. Suppose first that K is the splitting field of a set $\mathbf{P}$ of irreducible polynomials in $k[X]$. Then $K = k(\text{Roots}_{k_{\text{Alg}}} \mathbf{P})$. Since k and $\text{Roots}_{k_{\text{Alg}}} \mathbf{P}$ are stable under $\text{Aut}_k\, k_{\text{Alg}}$, K is also. Thus, K/k is normal. Suppose conversely that K/k is normal. Let $\mathbf{P} = \{f_x(X) \mid x \in K\}$ where $f_x(X)$ denotes the minimum polynomial of x over k. Since K is stable under $\text{Aut}_k\, k_{\text{Alg}}$, the orbit $\text{Roots}_{k_{\text{Alg}}} f_x(X)$ of x under $\text{Aut}_k\, k_{\text{Alg}}$ is contained in K (see 1.4.9). Thus, $f_x(X)$ splits in K. Since $K = \text{Roots}\, \mathbf{P}$, K/k is a splitting field for $\mathbf{P}$. □

2.3.5 Corollary. If K/k is normal, then K/k' is normal for any subfield k' of K containing k.

Proof. If K is the splitting field of $\mathbf{P}$ over k, then K is the splitting field for $\mathbf{P}$ over k'. □

2.3.6 Corollary. If K is a finite dimensional extension of k, then there is a finite dimensional normal extension K' of k containing K as subfield.

Proof. Let $x_1, \ldots, x_n$ be a basis for K/k. Then a splitting field K'/k of $\mathbf{P} = \{f_{x_1}(X), \ldots, f_{x_n}(X)\}$ is a normal extension of K, and K'/k is finite dimensional since $K' = k(\text{Roots}_{K'} \mathbf{P})$ and $\mathbf{P}$ is finite. □

Since the intersection of subfields K' of k_{Alg} which are normal extensions of k is a normal extension of k, the above proposition says that the minimal normal extension K^{norm} of k containing K is finite dimensional.

2.3.7 Definition. K^{norm} is the *normal closure* of K.

The next proposition is a generalization of 1.4.12 and 1.4.13. The proof is precisely the same, in view of 2.3.1.

2.3.8 Proposition. Let K/k be normal and $G = \text{Aut}_k K$. Then $K^G = k_{\text{rad}}$ and K/k_{rad} is separable.

2.3.9 Corollary. Let K/k be normal. Then $K = k_{\text{sep}}k_{\text{rad}}$.

Proof. This follows from 2.2.19 and 2.3.8. □

We now briefly introduce Galois extensions and then compare normal extensions and Galois extensions. The main results on Galois extensions are in Chapter 3.

2.3.10 Definition. K/k is *Galois* if k is the fixed field $K^G = \{x \in K \mid \sigma(X) = x \text{ for } \sigma \in G\}$ of some subgroup G of Aut K.

Finite dimensional extensions of a finite field are Galois, by 1.5.7. We state this for future reference.

2.3.11 Proposition. If K is a finite dimensional extension of a finite field k, then K/k is Galois.

2.3.12 Proposition. The following conditions are equivalent.

1. K/k is Galois;
2. K/k is the splitting field over k of a set $\mathbf{P}$ of separable irreducible polynomials in $k[X]$.
3. K/k is normal and separable.

Proof. Let $G = \text{Aut}_k K$. Suppose that $k = K^G$, let $x \in K$ and let $x_1, \ldots, x_n$ be the distinct roots of the minimum polynomial $f_x(X)$ of x over k. Then $g(X) = \prod_1^n (X - x_i)$ is fixed by each $\sigma \in G$, since

$$\sigma(g(X)) = \prod_1^n \sigma(X - x_i) = \prod_1^n (X - \sigma(x_i)) = \prod_1^n (X - x_i) = g(X).$$

It follows that $g(X) \in k[X]$. Thus, $f_x(X)$ divides $g(X)$ so that $f_x(X)$ is separable and splits in K. Thus, $\mathbf{P} = \{f_x(X) \mid x \in K\}$ is a set of separable irreducible polynomials in $k[X]$ and K is a splitting field of $\mathbf{P}$ over k, and 1 implies 2. Suppose next that K is the splitting field of a set $\mathbf{P}$ of separable irreducible polynomials in $k[X]$. Then K/k is normal, by 2.3.4. Since the elements of Roots $\mathbf{P}$ are separable over k and $K = k(\text{Roots}_K \mathbf{P})$, it follows from 2.2.15 that K/k is separable. Thus 2 implies 3. Finally, let K/k be normal and separable. Then $K^G = k_{\text{rad}} = k$, by 2.3.8, and K/k is Galois. □

2.3.13 Corollary. Let K/k be Galois and k' a subfield of K containing k. Then k'/k is Galois if and only if k' is stable under $\text{Aut}_k K$.

Proof. K/k is normal and separable. Since k'/k is then separable, the assertion follows from 2.3.3 and 2.3.12. □

The following two corollaries are counterparts of 2.3.5 and 2.3.6. The proofs are similar and we omit them.

2.3.14 Corollary. Let K/k be Galois. Then K/k' is Galois for every subfield k' of K containing k.

2.3.15 Corollary. Let K/k be finite dimensional and separable. Then K^{norm} is a finite dimensional Galois extension of K containing K as subfield (see 2.3.7).

The next theorem shows that an algebraic extension K of k is a normal extension of k if and only if it is k-isomorphic to a field extension of k of the form $k_{\text{Gal}} \otimes_k k_{\text{rad}}$ where k/k_{Gal} and k_{rad}/k are algebraic Galois and radical extensions of k respectively (see A.2 and 2.2.19).

2.3.16 Theorem. K/k is normal if and only if $K = k_{\text{sep}}k_{\text{rad}}$ and k_{sep}/k is Galois.

Proof. Suppose first that K/k is normal. Then $K = k_{\text{sep}}k_{\text{rad}}$, by 2.3.9. Now k_{sep} is stable under $\text{Aut}_k K$. Thus, k_{sep}/k is normal (see 2.3.3), hence Galois by 2.3.12. Suppose conversely that $K = k_{\text{sep}}k_{\text{rad}}$ and that k_{sep}/k is Galois. Then $k_{\text{sep}}, k_{\text{rad}}$ are stable under $\text{Aut}_k k_{\text{Alg}}$ (see 2.2.7), so that K/k is normal. □

2.4 Composites

Let k be a field, K an extension field of k. For subrings A, B of K containing k, we let AB be the set of sums $\sum_1^n x_i y_i$ where $x_i \in A$, $y_i \in B$ for $1 \leq i \leq n$. Then AB is a subring of K containing k.

2.4.1 Proposition. Let A and B be subfields of K containing k and suppose that x is algebraic over B for $x \in A$. Then AB is a subfield of K and AB/B is algebraic.

Proof. Let $z = \sum_1^n x_i y_i$ be a nonzero element of AB where $x_i \in A$, $y_i \in B$ for $1 \leq i \leq n$. Then $z, z^{-1} \in B(x_1, \ldots, x_n) = B[x_1, \ldots, x_n] \subset AB$ (see 1.2.5). Thus, AB is a subfield and AB/B is algebraic. □

2.4.2 Proposition. Let A and B be subfields of K containing k. If A/k and B/k are finite dimensional/algebraic/algebraic and normal/algebraic and Galois/algebraic and separable/algebraic and radical, then so is AB/k.

Proof. If a_i $(1 \leq i \leq m)$ span A over k and b_j $(1 \leq j \leq n)$ span B over k, then $a_i b_j$ $(1 \leq i \leq m, 1 \leq j \leq n)$ span AB over k. Thus, AB/k is finite dimensional if A/k and B/k are. Clearly $A \subset k_{\text{alg}}$, $B \subset k_{\text{alg}}$ imply $AB \subset k_{\text{alg}}$. The same is true for k_{sep} and for k_{rad}. If A/k and B/k are algebraic and normal/Galois, take sets $\mathbf{P}_A$, $\mathbf{P}_B$ of irreducible polynomials/separable polynomials in $k[X]$ such that A/k, B/k are splitting fields for $\mathbf{P}_A$, $\mathbf{P}_B$ respectively. Then AB/k is the splitting field for $\mathbf{P}_A \cup \mathbf{P}_B$, and AB/k is an algebraic normal/Galois extension.

2.4.3 Proposition. Let A and B be subfields of K containing k. If A/k is finite dimensional/algebraic/algebraic and separable/algebraic and radical, then so is AB/B.

Proof. Let S be a set spanning A over k and note that AB now is $B(S)$. Suppose that the elements of S are algebraic over k. Then $B(S)/B$ is algebraic (finite dimensional if S is finite), by 1.2.5. If the elements of S are separable/radical over k, then they are separable/radical over B and $B(S)/B$ is separable/radical by 2.2.15. ▯

2.4.4 Proposition. Let A, B be subfields of K containing k. Suppose that A/k is algebraic and normal/Galois. Then AB/B is normal/Galois and A is stable under $\mathrm{Aut}_B\, AB$. Furthermore, the groups $\mathrm{Aut}_B\, AB$ and $\mathrm{Aut}_{A\cap B}\, A$ are isomorphic under the restriction mapping $\sigma \mapsto \sigma|_A$ $(\sigma \in \mathrm{Aut}_B\, AB)$.

Proof. Consider the diagram

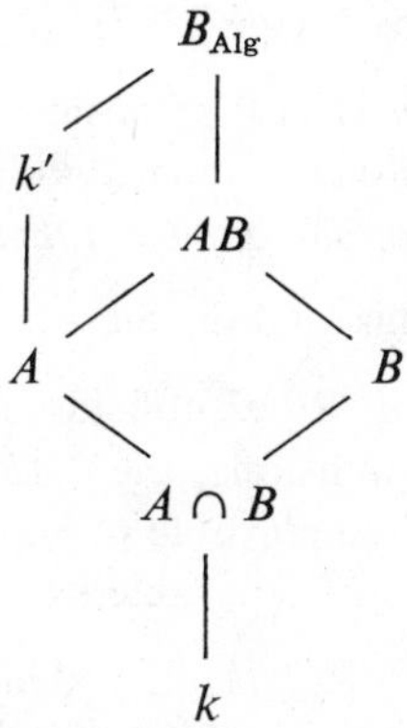

where B_{Alg} is an algebraic closure of B containing AB and k' is the algebraic closure of k in B_{Alg}. Then k' is stable under $\mathrm{Aut}_B\, B_{\mathrm{Alg}}$. But A/k is normal and stable under $\mathrm{Aut}_k\, k'$ (see 2.3.3). It follows that A is stable under $\mathrm{Aut}_B\, B_{\mathrm{Alg}}$. Thus, AB is also stable under $\mathrm{Aut}_B\, B_{\mathrm{Alg}}$ and AB/B is normal. If A/k is also separable, then AB/B is also separable, by 2.2.8, so that AB/B is Galois if A/k is Galois. The homomorphism $f(\sigma) = \sigma|_A$ is obviously injective on $\mathrm{Aut}_B\, AB$. We now show that f is surjective and assume first that $A:k < \infty$. The groups $H_1 = f(\mathrm{Aut}_B\, AB)$ and $H_2 = \mathrm{Aut}_{A\cap B}\, A$ are then finite subgroups of $\mathrm{Aut}\, A$. Thus, we can show that they are equal by showing the fields A^{H_1}, A^{H_2} to be equal (see 2.2.17). But $A^{H_1} = \{x \in A \mid x \text{ is radical over } B\} = A^{H_2}$, by 2.2.7. Thus, $f(\mathrm{Aut}_B\, AB) = H_1 = H_2 = \mathrm{Aut}_{A\cap B}\, A$ and f is surjective. We now drop the assumption that $A:k < \infty$ and let $A_S = k(S)$ for $S \in \mathbf{S}$ where $\mathbf{S}$ is the collection of finite subsets S of A stable under $\mathrm{Aut}_k\, A$. Then A_S/k is finite dimensional and normal, and $A = \bigcup_{S\in\mathbf{S}} A_S$. For $\sigma \in \mathrm{Aut}_{A\cap B}\, A$, let $\sigma_S = \sigma|_{A_S}$ and let τ_S be the element of $\mathrm{Aut}_B\, A_S B$ such that $\tau_{S|A_S} = \sigma_S$. For $S_1 \subset S_2$, we have $\sigma_{S_2|A_{S_1}} = \sigma_{S_1}$, therefore also $\tau_{S_2|A_{S_1}B} = \tau_{S_1}$. Thus, $\tau = \bigcup_{S\in\mathbf{S}} \tau_S$ is a B-automorphism of $\bigcup_{S\in\mathbf{S}} A_S B = AB$. Clearly, $f(\tau) = \tau|_A = \sigma$. Thus, f is surjective. ▯

2.4.5 Proposition. Let A, B be subfields of K containing k and suppose that A/k and B/k are algebraic and normal. Then the homomorphism $f\colon \operatorname{Aut}_k AB \to \operatorname{Aut}_k A \times \operatorname{Aut}_k B$ (outer direct product) defined by $f(\sigma) = \sigma|_A \times \sigma|_B$ is injective.

Proof. Let $\sigma \in \operatorname{Kernel} f$. Then $\sigma(x) = x$ for $x \in A \cup B$, thus for $x \in AB$. Thus, f has trivial kernel and f is injective. □

2.4.6 Definition. A property **P** of groups is *injective* if

1. any subgroup H of a group with **P** has **P**;
2. If G and H are groups having **P**, then $G \times H$ (outer direct product) has **P**.

The properties "G is Abelian," "G is solvable," "Exp G divides n," n being a fixed positive integer, are injective properties of groups G.

2.4.7 Definition. An extension K/k is a **P**-*extension* if K/k is separable algebraic and Aut K^{norm}/k has **P** (see 2.3.7).

2.4.8 Definition. If K/k is a **P**-extension where **P** is the property of being cyclic/Abelian/solvable/of exponent dividing n, then K/k is said to be *cyclic/Abelian/solvable/of separable exponent dividing n.*

The following proposition is an immediate consequence of 2.4.5.

2.4.9 Proposition. Let A, B be subfields of K containing k. If **P** is injective and A/k, B/k are **P**-extensions, then AB/k is a **P**-extension. In particular, if A/k, B/k are Abelian/solvable/of separable exponent dividing n, then AB/k is Abelian/solvable/of separable exponent dividing n.

2.4.10 Definition. We let $k_{\text{abel}}/k_{\text{solv}}/{}_nk$ denote the union of all subfields A of K containing k such that A/k is Abelian/solvable/of separable exponent dividing n. If $K = k_{\text{Alg}}$, these sets are denoted $k_{\text{Abel}}/k_{\text{Solv}}/{}_nk$ and we let ${}_nk_{\text{Abel}} = {}_nk \cap k_{\text{Abel}}$.

The sets k_{abel}, k_{solv}, ${}_nk$ are subfields of K containing k, by 2.4.9. The field extensions $k_{\text{Abel}}/k_{\text{Solv}}/{}_nk/{}_nk_{\text{Abel}}$ of k are normal and every finite dimensional subextension is Abelian/solvable/of separable exponent dividing n/Abelian of separable exponent dividing n.

2.4.11 Proposition. The extensions $k_{\text{Abel}}/{}_nk/{}_nk_{\text{Abel}}$ are Abelian/of separable exponent dividing n/Abelian of separable exponent dividing n.

Proof. Let $\sigma, \tau \in \operatorname{Aut}_k k_{\text{Abel}}$ and let k' be a subfield of k_{Abel} containing k such that k'/k is normal and Abelian. Then $\sigma\tau|_{k'} = \sigma|_{k'}\tau|_{k'} = \tau|_{k'}\sigma|_{k'} = \tau\sigma|_{k'}$. Since k_{Abel} is the union of such k', $\sigma\tau = \tau\sigma$ and k_{Abel}/k is Abelian. Next let $\sigma \in \operatorname{Aut}_k k_{\text{Alg}}$ and let k' be a subfield of ${}_nk$ containing k such that k'/k is normal and of separable exponent dividing n. Then $\sigma^n|_{k'} = (\sigma|_{k'})^n$ is the identity mapping on k'. Since ${}_nk$ is the union of such k', σ^n is the identity on k and ${}_nk$ is of separable exponent dividing n. Since ${}_nk_{\text{Abel}} = {}_nk \cap k_{\text{Abel}}$ is normal over k, we have

$$\operatorname{Aut}_k ({}_nk_{\text{Abel}}) = \operatorname{Aut}_k ({}_nk)|_{{}_nk_{\text{Abel}}} \quad \text{and} \quad \operatorname{Aut}_k ({}_nk_{\text{Abel}}) = \operatorname{Aut}_k (k_{\text{Abel}})|_{{}_nk_{\text{Abel}}},$$

by 2.3.2. Thus, ${}_nk_{\text{Abel}}$ is Abelian of separable exponent dividing n over k. □

We can blow up an arbitrary field k as follows in terms of these constructions:

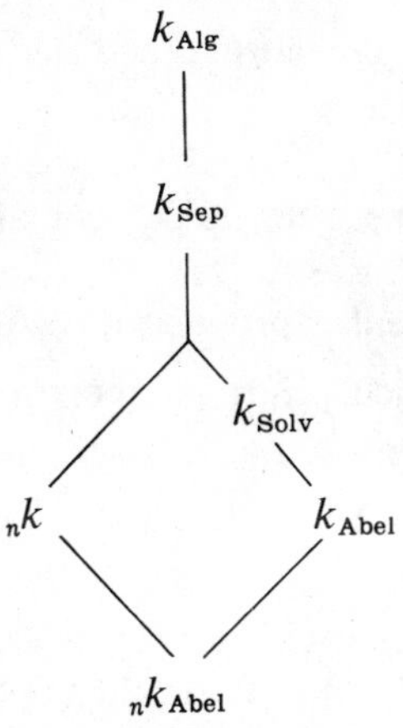

The bottom extension ${}_nk_{\text{Abel}}/k$ is an important part of the extension k_{Sep}/k and is discussed in 3.10.

E.2 Exercises to Chapter 2

E.2.1 (Fundamental Theorem of Algebra). Let $f(X)$ be an element of $\mathbb{R}[X]$ of odd degree. Show that there exist $a, b \in \mathbb{R}$ with $a < b$ such that $f(a)$ and $f(b)$ are nonzero of opposite sign. Show that $f(c) = 0$ for some c such that $a < c < b$. (Hint: Otherwise, there is a sequence of pairs (a_i, b_i) with the same property as (a, b) such that $a < a_1 < a_2 \cdots$, $b > b_1 > b_2 \cdots$, and $b_i - a_i$ converges to 0. Show that a_i and b_i are Cauchy sequences and $f(c) = 0$ where c is their limit).

E.2.2 (Fundamental Theorem of Algebra). Let $f(X) = \prod_1^n (X - x_i)$ be an element of $\mathbb{R}[X]$, the x_i being elements of an algebraic closure of $\mathbb{C}$. Show that

1. $g(X) = \prod_{i<j} (X - (x_i + x_j))$ and $h(X) = \prod_{i<j} (X - x_ix_j)$ are elements of $\mathbb{R}[X]$ of degree $n(n - 1)/2$;
2. if $x_i + x_j$ and x_ix_j are elements of $\mathbb{C}$, then x_i and x_j are elements of $\mathbb{C}$.

E.2.3 (Fundamental Theorem of Algebra). Using the preceding two exercises, prove by induction on the highest power of 2 dividing $\operatorname{Deg} f(X)$ that every nonconstant $f(X)$ in $\mathbb{R}[X]$ has all its roots in $\mathbb{C}$. Show that $\mathbb{C}$ is algebraically closed. (Hint: the highest power of 2 dividing $n(n - 1)/2$ is less than that for n).

E.2.4. Show that $X^{10} + X^5 + 1$ is not irreducible over $\mathbb{Z}_5$. What is the radical exponent of $X^{10} + X^5 + 1$ over $\mathbb{Z}_5$?

E.2.5. Let $K = \mathbb{Z}_5(X)$ be the field of rational expressions in an indeterminant X over a field of five elements. Let $k = \mathbb{Z}_5(X^{10} + X^5 + 1)$. Show that

(a) $K{:}k = 10$;
(b) K/k is neither separable nor radical. Describe k_{rad}.

E.2.6. Let $K = \mathbb{Q}(x)$ where x is a complex root of $X^4 + X^2 + 1$. How many homomorphisms are there from K into $\mathbb{C}$?

E.2.7. Show that every quadratic extension K/k is normal.

E.2.8. A *cubic* extension is an extension K/k of degree 3. Show that if K/k is a cubic extension which is not normal and if L is the normal closure of K over k, then

(a) $L{:}k = 6$;
(b) L has precisely one subfield k' containing k such that k'/k is quadratic;
(c) $L = Kk'$ (internal tensor product) (See Appendix A).

E.2.9. Determine for which prime numbers p the polynomial $X^2 + 1$ in $\mathbb{Z}_p[X]$ is irreducible.

E.2.10. Show that $X^2 + X - c$ is irreducible in $K[X]$ if and only if $c = d^2 + d$ for some $d \in K$.

E.2.11. Let $f(X)$ be a monic polynomial in $\mathbb{Z}[X]$. Show that if $x \in \mathbb{Q}$ is a root of $f(X)$, then $x \in \mathbb{Z}$. If $f(0)$ is prime, then $f(X)$ has at most three different roots in $\mathbb{Q}$. What can be said if $f(0) = p^2q$ where p and q are prime?

E.2.12 (Artin). Let K/k be a finite dimensional field extension and suppose that k is infinite. Show that

(a) if $K = k(x)$ and k' is a subfield of K containing k, then $k' = k(S)$ where S is the set of coefficients of the minimum polynomial of x over k';
(b) if K/k is simple, then K has only finitely many subfields containing k;
(c) if $x, y \in K$ and $k(x, y)$ has only finitely many subfields containing k, then $k(x, y) = k(x + ay)$ for infinitely many elements a of k;
(d) if K has only finitely many subfields containing k, then K is a simple extension of k.

E.2.13. Let K/k be a finite dimensional separable extension. Show that K has only finitely many subfields containing k.

E.2.14. Construct a splitting field K for $(X^2 - 2)(X^2 + X + 1)$ over $\mathbb{Q}$ and find $s \in K$ such that $K = \mathbb{Q}(s)$.

E.2.15. Construct a splitting field K for $(X^2 + 2)(X - 1)(X^2 + X - 1)$ over $\mathbb{Z}_3$ and find $s \in K$ such that $K = \mathbb{Z}_3(s)$.

E.2.16. Construct x such that $K = \mathbb{Q}(x)$ where K is the splitting field over $\mathbb{Q}$ of $X^2 - 3$ and $X^2 - 5$.

E.2.17. Let K/k be an extension of degree p^2 (p being the characteristic of K and p being nonzero). Show that K/k is not a simple extension if and only if K/k is radical of exponent one (that is, $x^p \in k$ for all $x \in K$).

E.2.18 (Kaplanski). Let K/k be a finite dimensional extension and suppose that $K = k_{\text{sep}}(y)$ for some $y \in K$. Show that K/k is a simple extension. (Hint: Show that K has only finitely many subfields k' containing k by considering all k' with a fixed k'_{sep} and replacing k by k'_{sep}).

E.2.19. Let K/k be an algebraic extension. Show that k_{rad} is a subfield of K containing k. (Use 2.2.5 and the equations $(st)^p = s^p t^p$, $(s + t)^p = s^p + t^p$ for $s, t \in K$).

E.2.20. Let $K = k_0(X_1, \ldots, X_n)$ be the field of rational expressions in the algebraically independent variables $X_1, \ldots, X_n$ over the field k_0 and let $k = k_0(X_1{}^p, \ldots, X_n{}^p)$.

Show that

(a) K/k is a radical extension of degree p^n;

(b) $K = k(X_1) \cdots k(X_n)$ (internal tensor product of k-algebras) (see Appendix A).

E.2.21. Let k be a field of characteristic $p > 0$ and let $a \in k$. Show that either $X^p - X - a$ is irreducible over k, or $X^p - X - a$ has p distinct roots in k.

E.2.22 (Perfect fields). Show that

(a) a field k is perfect if and only if every finite dimensional extension K of k is separable;

(b) k is perfect if and only if $k = k^p$ (where p is the exponent characteristic of k and $k^p = \{x^p \mid x \in k\}$);

(c) if k is perfect and k' is a finite dimensional extension of k, then k' is perfect;

(d) if k' is perfect and k is a subfield such that $k'{:}k < \infty$, then k is perfect.

E.2.23. Let K/k be finite dimensional of characteristic p. Show that if K/k is not separable, then K has subfields k_n such that K/k_n is radical of degree p^n for every positive integer n.

E.2.24. Show that the hypothesis that k be infinite can be dropped in Theorem 2.2.10 without affecting the validity of the conclusion.

E.2.25. Show that for a finite dimensional extension K/k to be normal, it is necessary and sufficient that for any monic irreducible element $f(X)$ of $k[X]$ having a root in $K, f(X) = \prod_1^n (X - x_i)$ with $x_1, \ldots, x_n$ in K.

E.2.26. Give an example of a finite dimensional normal extension K of k having a subfield k' containing k such that k'/k is not normal.

E.2.27. Describe a finite dimensional field extension K of k which is not normal over k but which has a subfield k' containing k such that K/k' and k'/k are normal.

E.2.28. Give an example of a Galois extension K of k of degree 6 having a subfield k' containing k such that k'/k is not Galois.

E.2.29. Prove the Chain Rule $g(f(X))' = g'(f(X))f'(X)$ for $f(X)$, $g(X) \in K[X]$.

E.2.30. Prove Taylor's Theorem $f(X) - f(x) = \sum_0^n 1/i!\, f^{(i)}(x)(X - x)^i$ for $f(X) \in K[X]$, K being of characteristic 0 and $f^{(i)}(X)$ being appropriately defined.

E.2.31. Prove the formula

$$\frac{(f_1 \cdots f_n)'}{f_1 \cdots f_n} = \frac{f_1'}{f_1} + \cdots + \frac{f_n'}{f_n}$$

for Logarithmic Differentiation, $f_1, \ldots, f_n$ being nonzero elements of $K[X]$.

E.2.32. For $f(X) = \prod_1^n (X - x_i)$ in $K[X]$, show that

$$f(X)' = \sum_1^n (f(X)/(X - x_i)).$$

E.2.33. Show that the derivative $(f(X)/g(X))'$ is well defined by

$$\left(\frac{f(X)}{g(X)}\right)' = \frac{1}{g(X)^2}\,(f(X)'g(X) - f(X)g(X)')$$

for $f(X)/g(X) \in K(X)$ (field of fractions of $K[X]$). Derive the formulas for derivatives of sums, products and quotients in $K(X)$.

E.2.34. To what degree of generality can one prove the following in $K(X)$:
Chain Rule
Taylor's Theorem
Formula for Logarithmic Differentiation.

E.2.35. Let $(f(X)')' = 0$, where $f(X) \in K[X]$ and K is of characteristic $p > 0$. Show that $f(X) = g(X^p) + Xh(X^p)$ for suitable $g(X), h(X) \in K[X]$. What can be said if $f^{(m)}(X) = 0$?

E.2.36. Show that $Q(\sqrt{2} + \sqrt[3]{3})$ contains $\sqrt{2}$ and $\sqrt[3]{3}$.

E.2.37. Show that "G is nilpotent" is an injective property of groups. Show that if A and B are subfields of K containing k and A/k, B/k are finite dimensional and nilpotent, then AB/k is nilpotent.

3 Classical Galois theory

We now turn to the study of Galois extensions K/k. These are studied by comparing the structure of K/k with that of its Galois group $\text{Aut}_k K$. We begin with some general material on isomorphisms and groups of automorphisms of fields which leads to the Galois Correspondence Theorem (see 3.3) and the Normal Basis Theorem (see 3.4). We then specialize to the study of cyclotomic extensions and extensions K/k whose Galois group $\text{Aut}_k K$ is cyclic, Abelian or solvable.

3.1 Linear independence of homomorphisms

Let A be a group, L a field and V a vector space over L. Let $\sigma_1, \ldots, \sigma_n$ be distinct elements of the set Hom (A, L^*) of homomorphisms from A to the group $L^* = L - \{0\}$ of units of L and let $v_1, \ldots, v_n \in V$. Let $\sum_1^n \sigma_i v_i$ denote the function from A to V which maps $a \in A$ to $\sum_1^n \sigma_i(a) v_i \in V$ and let 0 denote the function mapping each element $a \in A$ to $0 \in V$. Then $\sigma_1, \ldots, \sigma_n$ are linearly independent over V in the following sense.

3.1.1 Theorem. If $\sum_1^n \sigma_i v_i$, then $v_1 = \cdots = v_n = 0$.

Proof. Suppose not and take $\sigma_1, \ldots, \sigma_n$ and $v_1, \ldots, v_n$ (not all v_i zero) with n minimal such that $\sum_1^n \sigma_i v_i = 0$. By the minimality of n, $v_i \neq 0$ for $1 \leq i \leq n$. For $a, b \in A$, we then have

$$0 = \sum_1^n \sigma_i(ab) v_i = \sum_1^n \sigma_i(a)\sigma_i(b) v_i$$

$$0 = \sigma_n(a) \sum_1^n \sigma_i(b) v_i = \sum_1^n \sigma_n(a)\sigma_i(b) v_i.$$

It follows that

$$0 = \sum_1^{n-1} (\sigma_i(a) - \sigma_n(a))\sigma_i(b) v_i \qquad \text{for all } b \in A,$$

therefore that $0 = \sum_1^{n-1} \sigma_i v_i'$ where $v_i' = (\sigma_i(a) - \sigma_n(a)) v_i$. By the minimality of n, the v_i' are 0. Since the v_i are nonzero, $\sigma_i(a) = \sigma_n(a)$, for all i and $a \in A$. Thus, $\sigma_i = \sigma_n$ for all i, contradicting the distinctness of the σ_i. Thus, the σ_i are linearly independent over V. □

3.1.2 Corollary. Let $\sigma_1, \ldots, \sigma_n$ be distinct homomorphisms from a field K to a field L and let $v_1, \ldots, v_n$ be elements of a vector space V over L. Then $\sum_1^n \sigma_i v_i = 0 \Rightarrow v_1 = \cdots = v_n = 0$.

Proof. Taking A to be the group $K^* = K - \{0\}$ of units of K, $\sigma_{1|A}, \ldots, \sigma_{n|A}$ are distinct elements of Hom (A, L^*). Now apply 3.1.1. □

3.1.3 Corollary (Dedekind Independence Theorem). Let $\sigma_1, \ldots, \sigma_n$ be distinct homomorphisms from a field K to a field L. Then $\sigma_1, \ldots, \sigma_n$ are linearly independent over L in the vector space $\mathbf{F}(K, L)$ of functions from K to L.

Proof. Take $V = L$ in 3.1.2. □

3.2 Galois Descent

Let K be a field, G a subgroup of the group Aut K of automorphisms of K. We are concerned here with the passage (called *descent*) from K to K^G. At times, it is necessary to assume that the orbits of G in K are finite, that is, that G acts continuously on K (see 0.4.5). This is equivalent to the assumption that K/K^G is algebraic, as we showed in 1.2.8.

3.2.1 Definition. The subgroup G of Aut K is *algebraic* if the orbits of G in K are finite.

3.2.2 Definition. A *G-product* on a vector space V over K is a mapping from $G \times V$ to V, denoted $(\sigma, v) \mapsto \sigma(v)$, such that

1. V with $\sigma(v)$ is a G-space, that is, $\epsilon_G(v) = v$ and $\sigma(\tau(v)) = (\sigma\tau)(v)$ for $\sigma, \tau \in G, v \in V$, where ϵ_G is the identity element of G;
2. $\sigma(v + w) = \sigma(v) + \sigma(w)$ and $\sigma(xv) = \sigma(x)\sigma(v)$ for $v, w \in V, x \in K$.

A G-product on V over K is *continuous* if the G-orbits in V are finite (see 0.4.5).

The G-products $\sigma(v)$ on a vector space V correspond bijectively to those homomorphisms f from G to the group Aut $(V, +)$ of automorphisms of the additive group $(V, +)$ of V such that $f(\sigma)$ is *σ-linear* for $\sigma \in G$, that is, such that $f(\sigma)(xv) = \sigma(x)f(\sigma)(v)$ for $x \in K, v \in V$. The correspondence is given by $f(\sigma)(v) = \sigma(v)$.

Starting from a vector space V^k over a subfield k of K^G, we get a vector space $V^K = K \otimes_k V^k$ over K with scalar product such that $x(y \otimes v) = (xy) \otimes v$ for $x, y \in K$ and $v \in V$. We also get a G-product on V^K such that $\sigma(x \otimes v) = \sigma(x) \otimes v$ for $x \in K, v \in V^k$ (see T.4). Note that if the orbits of G in K are finite, then the orbits of G in V^K are finite. The vector space V^K/G-product $\sigma(x \otimes v)$ is called the *vector space/G-product obtained from V^k by ascent from k to K*. The k-subspace $1 \otimes V^k$ is isomorphic to V^k and is a k-form of V^K in the sense of the following definition (see T.4).

3.2.3 Definition. Let V be a vector space over K and let k be a subfield of K. Then a *k-form* of V is a k-subspace V^k of V such that a k-basis for V^k is a K-basis for V.

We now proceed to prove that if V is a vector space over K with continuous G-product $\sigma(v)$, $V^G = \{v \in V \mid \sigma(v) = v \text{ for } \sigma \in G\}$ is a K^G-form of V. Roughly speaking, this amounts to saying that V and the given G-product on V are obtained from the K^G-space V^G by ascent from K^G to K. The K^G-space V^G is called the *vector space obtained* from V and the given G-product on V *by Galois descent from K to K^G*. A first step in the proof is the following proposition.

3.2.4 Proposition. Let G be finite, let V be a vector space over K and let $\sigma(v)$ be a G-product on V. For $v \in V$, let $\bar{v} = \sum_{\sigma \in G} \sigma(v)$. Then the $\bar{v}$ are elements of the set $V^G = \{v \in V \mid \sigma(v) = v \text{ for } \sigma \in G\}$ of fixed points of G in V, and for each nonzero $v \in V$, there exists $a \in K$ such that $\overline{av} \neq 0$.

Proof. Since

$$\tau(\bar{v}) = \sum_{\sigma \in G} \tau(\sigma(v)) = \sum_{\sigma \in G} \sigma(v) = \bar{v},$$

the $\bar{v}$ are in V^G. Now suppose that $\overline{av} = 0$ for all $a \in K$. Then

$$0 = \sum_{\sigma \in G} \sigma(av) = \sum_{\sigma \in G} \sigma(a)\sigma(v) \qquad \text{for } a \in K,$$

so that $0 = \sum_{\sigma \in G} \sigma[\sigma(v)]$ and $\sigma(v) = 0$ for $\sigma \in G$ by 3.1.2. But then $v = 0$, a contradiction. Thus, $\overline{av} \neq 0$ for some $a \in K$. ☐

We can now prove the following Galois Descent Theorem.

3.2.5 Theorem (Speiser). Let V be a vector space over K, $\sigma(v)$ a continuous G-product on V. Then V^G is a K^G-form of V.

Proof. We first show that if $v_1, \ldots, v_n$ are linearly independent elements of V^G over K^G, then $v_1, \ldots, v_n$ are linearly independent over K. Suppose not and let $v_1, \ldots, v_n$ be K^G-linearly independent elements of V^G with n minimal such that $\sum_1^n c_i v_i = 0$ for suitable c_i (not all zero) in K. Then the c_i are all nonzero, by the minimality of n, and we may take $c_n = 1$ (by dividing each c_i by the original c_n). For $\sigma \in G$, we then have $\sum_1^n \sigma(c_i)v_i = 0$, since the v_i are in V^G, so that $\sum_1^{n-1} (c_i - \sigma(c_i))v_i = 0$. By the minimality of n, $c_i = \sigma(c_i)$ for $1 \leq i \leq n - 1$ and $\sigma \in G$. Thus, $c_1, \ldots, c_{n-1}$ and $c_n = 1$ are elements of K^G, contradicting the linear independence of $v_1, \ldots, v_n$ over K^G. We next show that V^G spans V over K, thereby completing the proof. We suppose first that G is finite and let W be the span of V^G over K. Then we have a G-product on V/W, defined by $\sigma(v + W) = \sigma(v) + W$ for $\sigma \in G$, $v \in V$. If $v + W$ is a nonzero element of V/W, then $\overline{a(v + W)}$ is nonzero for some $a \in K$, by 3.2.4. But $\overline{a(v + W)} = \overline{av} + W$ and $\overline{av} \in V^G \subset W$, so that $\overline{a(v + W)}$ is zero for $a \in K$. Thus, V/W has no nonzero $v + W$ and $V = W$. Now we drop the assumption that G is finite and let $v \in V$. We claim that v is contained in the K-span of V^G. The orbit v^G of v under G is finite. Thus, the normal subgroup

$$H = \{\sigma \in G \mid \sigma(w) = w \qquad \text{for } w \in V^G\}$$

is of finite index in G. We may regard the finite group $G' = G/H$ as a subgroup of the group Aut K' of automorphisms of the field $K' = K^H$. The given G-product on V now induces a G'-product on the K'-span V' of V^G. Since G' is finite, V' is the K'-span of $V'^{G'}$. Since $v \in V'$ and $V'^{G'} \subset V^G$, v is therefore in in the K-span of V^G. Thus, V is the K-span of V^G. □

If V^{K^G} is a K^G-form of a vector space V over K, then we have a corresponding G-product in V such that $V^G = V^{K^G}$, namely, $\sigma(\sum_1^n x_i v_i) = \sum_1^n \sigma(x_i) v_i$ where the v_i form a basis for V^{K^G} over K^G and the x_i are in K. (This G-product is essentially the G-product in $V^K = K \otimes_{K^G} V^{K^G}$ obtained by ascent from K^G to K as described earlier.) If the orbits of G in K are finite, then the orbits of G in V with respect to this G-product are finite. The following proposition is an immediate consequence of these observations and the above Galois Descent Theorem.

3.2.6 ***Proposition.*** Let G be algebraic and let V be a vector space over K. Then the mapping defined above maps the set of K^G-forms of V bijectively onto the set of G-products on V such that the orbits of G in V are finite.

We now apply the Galois Descent Theorem to the K-vector space $\mathbf{F}(X, K)$ of functions from a G-space X to K. This is then used in the next section in giving an otherwise self-contained proof of the Galois Correspondence Theorem for algebraic extensions. (See 2.2.17 for a more classical proof for finite dimensional extensions.)

3.2.7 ***Definition.*** For $f \in \mathbf{F}(X, K)$ and $\sigma \in G$, let $\sigma(f)$ be the element of $\mathbf{F}(X, K)$ defined by the diagram

$$\begin{array}{ccc} X & \xrightarrow{\ \sigma\ } & X \\ {\scriptstyle f}\downarrow & & \downarrow{\scriptstyle \sigma(f)} \\ K & \xrightarrow[\ \sigma\ ]{} & K \end{array}$$

that is, the function on X defined by $\sigma(f)(x) = \sigma(f(\sigma^{-1}(x)))$.

3.2.8 ***Proposition.*** The mapping $G \times \mathbf{F}(X, K) \to \mathbf{F}(X, K)$ sending (σ, f) to $\sigma(f)$ for $\sigma \in G, f \in \mathbf{F}(X, K)$ is a G-product on $\mathbf{F}(X, K)$. If X is finite and G is algebraic, this G-product is continuous and the set $\mathbf{F}^G(X, K) = \{f \in \mathbf{F}(X, K) \mid \sigma(f(x)) = f(\sigma(x))$ for $\sigma \in G, x \in K\}$ of G-linear functions from X to K is a K^G-form of $\mathbf{F}(X, K)$.

Proof. We have $\sigma(f) = \sigma_L f \sigma_L^{-1}$, where $\sigma_L : X \to X$ is defined by $\sigma_L(x) = \sigma(x)$ for $x \in X$ and the product is composition of functions. The equations $\epsilon_L f \epsilon_L^{-1} = f$, $(\sigma\tau)_L f (\sigma\tau)_L^{-1} = \sigma_L(\tau_L f \tau_L^{-1})\sigma_L^{-1}$, $\sigma_L(f_1 + f_2)\sigma_L^{-1} = \sigma_L f_1 \sigma_L^{-1} + \sigma_L f_2 \sigma_L^{-1}$, $\sigma_L(xf)\sigma_L^{-1} = \sigma(x)\sigma_L f \sigma_L^{-1}$ are then easily verified and show that $\sigma(f)$ is a G-product on $\mathbf{F}(X, K)$. Next, let X be finite and let the orbits of G in K be finite. Then for any $f \in \mathbf{F}(X, K)$, the possible values for $\sigma(f)(x) = \sigma(f(\sigma^{-1}(x)))$ are finitely many, being contained in the orbits of the finite set Image f. Since X is finite, it follows that the orbits Gf of G in $\mathbf{F}(X, K)$ are finite. Now the Galois Descent Theorem applies, so that $\mathbf{F}^G(X, K) = \mathbf{F}(X, K)^G$ is a K^G-form of $\mathbf{F}(X, K)$. □

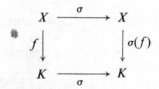

3.3 The Galois Correspondence Theorem

In this section, K is a field and G a subgroup of Aut K. We recall that for subfields k of K, K/k is Galois if $k = K^G$ for some subgroup G of Aut K. The analogous concepts for subgroups G of Aut K is the following one.

3.3.1 Definition. G is *Galois* if $G = \operatorname{Aut}_k K$ for some subfield k of K.

We now let $\mathbf{F} = \{k \mid k$ is a subfield of K and K/k is algebraic and Galois$\}$ and $\mathbf{G} = \{G \mid G$ is a subgroup of Aut K and G is algebraic and Galois$\}$. The mappings $k \mapsto \operatorname{Aut}_k K$ on $\mathbf{F}$ and $G \mapsto K^G$ on $\mathbf{G}$ map $\mathbf{F}$ to $\mathbf{G}$ and $\mathbf{G}$ to $\mathbf{F}$ (see 3.2.1, 1.2.8). Moreover, for $k \in \mathbf{F}$ and $G = \operatorname{Aut}_k K$, $k = K^G$ since K/k is Galois; and for $G \in \mathbf{G}$ and $k = K^G$, $G = \operatorname{Aut}_k K$ since G is Galois. We have now established that these mappings are bijective inverses tweeen $\mathbf{F}$ and $\mathbf{G}$. The mapping $\Gamma: \mathbf{F} \to \mathbf{G}$, defined by $\Gamma(k) = \operatorname{Aut}_k K$ for $k \in \mathbf{F}$, is called the *Galois Correspondence* between $\mathbf{F}$ and $\mathbf{G}$. We state the bijectivity of Γ for future reference.

3.3.2 Proposition. Γ is a bijection from $\mathbf{F}$ to $\mathbf{G}$.

The purpose of this section is to establish some fundamental properties of the Galois Correspondence between $\mathbf{F}$ and $\mathbf{G}$. The main tool is Galois Descent.

We begin by taking the set X of 3.2 to be the set $G/H = \{\tau H \mid \tau \in G\}$ of left cosets of a subgroup H of G and regard G/H as G-space with product $\sigma(\tau H) = (\sigma\tau)H$ $(\sigma \in G, \tau H \in G/H)$. Each $x \in K^H$ determines a function $\hat{x} \in \mathbf{F}(G/H, K)$, defined by $\hat{x}(\tau H) = \tau(x)$ for $\tau H \in G/H$. In fact, the set $\mathbf{F}^G(G/H, K)$ of *G-linear* functions from G/H to K is the set $\widehat{K^H} = \{\hat{x} \mid x \in K^H\}$. For on the one hand, $\hat{x}(\sigma(\tau H)) = \hat{x}((\sigma\tau)H) = (\sigma\tau)(x) = \sigma(\tau(x)) = \sigma(\hat{x}(\tau H))$. And on the other, if $f \in \mathbf{F}^G(G/H, K)$ and $x = f(\epsilon H)$ where ϵ is the identity of G, then $x \in K^H$ since $\sigma(x) = \sigma(f(\epsilon H)) = f(\sigma H) = f(\epsilon H) = x$ for $\sigma \in H$, and $f = \hat{x}$ since $f(\tau H) = f(\tau(\epsilon H)) = \tau(f(\epsilon H)) = \tau(x) = \hat{x}(\tau H)$ for $\tau H \in G/H$. If G is algebraic and G/H is finite, then it follows from 3.2.8 and 3.2.5 that $\widehat{K^H}$ is a K^G-form of $\mathbf{F}(G/H, K)$. We now have essentially proved the following generalization of 2.2.7, which is the main part of the Galois Correspondence Theorem for algebraic extensions.

3.3.3 Theorem. Suppose that G is algebraic and let H be a subgroup of G of finite index. Then $K^H : K^G = G : H$.

Proof. We have just seen that $\widehat{K^H}$ is a K^G-form of $\mathbf{F}(G/H, K)$. It follows that $K^H : K^G = \widehat{K^H} : K^G = \mathbf{F}(G/H, K) : K = G : H$. □

3.3.4 Corollary. Suppose that G is algebraic and let H, H' be subgroups of G of finite index. Then $K^H = K^{H'}$ if and only if $H = H'$.

Proof. Suppose that $K^H = K^{H'}$. Then $K^H = K^{H''}$ where H'' is the subgroup of G generated by H and H'. But then $H \subset H''$ and $G:H = K^H : K^G = K^{H''} : K^G = G : H''$. Thus, $H = H''$ so that $H \supset H'$. Similarly, one shows that $H' \supset H$, thus that $H = H'$. □

Recall that the closure $\overline{H}$ of H in G (in the Krull Topology of G) is $\overline{H} = \bigcap_{N \in \mathbf{N}} NH$ for $H \subset G$, where $\mathbf{N}$ is the set of normal subgroups of G of finite index, and H is closed if and only if $H = \overline{H}$. Recall also that the subgroups H of finite index in G are closed (see 0.4.1).

3.3.5 Proposition. Suppose that G is algebraic and let H, H' be subgroups of G. Then $K^H = K^{\overline{H}}$, and $K^H = K^{H'}$ if and only if $\overline{H} = \overline{H}'$.

Proof. Let $\sigma \in \overline{H}$ and $x \in K^H$. The orbit Gx of x is finite, so that $N = \{\tau \in G \mid \tau(y) = y \text{ for } y \in Gx\}$ is a normal subgroup of G of finite index. Thus, $\sigma \in NH$. But then $\sigma(x) = x$, since this is true of the elements of N and H. It follows that $K^H \subset K^{\overline{H}}$, hence that $K^H = K^{\overline{H}}$. In particular, $K^H = K^{H'}$ if $\overline{H} = \overline{H}'$. Suppose, finally, that $K^H = K^{H'}$ and let N be any normal subgroup of G of finite index. Then NH and NH' are subgroups of G of finite index and $K^{NH} = K^{NH'}$. Thus, $NH = NH'$ by 3.3.4. Since $NH = NH'$ for all such N, $\overline{H} = \overline{H}'$. ☐

3.3.6 Corollary. Let K/k be algebraic. Then $\mathrm{Aut}_k K$ is algebraic and the Galois subgroups of $\mathrm{Aut}_k K$ are the closed subgroups of $\mathrm{Aut}_k K$.

Proof. The subgroup $\mathrm{Aut}_k K$ is algebraic, by 1.2.8. Suppose that G is a Galois subgroup of $\mathrm{Aut}_k K$. Then $K^G = K^{\overline{G}}$, by 3.3.5, and we have $\overline{G} \subset \mathrm{Aut}_{K^G} K = G \subset \overline{G}$. Thus, $\overline{G} = G$ and G is closed. Suppose, conversely, that G is a closed subgroup of $\mathrm{Aut}_k K$. Then $K^G = K^{G'}$ where $G' = \mathrm{Aut}_{K^G} K$. Since G is closed and G' is closed by the preceding discussion, $G = G'$, by 3.3.5. That is, $G = \mathrm{Aut}_{K^G} K$ and G is Galois. ☐

We now let K/k be an algebraic Galois extension and recall that K/k' is Galois for any subfield k' of K containing k (see 2.3.14). Thus, letting $\mathbf{F}_k$ be the set $\{k' | k' \text{ is a subfield of } K \text{ containing } k\}$ of subfields of K/k and $\mathbf{G}_k$ the set of closed subgroups of $\mathrm{Aut}_k K$, the above corollary together with 3.3.2 says that the mapping $\Gamma_k = \Gamma\}_{\mathbf{F}_k}$ is a bijection from $\mathbf{F}_k$ to $\mathbf{G}_k$. The mapping Γ_k on $\mathbf{F}_k$ is called the *Galois Correspondence* between $\mathbf{F}_k$ and $\mathbf{G}_k$.

We next let $\sigma \in \mathrm{Aut}_k K$, $k' \in \mathbf{F}_k$ and note that $\sigma(\mathrm{Aut}_{k'} K)\sigma^{-1} = \mathrm{Aut}_{\sigma(k')} K$ (see E.1.42). Thus, $\sigma(\mathrm{Aut}_{k'} K)\sigma^{-1} = \mathrm{Aut}_{k'} K$ if and only if $\sigma(k') = k'$, by the injectivity of Γ_k. It follows that k'/k is normal (and therefore Galois) if and only if $\mathrm{Aut}_{k'} K$ is a normal subgroup of G, by 2.3.3.

For $k' \in \mathbf{F}_k$ and k'/k Galois, the restriction mapping $\sigma \mapsto \sigma|_{k'}$ is a surjective homomorphism from $\mathrm{Aut}_k K$ to $\mathrm{Aut}_k k'$, by 2.3.3. The kernel of this homomorphism is $\mathrm{Aut}_{k'} K$ so that it induces an isomorphism from $\mathrm{Aut}_k K/\mathrm{Aut}_{k'} K$ to $\mathrm{Aut}_k k'$.

We summarize the above observations in the following *Galois Correspondence Theorem* for algebraic extensions.

3.3.7 Theorem. Let K/k be an algebraic Galois extension. Then

1. the mapping Γ_k defined by $\Gamma_k(k') = \mathrm{Aut}_{k'} K$ is a bijection from the set $\mathbf{F}_k$ of subfields k' of K containing k to the set $\mathbf{G}_k$ of closed subgroups of $\mathrm{Aut}_k K$;

2. a subfield $k' \in \mathbf{F}_k$ is Galois over k if and only if the corresponding subgroup $G' = \Gamma_k(k')$ is normal in $\mathrm{Aut}_k K$;
3. if $k' \in \mathbf{F}_k$ is Galois over k, then the restriction mapping $\sigma \mapsto \sigma|_{k'}$ is a surjective homomorphism from $\mathrm{Aut}_k K$ to $\mathrm{Aut}_k k'$ with Kernel $\mathrm{Aut}_{k'} K$ and induces an isomorphism from $\mathrm{Aut}_k K/\mathrm{Aut}_{k'} K$ to $\mathrm{Aut}_k k'$.

Letting K/k be a finite dimensional Galois extension, the Galois Correspondence Γ_k is a bijection from the set $\mathbf{F}_k$ of all subfields of K/k to the set $\mathbf{G}_k$ of all subgroups of $\mathrm{Aut}_k K$. Since $\mathrm{Aut}_k K$ is then finite, it follows that K/k has only finitely many subfields. More generally, this is true of finite dimensional separable extensions.

3.3.8 Proposition. Let K/k be a finite dimensional separable extension. Then K/k has only finitely many subfields. (See 2.3.15.)

Proof. There is a finite dimensional Galois extension K^{norm} of k containing K and K^{norm}/k has only finitely many subfields. □

The condition that an algebraic extension K/k have only finitely many subfields is equivalent to the condition that K/k be simple (see E.2.12). This together with the simplicity of a finite dimensional separable extension K/k (see 2.2.13) provides an alternate proof of the above proposition. Conversely, this and the above proposition provide an alternate proof of the simplicity of a finite dimensional separable extension K/k.

We conclude this section with a comparison of decompositions of a Galois extension K/k with decompositions of its Galois group G. The kinds of decompositions that we have in mind are as follows.

3.3.9 Definition. Let $G_1, \ldots, G_n$ be subgroups of a group G. Denote $\{\sigma_1 \cdots \sigma_n \mid \sigma_1 \in G_1, \ldots, \sigma_n \in G_n\}$ by $G_1 \cdots G_n$. Then we say that G' is the *direct product* of $G_1, \ldots, G_n$ *over* G, written $G' = G_1 \cdots G_n$ (*direct product over* G), if

1. the G_i are closed normal subgroups of G;
2. $G' = G_1 \cdots G_n$;
3. For all i, $G_i \cap G'//G_i = \mathbf{1}$ where $G'//G_i = G_1 \cdots G_{i-1}G_{i+1} \cdots G_n$.

The condition $G' = G_1 \cdots G_n$ (direct product over G) is equivalent to the conditions $G' = G_1 \cdots G_n$ (internal direct product) and $G_1, \ldots, G_n$ are closed normal subgroups of G (see 0.2).

3.3.10 Definition. Let K/k be a Galois extension and let $k_1, \ldots, k_n$ be subfields of K containing k. Then k' is the *disjoint product* of $k_1, \ldots, k_n$ *over* k, written $k' = k_1 \cdots k_n$ (*disjoint product over* k), if

1. the k_i are normal extensions of k;
2. $k' = k_1 \cdots k_n$;
3. for all i, $k_i \cap k'//k_i = k$ where $k'//k = k_1 \cdots k_{i-1}k_{i+1} \cdots k_n$.

3.3.11 Proposition. Let K/k be Galois and $G = \mathrm{Aut}_k K$. Then

1. if $G = G_1 \cdots G_n$ (direct product over G) and $k_i = K^{G//G_i}$ for all i, then $K = k_1 \cdots k_n$ (disjoint product over k);
2. if $K = k_1 \cdots k_n$ (disjoint product over k) and $G_i = \mathrm{Aut}_{K//k_i} K$ for all i, then $G = G_1 \cdots G_n$ (direct product over G);
3. $\mathbf{1} \to G//G_i \to G \to \mathrm{Aut}_k k_i \to \mathbf{1}$ is exact, in 1 and 2, where $G//G_i \to G$ is inclusion and $G \to \mathrm{Aut}_k k_i$ is restriction $(1 \le i \le n)$;
4. $G \to \mathrm{Aut}_k k_1 \times \cdots \times \mathrm{Aut}_k k_n$, defined by $\sigma \mapsto \sigma|_{k_1} \times \cdots \times \sigma|_{k_n}$, is an isomorphism.

Proof. The proof is based on the Galois Correspondence Theorem. Let $G = G_1 \cdots G_n$ (direct product over G) and $k_i = K^{G//G_i}$ for all i. Let $\sigma = \sigma_1 \cdots \sigma_n$ be a typical element of G, $\sigma_i \in G_i$ for all i. Then $\sigma_{j|k_i}$ is the identity for $i \neq j$. Thus, $\sigma_{j|K//k_j}$ is the identity for all j. Letting $x \in k_i \cap K//k_i$, we have $\sigma_j(x) = x$ for $j \neq i$ since $x \in k_i$ and $\sigma_j(x) = x$ for $j = i$ since $x \in K//k_i$. Thus, $\sigma(x) = x$. It follows that $\mathrm{Aut}_{k_i \cap K//k_i} K = G = \mathrm{Aut}_k K$. Thus, $k_i \cap K//k_i = k$. Next, suppose that $\sigma|_{k_1 \cdots k_n}$ is the identity. Then $\sigma_{i|k_i}$ is the identity and $\sigma_i \in \mathrm{Aut}_{k_i} K = G//G_i$. Thus, $\sigma_i \in G_i \cap G//G_i = 1$ and σ_i is the identity on K. Thus, σ is the identity, that is, $\mathrm{Aut}_{k_1 \cdots k_n} K = \mathbf{1} = \mathrm{Aut}_K K$ and $k_1 \cdots k_n = K$. The k_i are normal over k since the $G//G_i$ are normal in G. Thus, $K = k_1 \cdots k_n$ (disjoint product over k).

Next, let $K = k_1 \cdots k_n$ (disjoint product over k) and $G_i = \mathrm{Aut}_{K//k_i} K$ for all i. Then the G_i are closed and normal in G and we prove that $G = G_1 \cdots G_n$ (direct product over G) by induction on n. Let $A = k_1$, $B = k_2 \cdots k_n$. Then $K = AB$ and $A \cap B = k$. Recall that $\mathrm{Aut}_B AB$ and $\mathrm{Aut}_{A \cap B} A$ are isomorphic under restriction to A (see 2.4.4). In the present case, $\mathrm{Aut}_B K$ and $\mathrm{Aut}_k A$ are isomorphic under restriction to A, and $\mathrm{Aut}_A K$ and $\mathrm{Aut}_k B$ under restriction to B. Thus, if $\sigma \in G$, there exists $\sigma_1 \in \mathrm{Aut}_B K$, $\sigma_2 \in \mathrm{Aut}_A K$ such that $\sigma|_A = \sigma_{1|A}$ and $\sigma|_B = \sigma_{2|B}$. Now $\sigma = \sigma_1\sigma_2$ and we have shown that $\mathrm{Aut}_k K = \mathrm{Aut}_B K \, \mathrm{Aut}_A K$. Next, let $\sigma \in \mathrm{Aut}_B K \cap \mathrm{Aut}_A K$. Then $\sigma(x) = x$ for $x \in B \cup A$, hence for $x \in BA = K$. Thus, $\mathrm{Aut}_B K \cap \mathrm{Aut}_A K = \mathbf{1}$ and $G = \mathrm{Aut}_B K \, \mathrm{Aut}_A K$ (internal direct product). Now $\mathrm{Aut}_B K = G_1$, and we claim that $\mathrm{Aut}_A K = G_2 \cdots G_n$ (direct product over $\mathrm{Aut}_A K$). In fact $\mathrm{Aut}_A K$ and $\mathrm{Aut}_k B$ are isomorphic under restriction to B and, by induction, $\mathrm{Aut}_k B = G_2' \cdots G_n'$ (direct product over $\mathrm{Aut}_k B$) where $G_i' = \mathrm{Aut}_{B//k_i} B$ for $2 \le i \le n$. The subgroup of $\mathrm{Aut}_A K$ corresponding to $G_i' = \mathrm{Aut}_{B//k_i} B$ under the restriction isomorphism is $G_i = \mathrm{Aut}_{K//k_i} K$ for $2 \le i \le n$. Thus, $\mathrm{Aut}_A K = G_2 \cdots G_n$ (direct product over $\mathrm{Aut}_A K$). It now follows that $G = G_1 G_2 \cdots G_n$ (direct product over G).

Now let $G = G_1 \cdots G_n$, $K = k_1 \cdots k_n$ be as in 1 and 2. By 1, $k_1 = K^{G//G_1}$. By the proof of 2, $G//G_1 = \mathrm{Aut}_{k_1} K$ and $G \to \mathrm{Aut}_k k_1$ is surjective. It follows that $\mathbf{1} \to G//G_1 \to G \to \mathrm{Aut}_k k_1 \to \mathbf{1}$ is exact. The same argument applies upon replacing 1 by i for $1 \le i \le n$.

Since $G = G_1 \cdots G_n$ (direct product over G), the exactness of the sequences $\mathbf{1} \to G//G_i \to G \to \mathrm{Aut}_k k_i \to \mathbf{1}$ imply that $G \to \mathrm{Aut}_k k_1 \times \cdots \times \mathrm{Aut}_k k_n$ is an isomorphism. For if $\tau_i \in \mathrm{Aut}_k k_i$, choose $\sigma_i \in G_i$ such that

$\sigma_{i|k_i} = \tau_i$ $(1 \leq i \leq n)$. Then $\sigma \mapsto \tau_1 \times \cdots \times \tau_n$, where $\sigma = \sigma_1 \cdots \sigma_n$. Thus, the homomorphism is surjective. And if σ is in the kernel, $\sigma(x) = x$ for $x \in k_1 \cup \cdots \cup k_n$, hence for $x \in k_1 \cdots k_n = K$. Thus, the kernel is $\mathbf{1}$ and the homomorphism is injective. ▯

3.3.12 **Corollary.** Let K/k be an algebraic Galois extension and $K = k_1 \cdots k_n$ (disjoint product over k). Then $k_1 \otimes_k \cdots \otimes_k k_n$ and K are k-isomorphic by the k-linear mapping f sending $x_1 \otimes \cdots \otimes x_n$ to $x_1 \cdots x_n$.

Proof. If K/k is finite dimensional, then

$$K{:}k = G{:}\mathbf{1} = (G_1{:}\mathbf{1}) \cdots (G_n{:}\mathbf{1}) = (k_1{:}k) \cdots (k_n{:}k) = (k_1 \otimes \cdots \otimes k_n){:}k$$

and f, being surjective, is therefore a k-isomorphism. Since for each i, $k_i = \bigcup_{k_i' \in \mathbf{S}_i} k_i'$ where $\mathbf{S}_i$ is the set of subfields k_i' of k_i containing k such that k_i'/k is finite dimensional Galois, f is a k-isomorphism in general. ▯

3.4 The Normal Basis Theorem

In the preceding section, we proved that for a finite dimensional Galois extension K/k, $K{:}k = \operatorname{Aut}_k K{:}\mathbf{1}$. We show here that such a K has a basis over k of the form $\sigma_1(y), \ldots, \sigma_n(y)$ where $\operatorname{Aut}_k K = \{\sigma_1, \ldots, \sigma_n\}$ and y is a suitable element of K. Such a basis is called a *normal basis* of K/k.

If K is finite, the proof goes as follows. The group $\operatorname{Aut}_k K$ then has the form $\operatorname{Aut}_k K = \{\epsilon, \tau, \tau^2, \ldots, \tau^{n-1}\}$ where $n = K{:}k$ (see 1.5.7). Since $\epsilon, \tau, \ldots, \tau^{n-1}$ are linearly independent (since distinct), the linear transformation τ of the n-dimensional vector space K over k is cyclic. Thus, $y, \tau(y), \ldots, \tau^{n-1}(y)$ is a basis for K over k for some y.

We now let K be any field, k a subfield of K.

3.4.1 **Theorem.** Let k be infinite and let $\sigma_1, \ldots, \sigma_n$ be distinct elements of $\operatorname{Aut}_k K$. Then there exists $y \in K$ such that $\operatorname{Det}(\sigma_i^{-1}\sigma_j(y)) \neq 0$.

Proof. Choose $z \in K$ such that $\sigma \mapsto \sigma(z)$ is injective on

$$T = \{\sigma_i^{-1}\sigma_j \mid 1 \leq j \leq n\}$$

(see 2.2.10). Let $f(X)$ be a polynomial in $K[X]$ such that $f(\sigma^{-1}(z)) = \delta_{\epsilon,\sigma}$ (Kronecker delta) for $\sigma \in T$ where ϵ is the identity of G (see E.1.30). Letting $\sigma(f)(X)$ denote the polynomial $\sigma(f(X))$, we then have $\sigma(f)(z) = \delta_{\epsilon,\sigma}$ for $\sigma = \sigma_i^{-1}\sigma_j \in T$. That is, $(\sigma_i^{-1}\sigma_j(f))(z) = \delta_{\epsilon,\sigma_i^{-1}\sigma_j} = \delta_{i,j}$. It follows that the polynomial $\operatorname{Det}((\sigma_i^{-1}\sigma_j)(f)(X))$ does not vanish at z, hence is nonzero. Since k is infinite, there consequently exists $x \in k$ such that

$$\operatorname{Det}((\sigma_i^{-1}\sigma_j)(f)(x)) \neq 0$$

(see E.1.41). But $(\sigma_i^{-1}\sigma_j)(f)(x) = \sigma_i^{-1}\sigma_j(f(x))$ since $x \in k$. Taking $y = f(x)$, we then have $\operatorname{Det}(\sigma_i^{-1}\sigma_j(y)) \neq 0$. ▯

3.4.2 **Theorem.** Let k be infinite or K/k algebraic. Let $\sigma_1, \ldots, \sigma_n$ be distinct elements of $\operatorname{Aut}_k K$. Then there exists $y \in K$ such that $\sigma_1(y), \ldots, \sigma_n(y)$ are linearly independent over k.

Proof. If k is infinite, take y as in the preceding theorem. A k-dependence of $\sigma_1(y), \ldots, \sigma_n(y)$ does not exist, for it would lead to a dependence of the columns of $(\sigma_i^{-1}\sigma_j(y))$. Suppose next that K/k is algebraic and k finite. Let S be a finite subset of K such that $\sigma_{1|S}, \ldots, \sigma_{n|S}$ are distinct. Since the orbits of $\mathrm{Aut}_k K$ are finite, we may take S to be $\mathrm{Aut}_k K$-stable. Let $k' = k(S)$. Then k' is finite and $\sigma_{1|k'}, \ldots, \sigma_{n|k'}$ are distinct elements of $\mathrm{Aut}_k k'$. But k' has a normal basis $\sigma(y)$ $(\sigma \in \mathrm{Aut}_k k')$, as we showed earlier in the section. Thus, $\sigma_1(y), \ldots, \sigma_n(y)$ are linearly independent over k for some $y \in K$. □

3.4.3 Theorem (Normal Basis Theorem). Let K/k be a finite dimensional Galois extension. Let $\mathrm{Aut}_k K = \{\sigma_1, \ldots, \sigma_n\}$. Then $\sigma_1(y), \ldots, \sigma_n(y)$ is a basis for K over k for some $y \in K$.

Proof. Choose y as in the preceding theorem. Then $\sigma_1(y), \ldots, \sigma_n(y)$ is a basis for K over k since $n = K{:}k$. □

3.5 Algebraic independence of homomorphisms

Let k be a field and let K and L be extension fields of k. Let L' be any extension field of L. Let $\sigma_1, \ldots, \sigma_n$ be distinct k-homomorphisms from K into L and let $f(X_1, \ldots, X_n)$ be an element of the ring $L'[X_1, \ldots, X_n]$ of polynomials in commuting indeterminants $X_1, \ldots, X_n$ with coefficients in L' (see E.0.57).

3.5.1 Definition. $f(\sigma_1, \ldots, \sigma_n)$ is the mapping from K to L' defined by $f(\sigma_1, \ldots, \sigma_n)(x) = f(\sigma_1(x), \ldots, \sigma_n(x))$ for $x \in K$.

For $M = (m_{ij})$ an element of the ring $M_n(L')$ of $n \times n$ matrices with entries in L', we let $XM = Y$ where $X = (X_1, \ldots, X_n)$, $Y = (Y_1, \ldots, Y_n)$ and $Y_j = \sum_{i=1}^n X_i m_{ij}$. We adopt the notation $f(X) = f(X_1, \ldots, X_n)$, and let $f(Y)$ be the element of $L'[X_1, \ldots, X_n]$ obtained from $f(X)$ by specializing X_i to Y_i for $1 \le i \le n$. We then let ${}^M f(X) = f(XM)$. We have ${}^{(MN)}f(X) = f(X(MN)) = f((XM)N) = {}^N f(XM) = {}^M({}^N f(X))$ and ${}^{(MN)}f(X) = {}^M({}^N f(X))$ for $M, N \in M_n(L')$. Letting I be the identity element of $M_n(L')$, we have ${}^I f(X) = f(X)$.

Assume next that k is infinite. By 3.1.3, there exist elements $a_1, \ldots, a_n \in K$ such that the matrix $M = (\sigma_j(a_i))$ is nonsingular. For $x_1, \ldots, x_n \in k$, let $x = \sum_{i=1}^n x_i a_i$. Then $\sigma_j(x) = \sum_{i=1}^n x_i \sigma_j(a_i)$ $(1 \le j \le n)$ and $(\sigma_1(x), \ldots, \sigma_n(x)) = (x_1, \ldots, x_n)M$, so that $f(\sigma_1(x), \ldots, \sigma_n(x)) = {}^M f(x_1, \ldots, x_n)$.

3.5.2 Lemma. Let k be an infinite field, L' a field extension of k, $f(X_1, \ldots, X_n)$ an element of $L'[X_1, \ldots, X_n]$. Then if $f(x_1, \ldots, x_n) = 0$ for all $x_1, \ldots, x_n \in k$, $f(X_1, \ldots, X_n) = 0$.

Proof. The proof is by induction on n and follows from E.1.41 if $n = 1$. Next, let $n > 1$ and suppose that $f(x_1, \ldots, x_n) = 0$ for all $x_1, \ldots, x_n \in k$. Let $f(X_1, \ldots, X_n) = \sum_i a_i(X_2, \ldots, X_n)X_1{}^i$ and consider $\bar{f}(X_1) = f(X_1, x_2, \ldots, x_n)$, $\bar{a}_i = a_i(x_2, \ldots, x_n)$ where $x_2, \ldots, x_n$ are fixed elements of k. Then $\bar{f}(X_1) = \sum_i \bar{a}_i X_1{}^i$ and $\bar{f}(x_1) = 0$ for all $x_1 \in k$. By the case $n = 1$, we have $\bar{f}(X_1) = 0$, and $\bar{a}_i = 0$ for all i. Thus, $a_i(x_2, \ldots, x_n) = 0$ for all i and all $x_2, \ldots, x_n \in k$. By the induction hypothesis, $a_i(X_2, \ldots, X_n) = 0$ for all i. Thus,

$$f(X_1, \ldots, X_n) = 0. \qquad \square$$

3.5.3 Theorem. Let k be an infinite field, and let K and L be extension fields of k and L' an extension field of L. Let $\sigma_1, \ldots, \sigma_n$ be distinct k-homomorphisms from K to L. Then for $f(X_1, \ldots, X_n) \in L'[X_1, \ldots, X_n]$, the function $f(\sigma_1, \ldots, \sigma_n)$ is 0 if and only if the polynomial $f(X_1, \ldots, X_n)$ is 0.

Proof. One direction is clear. For the other, suppose that the function $f(\sigma_1, \ldots, \sigma_n)$ is 0. Taking $a_1, \ldots, a_n$ and M as above, we then have $0 = f(\sigma_1(x), \ldots, \sigma_n(x)) = {}^M f(x_1, \ldots, x_n)$ for all $x_1, \ldots, x_n \in K$. But then ${}^M f(X_1, \ldots, X_n) = 0$, by 3.5.2. Since M is non-singular, we then have $f(X_1, \ldots, X_n) = {}^{(M^{-1}M)} f(X_1, \ldots, X_n) = {}^{M^{-1}}({}^M f(X_1, \ldots, X_n)) = {}^{M^{-1}}(0) = 0$. $\square$

The property of the k-homomorphisms $\sigma_1, \ldots, \sigma_n$ described in the above theorem is referred to as the *algebraic independence* of $\sigma_1, \ldots, \sigma_n$, for k infinite.

3.6 Norm and trace

Let k be a field, K a finite dimensional field extension of k and k_{Alg} an algebraic closure of k containing K. Let $\sigma_1, \ldots, \sigma_m$ be the distinct k-homomorphisms from K into k_{Alg} and note that $\sigma_{1|k_{\text{sep}}}, \ldots, \sigma_{m|k_{\text{sep}}}$ are the distinct k-homomorphisms from k_{sep} into k_{Alg} (see 2.2.13). For $x \in K$, the *conjugates of x over k relative to K* are $\sigma_1(x), \ldots, \sigma_m(x)$. If $\sigma \in G = \text{Aut}_k K^{\text{norm}}$ where K^{norm} is the normal closure of K over k in k_{Alg}, then σ permutes the $\sigma_1, \ldots, \sigma_m$ in the sense that $\{\sigma \circ \sigma_1, \ldots, \sigma \circ \sigma_m\} = \{\sigma_1, \ldots, \sigma_m\}$, so that σ permutes the conjugates $\sigma_1(x), \ldots, \sigma_m(x)$ of x. Consequently, $\prod_1^m \sigma_i(x)$ and $\sum_1^m \sigma_i(x)$ are elements of $(K^{\text{norm}})^G$. If K/k is separable, then K^{norm}/k is Galois and $(K^{\text{norm}})^G = k$, so that $\prod_1^m \sigma_i(x)$ and $\sum_1^m \sigma_i(x)$ are elements of k.

3.6.1 Definition. $N_{K/k}$ and $Tr_{K/k}$ are the mappings from K to k_{Alg} defined by $N_{K/k}(x) = \prod_1^m \sigma_i(x^{p^e})$ and $Tr_{K/k}(x) = \sum_1^m \sigma_i(x^{p^e})$ for $x \in K$ where $p^e = (K:k)_{\text{rad}}$. For $x \in K$, $N_{K/k}(x)$ and $Tr_{K/k}$ are the *norm* and *trace* of x.

3.6.2 Proposition. $N_{K/k}$ is a multiplication preserving mapping from K to k such that $N_{K/k}(x) = x^{K:k}$ for $x \in k_{\text{rad}}$ and $Tr_{K/k}$ is an addition preserving mapping from K to k such that $Tr_{K/k}(x) = (K:k)x$ for $x \in k_{\text{rad}}$.

Proof. Since $x^{p^e} \in k_{\text{sep}}$ (see 2.2.18) and $\sigma_{1|k_{\text{sep}}} \ldots, \sigma_{m|k_{\text{sep}}}$ are the distinct k-homomorphisms from k_{sep} to k_{Alg} (see 2.2.13), $N_{K/k}(x)$ and $Tr_{K/k}(x)$ are in k by the remarks at the beginning of this section. Clearly $N_{K/k}$, $Tr_{K/k}$ preserve multiplication and addition respectively. Since $m = (K:k)_{\text{sep}}$ (see 2.2.17), we have $K:k = mp^e$ (see 2.2.18) and consequently $N_{K/k}(x) = x^{K:k}$, $Tr_{K/k}(x) = (K:k)x$ for $x \in k_{\text{rad}}$. $\square$

3.6.3 Proposition. Let $K \supset k' \supset k$. Then $N_{k'/k} \circ N_{K/k'} = N_{K/k}$ and $Tr_{k'/k} \circ Tr_{K/k'} = Tr_{K/k}$.

Proof. Note that the above equations make sense, since $N_{K/k'}$ and $Tr_{K/k'}$ map K to k', $N_{k'/k}$ and $Tr_{k'/k}$ map k' to k and $N_{K/k}$ and $Tr_{K/k}$ map K to k. Now let $\sigma_1, \ldots, \sigma_m$ be the distinct k'-homomorphisms from K to k_{Alg} and $\sigma_1', \ldots, \sigma_{m'}'$ the distinct k-homomorphisms from k' to k_{Alg}. Then the $\sigma_1', \ldots, \sigma_{m'}'$ can be extended to k-homomorphisms from K^{norm} to k_{Alg} (see 1.3.5 and 2.3.7) also denoted $\sigma_1', \ldots, \sigma_{m'}'$, and the mm' distinct $\sigma_j'\sigma_i$ ($1 \leq i \leq m$, $1 \leq j \leq m'$) are all of the distinct k-homomorphisms from K to k_{Alg} (see E.3.4). Letting $p^a = (K:k')_{\text{rad}}$ and $p^b = (k':k)_{\text{rad}}$, we have $p^{a+b} = (K:k)_{\text{rad}}$ (see 2.2.21). Now

$$N_{k'/k}(N_{K/k'}(x)) = N_{k'/k}\left(\prod_1^m \sigma_i(x^{p^a})\right) = \prod_1^{m'} \sigma_j'\left(\left(\prod_1^m \sigma_i(x^{p^a})\right)^{p^b}\right)$$
$$= \prod_1^{m'}\prod_1^m \sigma_j'\sigma_i(x^{p^{a+b}}) = N_{K/k}(x) \qquad \text{for } x \in K.$$

Similarly,

$$Tr_{k'/k}(Tr_{K/k'}(x)) = \sum_1^{m'} \sigma_j'\left(\left(\sum_1^m \sigma_i(x^{p^a})\right)^{p^b}\right) = Tr_{K/k}(x) \qquad \text{for } x \in K. \quad \square$$

3.7 Galois cohomology

Let G be a group and let H be a G-group (see 0.3). We first regard G and H as discrete groups, that is, as groups with the discrete topology. Let ϵ_G, e_H be the identity elements of G, H respectively. An element h of H determines a function ∂h from G to H, defined by $\partial h(\sigma) = \sigma(h^{-1})h$ and called the *coboundary* (or 1-*coboundary*) determined by h. One verifies easily that $\partial h(\sigma\tau) = \sigma(\partial h(\tau))\partial h(\sigma)$ for $\sigma, \tau \in G$. A function f from G to H such that $f(\sigma\tau) = \sigma(f(\tau))f(\sigma)$ for $\sigma, \tau \in G$ is called a *cocycle* (or 1-*cocycle*) from G to H. The set of coboundaries ∂h ($h \in H$) is denoted $B^1(G, H)$, the set of cocycles of G in H denoted $Z^1(G, H)$. We have $Z^1(G, H) \supset B^1(G, H)$. Two cocycles f, f' are *cohomologous* if there exists $h \in H$ such that $f(\sigma) = \sigma(h^{-1})f'(\sigma)h$ for $\sigma \in G$. The relation "f is cohomologous to f'" on $Z^1(G, H)$ is an equivalence relation and $B^1(G, H)$ is the equivalence class containing $e = \partial e_H$. The set of equivalence classes in $Z^1(G, H)$ is denoted $H^1(G, H)$. If $Z^1(G, H) = B^1(G, H)$, there is only one equivalence class and we say that $H^1(G, H) = \mathbf{1}$. If H is Abelian, then $Z^1(G, H)$ together with the product $ff'(\sigma) = f(\sigma)f'(\sigma)$ ($\sigma \in G$) is an Abelian group with subgroup $B^1(G, H)$ and

$$H^1(G, H) = Z^1(G, H)/B^1(G, H).$$

We now determine $Z^1(G, H)$ for G a cyclic group of order n and generator σ. For this, let f be a function from H to H and let $f(\sigma) = h$. We claim that $f \in Z^1(G, H)$ if and only if $f(\sigma^{i+1}) = \sigma(f(\sigma^i))f(\sigma)$ for $i \geq 0$. The necessity of the condition is obvious. Now let $f(\sigma^{i+1}) = \sigma(f(\sigma^i)f(\sigma))$ for $i \geq 0$. Then $f(\sigma^0) = e_H$, since $f(\sigma^{0+1}) = \sigma(f(\sigma^0))f(\sigma)$. We now show by induction on i that $f(\sigma^i\sigma^j) = \sigma^i(f(\sigma^j))f(\sigma^i)$ for all $j \geq 0$. Note first that $f(\sigma^0\sigma^j) = f(\sigma^j) = \sigma^0(f(\sigma^j))f(\sigma^0)$. Next,

$$f(\sigma^{i+1}\sigma^j) = f(\sigma\sigma^i\sigma^j) = \sigma(f(\sigma^i\sigma^j))f(\sigma) = \sigma(\sigma^i(f(\sigma^j)f(\sigma^i)))f(\sigma)$$
$$= \sigma^{i+1}(f(\sigma^j))\sigma(f(\sigma^i))f(\sigma) = \sigma^{i+1}(f(\sigma^j))f(\sigma^{i+1})$$

verifies the assertion for $i + 1$ under the assumption of the assertion for i. Thus, $f \in Z^1(G, H)$. Now the condition that $f \in Z^1(G, H)$ is that

$$\begin{aligned} f(\sigma) &= h \\ f(\sigma^2) &= \sigma(h)h \\ &\vdots \\ f(\sigma^{n+1}) &= \sigma^n(h) \cdots \sigma(h)h \\ &\vdots \end{aligned}$$

Since $h = f(\sigma) = f(\sigma^{n+1})$, a necessary condition for the above equations is that $N_\sigma h = e_H$ where $N_\sigma h = \sigma^n(h) \cdots \sigma(h)$. The condition $N_\sigma h = e_H$ is also sufficient, since it insures that $f(\sigma^{m+1}) = \sigma^m(h) \cdots \sigma(h)$ is well defined for $m \geq 0$. For if $m \geq 0$, it insures that $f(\sigma^{m+1+n}) = \sigma^m(\sigma^n(h) \cdots \sigma(h))\sigma^m(h) \cdots \sigma(h)h = \sigma^m(h) \cdots \sigma(h)h = f(\sigma^{m+1})$. We have now proved the following.

3.7.1 Proposition. Let G be cyclic of order n and generator σ and let $N_\sigma(h) = \sigma^n(h) \cdots \sigma(h)$. Then $Z^1(G, H)$ is mapped bijectively to $\{h \in H \mid N_\sigma(h) = e_H\}$ by $f \mapsto f(\sigma)$. ▯

We now let G be a finite subgroup of Aut K, K a field. We let K^* be the multiplicative group of units of K and K^+ the group K with addition, and regard K^* and K^+ as G-groups. Our main objective is to show that $H^1(G, K^*) = \mathbf{1}$ and $H^1(G, K^+) = \mathbf{1}$.

3.7.2 Theorem. For G a finite subgroup of Aut K, $H^1(G, K^*) = \mathbf{1}$.

Proof. Let $f \in Z^1(G, K^*)$ and $y \in K$. For $x = \sum_{\tau \in G} f(\tau)\tau(y)$, we have

$$\begin{aligned} \sigma(x) &= \sum_{\tau \in G} \sigma(f(\tau))\sigma\tau(y) = \sum_{\tau \in G} f(\sigma\tau)f(\sigma)^{-1}\sigma\tau(y) \\ &= f(\sigma)^{-1}\left(\sum_{\sigma\tau \in G} f(\sigma\tau)\sigma\tau(y)\right) = f(\sigma)^{-1}x. \end{aligned}$$

Taking $y \in K$ such that $x \neq 0$ (see 3.1.3), we have $f(\sigma) = \sigma(x^{-1})x$ for $\sigma \in G$, and $f \in B^1(G, K^*)$. Thus, $H^1(G, K^*) = \mathbf{1}$. ▯

3.7.3 Theorem. For G a finite subgroup of Aut K, we have $H^1(G, K^+) = \mathbf{1}$.

Proof. Let $f \in Z(G, K^+)$ and $y \in K$. For $x = \sum_{\tau \in G} f(\tau)\tau(y)$, we have

$$\sigma(x) = \sum_{\tau \in G} \sigma(f(\tau))\sigma\tau(y) = \sum_{\sigma\tau \in G} (f(\sigma\tau) - f(\sigma))\sigma\tau(y).$$

Thus

$$x - \sigma(x) = f(\sigma) \sum_{\sigma\tau \in G} \sigma\tau(y) = f(\sigma) \sum_{\tau \in G} \tau(y).$$

Taking $y \in K$ such that $z = \sum_{\tau \in G} \tau(y)$ is nonzero (see 3.1.3), we have $z \in K^G$ and $f(\sigma) = xz^{-1} - \sigma(xz^{-1})$ for $\sigma \in G$. Thus, $f \in B^1(G, K^+)$ and $H^1(G, K^+) = \mathbf{1}$. ▯

We now derive "Hilbert's Theorem 90" as corollary to 3.7.2. This theorem is important in the study of cyclic extensions. We also derive a corollary analogous to 3.7.3. For this, we let $N_\sigma(x) = \prod_1^n \sigma^i(x)$ and $Tr_\sigma(x) = \sum_1^n \sigma^i(x)$ where $x \in K$, $\sigma \in \operatorname{Aut} K$ and σ has order n (see 3.6). Note that $N_\sigma(x)$, $Tr_\sigma(x) \in K^\sigma = \{y \in K \mid \sigma(y) = y\}$.

3.7.4 **Corollary.** Let σ be an element of Aut K of finite order and let $y \in K$ such that $N_\sigma(y) = 1$. Then $y = \sigma(x^{-1})x$ for some $x \in K$.

Proof. Let G be the cyclic group generated by σ. By 3.7.1, there exists $f \in Z^1(G, K^*)$ such that $f(\sigma) = y$. Since $H^1(G, K^*) = \mathbf{1}$, there exists $x \in K^*$ such that $y = f(\sigma) = \sigma(x^{-1})x$. □

3.7.5 **Corollary.** Let σ be an element of Aut K of finite order and let $y \in K$ such that $Tr_\sigma(y) = 0$. Then $y = x - \sigma(x)$ for some $x \in K$.

Proof. Let G be the cyclic group generated by σ. By 3.7.1, there exists $f \in Z^1(G, K^+)$ such that $f(\sigma) = y$. Since $H^1(G, K^+) = \mathbf{1}$, there exists $x \in K^+$ such that $y = f(\sigma) = x - \sigma(x)$. □

We now describe a simple application of Hilbert's Theorem 90. This application is proved directly later (see 3.9.3).

3.7.6 **Proposition.** Let $\sigma \in \operatorname{Aut} K$ have order d. Let $y \in K$, $\sigma(y) = y$ and $y^d = 1$. Then there exists $x \in K$ such that $\sigma(x) = xy$.

Proof. Since $N_\sigma(y^{-1}) = 1$, $y^{-1} = \sigma(x^{-1})x$ for some $x \in K$. □

The additive analogue of 3.7.6 is the following (see 3.9.4).

3.7.7 **Proposition.** Let K have characteristic $p \neq 0$ and let $\sigma \in \operatorname{Aut} K$ have order p. Then there exists $x \in K$ such that $\sigma(x) = x + 1$.

Proof. Since $Tr_\sigma(-1) = p(-1) = 0$, $-1 = x - \sigma(x)$ for some $x \in K$. □

We now let G be a group together with its Krull Topology and let H be a G-group with the discrete topology. We assume that G acts continuously on H, that is, that the G-orbits in H are finite (see 0.4.5).

3.7.8 **Definition.** For G and H as above, $Z^1(G, H)^{\text{cont}}$ is the set of continuous cocycles from G to H.

3.7.9 **Proposition.** Let f be a cocycle from G to H. Then the following conditions are equivalent.

1. f is continuous;
2. Kernel $f = \{\sigma \in G \mid f(\sigma)$ is the identity of $H\}$ is a subgroup of G of finite index;
3. Image f is finite.

Proof. One easily sees that Kernel f is a subgroup of G and that the coset σ Kernel f is the set of preimages of $f(\sigma)$ under f for $\sigma \in G$. Now Kernel f is open if and only if Kernel f is of finite index, by 0.4.3, and it is now immediate that 1–3 are equivalent. □

3.7.10 **Corollary.** Let f and f' be cohomologous cocycles from G to H. Then f is continuous if f' is continuous.

Proof. Let $f(\sigma) = \sigma(h^{-1})f'(\sigma)h$ for all $\sigma \in G$. Since the orbit of h^{-1} under G is finite, Image f is finite if Image f' is finite. □

3.7.11 Corollary. Every coboundary is continuous.

Proof. Every coboundary is cohomologous to the trivial cocycle, which is continuous. □

3.7.12 Definition. $H^1(G, H)^{\text{cont}}$ is the set of equivalence classes in $Z^1(G, H)^{\text{cont}}$. If $Z^1(G, H)^{\text{cont}} = B^1(G, H)$, we say that $H^1(G, H)^{\text{cont}} = \mathbf{1}$. If H is Abelian, we regard $Z^1(G, H)^{\text{cont}}$ as a group and $H^1(G, H)^{\text{cont}}$ as $Z^1(G, H)^{\text{cont}}/B^1(G, H)$.

We now generalize 3.7.2. It is instructive to invoke the Galois Descent Theorem 3.2.5 for this.

3.7.13 Theorem. Let G be an algebraic subgroup of Aut K. Then $H^1(G, K^*)^{\text{cont}} = \mathbf{1}$.

Proof. Let f be a continuous cocycle from G to K^*. Introduce the G-product $\sigma \cdot x = f(\sigma)^{-1}\sigma(x)$ on the K-space K (see E.3.9). The G-orbits under the new action $\sigma \cdot x$ are finite since the G-orbits under the old action $\sigma(x)$ are finite. It therefore follows from 3.2.5 that there exists a nonzero $x \in K$ such that $\sigma \cdot x = x$ for all $\sigma \in G$. But then $f(\sigma)^{-1}\sigma(x) = x$ and $f(\sigma) = \sigma(x)x^{-1}$ for $\sigma \in G$. Thus, $f = \partial x^{-1}$ and $Z^1(G, H)^{\text{cont}} = B^1(G, H)$. □

We also have the corresponding generalization of 3.7.3.

3.7.14 Theorem. Let G be an algebraic subgroup of Aut K. Then $H^1(G, K^+)^{\text{cont}} = \mathbf{1}$.

Proof. Let f be a continuous cocycle from G to K^* and let $H =$ Kernel f. Then H is open and therefore contains a normal subgroup N of finite index. We let $\bar{f}(\sigma N) = f(\sigma)$ define the cocycle $\bar{f}$ from $\bar{G} = G/N$ to K^+. Regarding $\bar{G}$ as a finite subgroup of Aut K^N, we know from 3.7.3 that there exists $x \in K^N$ such that $\bar{f}(\bar{\sigma}) = x - \bar{\sigma}(x)$ for $\bar{\sigma} \in \bar{G}$. But then $f(\sigma) = x - \sigma(x)$ for $\sigma \in G$ and $f = \partial x$. □

3.8 Cyclotomic extensions

Let k be a field of exponent characteristic p, k_{Alg} an algebraic closure of k and $n > 1$ an integer.

3.8.1 Definition. An element x of k_{Alg} is a *primitive nth root of unity* if the cyclic group $\langle x \rangle$ generated by x is of order n. The *nth roots of unity* of k_{Alg} are the elements of the group $E_n = \{x \in k_{\text{Alg}} | x^n = 1\}$.

Note that k_{Alg} has a primitive nth root of unit only if n and p are relatively prime, since 1 is the only root of $X^p - 1$ in k_{Alg}.

3.8.2 Definition. The splitting field for $x^n - 1$ over k is called the *cyclotomic extension of* $X^n - 1$ *of* k.

Note that the cyclotomic extensions of $X^n - 1$ and $X^{np^e} - 1$ of k are the same.

3.8.3 Proposition. Let n and p be relatively prime and let K be the cyclotomic extension of $X^n - 1$ of k. Then K contains $\varphi(n)$ primitive nth roots of unity, where $\varphi(n)$ is the number of positive integers less than n which are relatively prime to n (see E.0.47). If x is any primitive nth root of unity in K, then $K = k(x)$. Finally, the extension K/k is Galois, $\mathrm{Aut}_k K$ is Abelian (cyclic if n is prime) and $\mathrm{Aut}_k K$ is isomorphic to a subgroup of the group $\mathbb{Z}_n^*$ of units of the ring of integers modulo n.

Proof. The set E_n of roots of $X^n - 1$ in K has n distinct elements, since $X^n - 1$ and its derivative nX^{n-1} are relatively prime. Thus, E_n is a subgroup of K^* of order n. By 1.5.5, E_n is cyclic. A generator x for E_n is now a primitive nth root of unity (see 3.8.1). The nth roots of unity are therefore $1, x, \ldots, x^{n-1}$. Among these, x^m is primitive if and only if m and n are relatively prime. Thus, the number of primitive nth roots of unity in K is $\varphi(n)$. Since $\langle x \rangle = E_n$, we have $K = k(x)$. Now K/k is normal, as a splitting field. Since $X^n - 1$ has n distinct roots, x is separable over k. Thus, K/k is Galois. For $\sigma \in \mathrm{Aut}_k K$, we have $\sigma(x) = x^{m_\sigma}$ where $1 \leq m_\sigma < n$. The mapping $\sigma \mapsto m_\sigma + n\mathbb{Z}$ is an injective homomorphism from $\mathrm{Aut}_k K$ to the group $\mathbb{Z}_n^*$ of units of $\mathbb{Z}_n$. Thus, $\mathrm{Aut}_k K$ is isomorphic to a subgroup of $\mathbb{Z}_n^*$. In particular, $\mathrm{Aut}_k K$ is Abelian (cyclic if n is prime). □

3.8.4 Theorem. Let n and p be relatively prime and let K be a cyclotomic extension of $X^n - 1$ of k. Then $\mathrm{Aut}_k K$ is isomorphic to $\mathbb{Z}_n^*$ if and only if $\prod_1^{\varphi(n)} (X - x_i)$ is irreducible over k, where $x_1, \ldots, x_{\varphi(n)}$ are the primitive nth roots of unity in K.

Proof. Note that $\{x_1, \ldots, x_{\varphi(n)}\}$ is stable under $G = \mathrm{Aut}_k K$, so that $\prod_1^{\varphi(n)} (X - x_i) \in K[X]^G = K^G[X] = k[X]$. Since $K = k(x_1)$, $\prod_1^{\varphi(n)} (X - x_i)$ is therefore irreducible if and only if its degree $\varphi(n)$ equals $K{:}k$, that is, if and only if $\mathbb{Z}_n^*{:}1 = \mathrm{Aut}_k K{:}\mathbf{1}$. Now apply the preceding proposition.

If $n = p - 1$, then $\mathbb{Z}_p^*{:}\mathbf{1} = p - 1$ and $\mathbb{Z}_p$ is the cyclotomic extension of $X^{p-1} - 1$ over $\mathbb{Z}_p$. Thus, $\prod_1^{\varphi(n)} (X - x_i)$ is not irreducible over the prime field in this case. However, if the prime field is $k = \mathbb{Q}$, $\prod_1^{\varphi(n)} (X - x_i)$ is irreducible (see E.3.18). Thus, we have the following corollary. □

3.8.5 Corollary. Let K be a cyclotomic extension of $X^n - 1$ of $\mathbb{Q}$. Then $\mathrm{Aut}_{\mathbb{Q}} K$ is isomorphic to $\mathbb{Z}_n^*$. □

The above results relate the Galois group $\mathrm{Aut}_k K$ of a cyclotomic extension K of $X^n - 1$ of k to the group $\mathbb{Z}_n^*$. The structure of the latter group is easy to determine. In fact if $n = p_1^{e_1} \cdots p_m^{e_m}$ (prime decomposition of n), then $\mathbb{Z}_n^*$ is isomorphic to $\prod_{i=1}^m \mathbb{Z}_{p_i^{e_i}}^*$ (outer direct product) and $\mathbb{Z}_{p_i^{e_i}}$ has order $\varphi(p_i^{e_i}) = p_i^{e_i - 1}(p_i - 1)$ and is cyclic for $p_i \neq 2$. If $p_i = 2$, then $\mathbb{Z}_{p_i^{e_i}}^*$ has a cyclic direct factor of index 2 (except, of course, when $e_i = 1$). (See E.0.45 and E.3.19.)

3.9 Cyclic extensions

Let K be an algebraic extension of a field k of exponent characteristic p and let k_{Alg} be an algebraic closure of k containing K. We begin with some simple general observations about K/k when K/k is Abelian. (See 2.4.8.)

3.9.1 Proposition. An Abelian extension K/k is Galois.

Proof. Let K/k be Abelian. Then K/k is separable and $\text{Aut}_k K^{\text{norm}}$ is Abelian. Now K^{norm}/k is Galois (see 2.3.15) and every subgroup of $\text{Aut}_k K^{\text{norm}}$ is normal. Thus, K/k is Galois by the Galois Correspondence Theorem. □

3.9.2 Proposition. Let K/k be finite dimensional and Abelian. Then $K = k_1 \cdots k_n$ (disjoint product over k) where, for each i, k_i is a cyclic extension of k of prime power degree $p_i^{e_i} = k_i{:}k$.

Proof. The group $G = \text{Aut}_k K$ is a finite Abelian group and $G = G_1 \cdots G_n$ (direct product over G) where, for each i, G_i is cyclic of prime power order $p_i^{e_i} = G{:}\mathbf{1}$. Now apply 3.3.11. □

The above proposition reduces the problem of determining the finite dimensional Abelian extensions of k to that of determining the cyclic extensions of k of prime power degree. In general, this is a difficult problem. However, we do describe the cyclic extensions K of k of prime power degree n in the case that k contains the group E_n of nth roots of 1 in k_{Alg}. The cases $n = p^e$ and $n = q^e$ (q prime, $q \neq p$) are treated separately. The following two propositions, treated less concretely in 3.7, are fundamental to the discussion.

3.9.3 Proposition. Let $\sigma \in \text{Aut}\, K$ have order d. Let $y \in K$, $\sigma(y) = y$ and $y^d = 1$. Then there exists $x \in K$ such that $\sigma(x) = xy$.

Proof. Choose $w \in K$ such that $x = \sum_{i=1}^{d} y^{-i}\sigma^i(w)$ is nonzero, by 3.1.3. Then $\sigma(x) = \sum_{i=1}^{d} y^{-i}\sigma^{i+1}(w) = xy$. □

3.9.4 Proposition. Let $\sigma \in \text{Aut}\, K$ have order p and assume that $p > 1$. Then there exists $x \in K$ such that $\sigma(x) = x + 1$.

Proof. Choose $w \in K$ such that $z = \sum_1^p \sigma^i(w)$ is nonzero. Then $\sigma(z) = z$ and $1 = \sum_1^p \sigma^i(wz^{-1})$. Now let $x = -\sum_1^p i\sigma^i(wz^{-1})$. Then

$$\begin{aligned}\sigma(x) &= -\sum_1^p i\sigma^{i+1}(wz^{-1}) = \sum_1^p \sigma^{i+1}(wz^{-1}) - \sum_1^p (i+1)\sigma^{i+1}(wz^{-1})\\ &= 1 + x.\end{aligned}$$

□

The next theorem treats the case of cyclic extensions K of degree n relatively prime to p of a field k containing E_n.

3.9.5 Theorem. Let n be relatively prime to p and suppose that k contains the group E_n of nth roots of unity in k_{Alg}. Then K/k is cyclic of degree dividing n if and only if $K = k(x)$ for some $x \in K$ such that $x^n \in k$. The minimum polynomial of any $x \in K$ with $x^n \in k$ is $f_x(X) = X^d - x^d$ where d divides n.

Proof. Let $x \in K$ and $x^n \in k$. Then the set of roots of $X^n - x^n$ in k_{Alg} is $\text{Roots}(X^n - x^n) = \{xa \mid a \in E^n\}$, so $\text{Roots}(X^n - x^n) \subset k(x)$ and $k(x)$ is the splitting field for $X^n - x^n$ over k. Since $X^n - x^n$ has n distinct roots, x is separable and $k(x)/k$ is Galois. We have an injective homomorphism from $\text{Aut}_k\, k(x)$ to E_n sending each $\sigma \in \text{Aut}_k\, k(x)$ to $a_\sigma \in E_n$ where $\sigma(x) = xa_\sigma$, since $\text{Aut}_k\, k(x)$ stabilizes $\text{Roots}(X^n - x^n)$. Since E_n is cyclic of order n, $\text{Aut}_k\, k(x)$ is therefore cyclic of order dividing n. Letting σ be a generator of $\text{Aut}_k\, k(x)$ and d its order, we have $\sigma(x) = xa_\sigma$ and $\sigma(x^d) = x^d(a_\sigma)^d = x^d$. Thus, x^d is fixed by $\text{Aut}_k\, k(x)$ and $x^d \in k$. Thus, $X^d - x^d$ is a polynomial in $k[X]$ vanishing at x of degree $d = k(x){:}k$ and $f_x(X) = X^d - x^d$. Suppose conversely, that K/k is cyclic of degree d dividing n. Let σ be a generator for $\text{Aut}_k\, K$, so that σ has order d. Let $a \in E_n \subset k$ be a primitive dth root of 1 and take $x \in K$ such that $\sigma(x) = xa$ (see 3.9.3). Then $f_x(X)$ has d distinct roots x, $\sigma(x) = xa, \ldots, \sigma^{d-1}(x) = xa^{d-1}$ and, consequently, $K = k(x)$. Moreover, $\sigma(x^d) = (\sigma(x))^d = (xa)^d = x^d a^d = x^d$ so that $x^d \in k$, hence $x^n \in k$. □

3.9.6 Corollary. Let n be any positive integer. Suppose that k contains the group E_n of nth roots of unity of k_{Alg}, and that $x \in k_{\text{Sep}}$ and $x^n \in k$. Then $k(x)$ is cyclic.

Proof. Let $n = mp^e$ where m and p are relatively prime, and let $y = x^{p^e}$. Then $y^m \in k$, so that $k(y)/k$ is cyclic, by 3.9.5. But $k(x) = k(y)$ since x is separable (see 2.2.5). □

We finally turn to the case of cyclic extensions K of degree $n = p^e$ of a field k. The case $e = 1$ is particularly simple, and it is instructive to discuss it before considering the general case.

3.9.7 Theorem. Let $p > 1$. Then K/k is cyclic of degree p if and only if $K = k(x)$ for some $x \in K$ such that $x^p - x \in k$ and $x \notin k$. The minimum polynomial $f_x(X)$ of any $x \in K$ with $x^p - x \in k$ and $x \notin k$ is $f_x(X) = X^p - X - (x^p - x)$.

Proof. Let $x^p - x \in k$ and $x \notin k$. The set of roots of $X^p - X - (x^p - x)$ in k_{Alg} is $\text{Roots}(X^p - X - (x^p - x)) = \{x + i \mid i \in \pi_p\}$, where $\pi_p = \{\bar{0}, \bar{1}, \ldots, \overline{p-1}\}$ is the prime field of k, so that $\text{Roots}(X^p - X - (x^p - x)) \subset k(x)$ and $k(x)$ is the splitting field of $X^p - X - (x^p - x)$ over k. Since $X^p - X - (x^p - x)$ has distinct roots, x is separable over k and $k(x)/k$ is Galois. We have an injective homomorphism from $\text{Aut}_k\, k(x)$ to π_p sending each $\sigma \in \text{Aut}_k\, k(x)$ to $i_\sigma \in \pi_p$ where $\sigma(x) = x + i_\sigma$, since $\text{Aut}_k\, k(x)$ stabilizes $\text{Roots}(X^p - X - (x^p - x))$. Since π_p is cyclic of order p, $\text{Aut}_k\, k(x)$ is therefore cyclic of order p. Thus, $k(x)/k$ is cyclic of degree p and $f_x(X) = X^p - X - (x^p - x)$. Suppose conversely that K/k is cyclic of degree p. Let σ be a generator for $\text{Aut}_k\, K$, so that σ has order p. Take $x \in K$ such that $\sigma(x) = x + 1$ (see 3.9.4). Then $\sigma(x^p) = \sigma(x)^p = (x + 1)^p = x^p + 1$ and $\sigma(x^p - x) = (x^p + 1) - (x + 1) = x^p - x$, so that $x^p - x \in k$. But $x \notin k$, since $\sigma(x) = x + 1$. Thus, $k(x)/k$ is cyclic of degree p, by the first part of the proof. But then $K = k(x)$, since $K \supset k(x)$ and $K{:}k = p$. □

The characterization of cyclic extension K/k of degree dividing n was established, under the assumption that k contain a multiplicative cyclic group E_n of order n, by analyzing conditions for the existence of an injective homomorphism from $\mathrm{Aut}_k K$ to E_n. We now characterize the cyclic extensions K/k of degree p^n by analyzing conditions for the existence of an injective homomorphism from $\mathrm{Aut}_k K$ to an additive cyclic group π_{p^e} of order p^e. Precisely this was done in 3.9.7 in the case $e = 1$, π_p then being the prime field of k. For $e \geq 1$, we pass from k and its prime field π_p to the ring $W_e k$ of Witt vectors and its prime subring π_{p^e}.

We begin by describing $W_e K$ and properties of $W_e K$ analogous to properties of K. The more technical parts of the discussion are given in Appendix W, where a complete discussion of Witt vectors is given.

3.9.8 Definition. $W_e K = \{x = (x_0, \ldots, x_{e-1}) \mid x_i \in K \text{ for } 0 \leq i \leq e - 1\}$.

We give $W_e K$ the operations of addition and multiplication constructed in W.2 and note that $W_e K$ is then a commutative ring with zero element $\mathbf{0} = (0, 0, \ldots, 0)$ and identity element $\mathbf{1} = (1, 0, \ldots, 0)$ (see W.9). It is clear that for any subfield k of K, $W_e k$ is a subring of $W_e K$. Letting $\mathbf{n} = \mathbf{1} + \cdots + \mathbf{1}$ (n times) so that $\mathbf{n}x = x + \cdots + x$ (n times) for n a positive integer and $x \in W_e K$, we have $\mathbf{p}x = (0, x_0{}^p, \ldots, x_{e-2}^p)$ for $x = (x_0, x_1, \ldots, x_{e-1})$ (see W.17). One then easily shows that the additive order of $\mathbf{1}$ is p^e. Since the subring $W_e\pi_p$ has p^e elements and contains the additive cyclic group with generator $\mathbf{1}$, $W_e\pi_p$ is generated as additive group by $\mathbf{1}$. Thus, $W_e\pi_p$ is the prime subring of $W_e k$ and $W_e\pi_p$ is additively cyclic. We let $\pi_{p^e} = W_e\pi_p$.

For any homomorphism σ from K to K, we let

$$\sigma(x) = x^\sigma = (\sigma(x_0), \sigma(x_1), \ldots, \sigma(x_{e-1}))$$

for $x = (x_0, x_1, \ldots, x_{e-1})$. Letting $\pi(a) = a^p$ for $a \in K$, we have, in particular, $\pi(x) = x^\pi = (x_0{}^p, x_1{}^p, \ldots, x_{e-1}^p)$. Since addition and multiplication are defined by polynomials with coefficients in π_p (see W.14) and since $\sigma(a) = a$ for $a \in \pi_p$, σ is a homomorphism from $W_e K$ to $W_e K$. In particular, π is a homomorphism from $W_e K$ to $W_e K$ whose subring of fixed points is the prime subring $\pi_{p^e} = W_e\pi_p$. If σ is a k-automorphism of K, the induced σ on $W_e K$ is an automorphism leaving fixed the elements of the subring $W_e k$.

The mapping on $W_e(K)$ defined by $V(x) = (0, x_0, \ldots, x_{e-2})$ for $x = (x_0, \ldots, x_{e-1})$ preserves addition (see W.5). Moreover, $V^i(x)V^j(y) = V^{i+j}(x^{p^j}y^{p^i})$ for $x, y \in W_e K$ (see W.17). In particular, $V(x)^e = 0$ for all x. Letting $y = (y_0, \ldots, y_{e-1})$, we have $y^e = 0$ if $y_0 = 0$, for we can then express y as $y = V(x)$. We claim that y is a unit if $y_0 \neq 0$. To see this, let $z = (y_0{}^{-1}, 0, \ldots, 0)$ and note that the zeroth coordinate of $\mathbf{1} - yz$ is 0 so that $(\mathbf{1} - yz)^e = 0$. Thus, $(yz) \sum_{i=0}^{e-1} (\mathbf{1} - yz)^i = \mathbf{1}$ and y is a unit.

3.9.9 Proposition. Let $\sigma \in \mathrm{Aut}\, K$ have finite order d. Then there exists $y \in W_e K$ such that $\sum_1^d \sigma^i(y) = -\mathbf{1}$.

Proof. Choose $w_0 \in K$ such that $z_0 = \sum_1^d \sigma^i(w_0)$ is nonzero. Letting $w = (w_0, 0, \ldots, 0)$, the zeroth coordinate of $z = \sum_1^n \sigma^i(w)$ is z_0, so that z is a unit. Since $\sigma(z) = z$, we therefore have $1 = \sum_1^d \sigma^i(wz^{-1})$ and we may take $y = -wz^{-1}$. ☐

For $f \le e$ and $y \in W_e K$, we let $y_f = V^{e-f}(y)$. In particular, $1_f = (0, \ldots, 0, 1, 0, \ldots, 0)$ where 1 is the $(e-f)$th entry of 1_f. Note that $1_e = 1$ and $1_0 = 0$. Note also that the additive order of 1_f is p^f and that $p^f y_f = 0$ for all $y \in W_e K$.

3.9.10 Proposition. Let $\sigma \in \operatorname{Aut} K$ have finite order p^f dividing p^e. Then there exists $x \in W_e K$ such that $\sigma(x) = x + 1_f$.

Proof. Choose y such that $\sum_{i=1}^{p^f} \sigma^i(y) = -1$. Applying V^{e-f}, we have

$$-1_f = V^{e-f}\left(\sum_{i=1}^{p^f} \sigma^i(y)\right) = \sum_{i=1}^{p^f} V^{e-f}\sigma^i(y) = \sum_{i=1}^{p^f} \sigma^i(V^{e-f}(y)) = \sum_{i=1}^{p^f} \sigma^i(y_f).$$

Letting $x = \sum_{i=1}^{p^f} i\sigma^i(y_f)$, we then have

$$\sigma(x) = \sum_{i=1}^{p^f} i\sigma^{i+1}(y_f) = \sum_{i=1}^{p^f} (i+1)\sigma^{i+1}(y_f) - \sum_{i=1}^{p^f} \sigma^{i+1}(y_f)$$

$$= \sum_{j=1}^{p^f} j\sigma^j(y_f) - \sum_{j=1}^{p^f} \sigma^j(y_f) = x + 1_f,$$

since $p^f y_f = 0$. ☐

The following related theorem is needed in the next section.

3.9.11 Theorem. Let G be an algebraic subgroup of $\operatorname{Aut} K$. Then $H^1(G, (W_e K)^+)^{\text{cont}} = 1$.

Proof. It is understood here that $(W_e K)^+$ is the additive group of $W_e K$ with the discrete topology. Since G acts continuously on K, it acts continuously on $(W_e K)^+$ (see 0.4.5). Now let f be a continuous cocycle from G to $(W_e K)^+$. Suppose first that G is finite and take $y = (y_0, 0, \ldots, 0) \in W_e K$. The zeroth coordinate of $z = \sum_{\tau \in G} \tau(y)$ is $z_0 = \sum_{\tau \in G} \tau(y_0)$ and we take y_0 such that z_0 is nonzero. Then z is a unit, by the discussion preceding 3.9.9. Letting $x = \sum_{\tau \in G} f(\tau)\tau(y)$, we have $x - \sigma(x) = f(\sigma)z$, as in 3.7.7, so that $f(\sigma) = xz^{-1} - \sigma(xz^{-1})$ for $\sigma \in G$. Now drop the assumption that G be finite. Since f is continuous, Kernel f is open and G has a normal subgroup N of finite index such that $N \subset$ Kernel f. Let $\bar{f}(\sigma N) = f(\sigma)$ define a cocycle $\bar{f}$ from $\bar{G} = G/N$ to $(W_e K)^+$. Since we can regard $\bar{G}$ as a finite subgroup of $\operatorname{Aut} K^N$, we know that there exists $x \in W_e K^N$ such that $\bar{f}(\bar{\sigma}) = x - \bar{\sigma}(x)$ for $\bar{\sigma} \in \bar{G}$. But then $f(\sigma) = x - \sigma(x)$ for $\sigma \in G$ and $f = \partial x$. ☐

We now can describe the cyclic extensions of degree dividing p^e. The following theorem provides this description. In the theorem, the condition that K/k be separable algebraic can be dropped, as we note at the end of this section. We use the notation $k(x)$ for $k(x_0, \ldots, x_{e-1})$ when $x = (x_0 \cdots x_{e-1})$.

3.9.12 Theorem. A (separable) algebraic extension K/k is cyclic of degree p^f dividing p^e if and only if $K = k(x)$ for some $x \in W_e K$ such that $x^\pi - x \in W_e k$.

Proof. Let the expression $x \in W_eK$ be such that $x^\pi - x \in W_eK$. The set Roots $(X^\pi - X - (x^\pi - x))$ of roots of $X^\pi - X - (x^\pi - x)$ in $W_e k_{\text{Alg}}$ is $\{x + i \mid i \in \pi_{p^e}\}$ where $\pi_{p^e} = W_e\pi_p$, since $(y^\pi - y) - (x^\pi - x) = 0$ if and only if $(y - x)^\pi = y - x$ that is, if and only if $y - x \in W_e\pi_p$. Since $x^\pi - x \in W_ek$, Roots $(X^\pi - X - (x^\pi - x))$ is stable under $\text{Aut}_k\, k_{\text{Alg}}$. Thus, $k(x)$ is stable under $\text{Aut}_k\, k_{\text{Alg}}$, for $k(x)$ contains the coordinates of the roots $x + i$ $(i \in \pi_{p^e})$ of Roots $(X^\pi - X - (x^\pi - x))$. Since $k(x)/k$ is separable algebraic, $k(x)/k$ is therefore Galois. We have an injective homomorphism from $\text{Aut}_k\, k(x)$ to π_{p^e} sending each $\sigma \in \text{Aut}_k\, k(x)$ to $i_\sigma \in \pi_{p^e}$ where $\sigma(x) = x + i_\sigma$. Since π_{p^e} is cyclic of order p^e, $\text{Aut}_k\, k(x)$ is therefore cyclic of order p^f dividing p^e. Thus, $k(x)/k$ is cyclic of degree p^f dividing p^e. Suppose, conversely, that K/k is cyclic of degree p^f dividing p^e. Let σ be a generator for $\text{Aut}_k\, K$, so that σ has order p^f. Take $x \in W_eK$ such that $\sigma(x) = x + \mathbf{1}_f$ (see 3.9.10). Then $\sigma(x^\pi) = \sigma(x)^\pi = (x + \mathbf{1}_f)^\pi = x^\pi + \mathbf{1}_f$ and $\sigma(x^\pi - x) = (x^\pi + \mathbf{1}_f) - (x + \mathbf{1}_f) = x^\pi - x$, so that $x^\pi - x \in W_ek$. Thus, $k(x)/k$ is cyclic of degree $p^{f'}$ dividing p^e, by the first part of the proof. Since $\mathbf{1}_f$ has additive order p^f, $\sigma|_{k(x)}$ has order p^f. Thus, $p^{f'} = p^f$ and $K = k(x)$. □

The above theorem together with the following proposition determine the cyclic extensions $k(x)$ of k of degree dividing p^e. In the proposition, we use the notation $(\pi - 1)x = x^\pi - x$.

3.9.13 Proposition. For each $a \in W_ek$, there exists $x \in W_e({}_{p^e}k_{\text{Abel}})$ such such that $(\pi - 1)x = a$.

Proof. We prove by induction on e that there exists $x = (x_0, \ldots, x_{e-1})$ such that $(\pi - 1)(x) = a$ and $x_i \in k_{\text{Sep}}$ for $0 \le i \le e - 1$. If $e = 1$, we simply take $x_0 \in k_{\text{Sep}}$ such that $x_0{}^p - x_0 = a$ (see 3.9.7). Next, let $e > 1$ and again choose $x_0 \in k_{\text{Sep}}$ such that $x_0{}^p - x_0 = a_0$. Then $(a_0, a_1, \ldots, a_{e-1}) - (\pi - 1)(x_0, 0, \ldots, 0) = (0, b_1, \ldots, b_{e-1})$ for suitable $b_k \in k_{\text{Sep}}$. Letting $k' = k(b_1, \ldots, b_{e-1})$, we can find, by induction, $y_1, \ldots, y_{e-1} \in k'_{\text{Sep}} = k_{\text{Sep}}$ such that $(b_1, \ldots, b_{e-1}) = (\pi - 1)(y_1, \ldots, y_{e-1}) = (y_1{}^p, \ldots, y_{e-1}^p) - (y_1, \ldots, y_{e-1})$. But then $(0, b_1, \ldots, b_{e-1}) = (0, y_1{}^p, \ldots, y_{e-1}^p) - (0, y_1, \ldots, y_{e-1}) = (\pi - 1) \times (0, y_1, \ldots, y_{e-1})$ (see W.11). We now have $(a_0, a_1, \ldots, a_{e-1}) = (\pi - 1) \times ((x_0, 0, \ldots, 0) + (0, y_1, \ldots, y_{e-1})) = (\pi - 1)(x_0, x_1, \ldots, x_{e-1})$ for suitable $x_i \in k_{\text{Sep}}$ $(1 \le i \le e - 1)$. Thus, $(\pi - 1)(x) = a$, for $x = (x_0, \ldots, x_{e-1})$ as asserted. Since the extension $k(x)/k$ is separable, it is cyclic of degree dividing p^e, by 3.9.12. Thus, $k(x) \subset {}_{p^e}k_{\text{Abel}}$ and $x \in W_e({}_{p^e}k_{\text{Abel}})$. □

It is appropriate at this point to note that the part of the hypothesis in 3.9.12 that K/k be separable algebraic can be dropped. For this condition is used only to insure that $k(x)/k$ is separable algebraic. However, that $k(x)/k$ is separable algebraic follows from the condition that $x^\pi - x \in W_ek$. For, by 3.9.13, there exists $y = (y_0, \ldots, y_{e-1})$ such that $y^\pi - y = x^\pi - x$ and such that $y_i \in k_{\text{Sep}}$ for $0 \le i \le e - 1$. But then $(y - x)^\pi = y - x$ and $y - x \in W_e\pi_p$. Thus, $x = y - (y - x) \in W_ek_{\text{Sep}}$ and $k(x)/k$ is separable algebraic.

3.10 Abelian extensions

In 3.9, we noted that an Abelian extension K of k of finite degree has the form $K = k_1 \cdots k_m$ (disjoint product over k), where the extension k_i/k is cyclic of prime power degree $p_i^{e_i}$ for $1 \leq i \leq m$ (see 3.9.2). Thus, the problem of determining the structure of an Abelian extension K of k of finite degree n reduces to the problem of determining the structure of the cyclic extensions of k of prime power degree dividing n. In the case that k contains the multiplicative group E_n of nth roots of unity in k_{Alg} and K/k has separable exponent dividing n, the solution has been given (see 3.9.5, 3.9.12). Without the assumption that k contain E_n, however, the problem is quite difficult.

In this section, we retain the assumption that k contain E_n and describe a bijective correspondence between the set of subfields K of k_{Alg} containing k such that K/k is a finite dimensional Abelian extension of separable exponent dividing n and the set of finite subgroups of the character group $\chi_n(G)$ of $G = \text{Aut}_k\,({}_nk_{\text{Abel}})$. We then determine $\chi_n(G)$.

We begin by noting that the set of subfields K of k_{Alg} containing k such that K/k is Abelian of separable exponent dividing n is the set of subfields of ${}_nk_{\text{Abel}}$ containing k (see 2.4.11). Next, let $G = \text{Aut}_k\,{}_nk_{\text{Abel}}$ and consider the group $\chi_n(G) = \text{Hom}\,(G, \mathbb{Z}_n)$ of homomorphisms from G to the additive group of $\mathbb{Z}_n$. If G is finite, G and $\chi_n(G)$ are *dual* in the sense that for $\sigma \in G$, $f(\sigma) = 0$ for all $f \in \chi_n(G)$ if and only if σ is the identity element of G (see E.3.20). Let $H^\perp = \{f \in \chi_n(G) \mid f(H) = \{0\}\}$ for $H \subset G$ and $S^\perp = \{\sigma \in G \mid f(\sigma) = 0 \text{ for } f \in S\}$ for $S \subset \chi_n(G)$. Note that for H a subgroup of G, the group G/H has exponent dividing n and that $\chi_n(G/H)$ and $H^\perp$ can be identified by letting any $f \in H^\perp$ be regarded also as the function on G/H mapping σH to $f(\sigma)$ for $\sigma \in G$. If H is a subgroup of G of finite index, then $H^\perp$ is a finite subgroup of $\chi_n(G)$ (see E.3.7), G/H and $\chi_n(G/H) = H^\perp$ are dual and, consequently, $H = (H^\perp)^\perp$. Note, conversely, that for S a finite subgroup of $\chi_n(G)$, $S^\perp$ is a subgroup of G of finite index and $S = (S^\perp)^\perp$ (see E.3.20). Thus, $H \mapsto H^\perp$ maps the set of subgroups of G of finite index bijectively to the set of finite subgroups of $\chi_n(G)$. From this and the Galois Correspondence Theorem we have the following proposition.

3.10.1 Proposition. The set of subfields K of k_{Alg} containing k such that K/k is a finite dimensional Abelian extension of separable exponent dividing n is mapped bijectively to the set of finite subgroups of $\chi_n(G)$ by the mapping sending K to $(G_K)^\perp$ where G_K denotes the subgroup

$$\{\sigma \in G \mid \sigma(x) = x \text{ for } x \in K\}. \qquad \square$$

We now assume that k contain the multiplicative group E_n of nth roots of unity in k_{Alg} and let $n = mp^e$ where m and p are relatively prime. Then $E_n = E_m$ and E_m is isomorphic to the additive group of $\mathbb{Z}_m$. Consequently, the additive group of $\mathbb{Z}_n$ is isomorphic to the direct product $E_m \times \mathbb{Z}_{p^e}$ of E_n and the additive group of $\mathbb{Z}_{p^e}$. Thus, $\chi_n(G) = \text{Hom}\,(G, \mathbb{Z}_n)$ is isomorphic to the direct product $\text{Hom}\,(G, E_m) \times \text{Hom}\,(G, \mathbb{Z}_{p^e})$ (see E.3.20). Thus, to determine

$\chi_n(G)$, it suffices to determine Hom (G, E_m) and Hom $(G, \mathbb{Z}_{p^e})$. This is now done in 3.10.2 and 3.10.3.

3.10.2 Theorem (Kummer). Let k contain the multiplicative group E_n of nth roots of unity in k_{Alg} and let $n = mp^e$ where m and p are relatively prime. Then Hom (G, E_m) and k^*/k^{*m} are canonically isomorphic, where k^{*m} denotes the subgroup $\{x^m \mid x \in k^*\}$ of the multiplicative group k^* of units of k.

Proof. For $a \in k^*$, we define $f_a \in$ Hom (G, E_m) as follows. Let x be an mth root of a in ${}_nk_{\text{Abel}}$ (see 3.9.5). For $\sigma \in G$, let $f_a(\sigma) = \sigma(x)x^{-1}$. Since any two such x differ multiplicatively by an element of $E_m \subset k$, $f_a(\sigma)$ is independent of the choice of x. Since $x^m = a \in k$, $f_a(\sigma)$ is an element of E_m. Since $f_a(\sigma) \in E_m \subset k$, we have $\sigma(x)x^{-1}\tau(x)x^{-1} = \sigma x\tau(x)x^{-1}x^{-1} = (\sigma\tau)(x)x^{-1}$ for $\sigma, \tau \in G$, so that $f_a \in$ Hom (G, E_m). Finally, $f_a = f_b$ if $ak^{*m} = bk^{*m}$. For let $b = ac^m$ with $c \in k^*$ and let y be an mth root of b in ${}_nk_{\text{Abel}}$. Then $y^m = ac^m = x^mc^m = (xc)^m$, so that $y^{-1}xc \in E_m \subset k$ and $y^{-1}x \in k$. Thus, $f_b(\sigma) = \sigma(y)y^{-1} = \sigma(x)x^{-1} = f_a(\sigma)$ for $\sigma \in G$ and $f_b = f_a$. We have now shown that $a \mapsto f_a$ induces a mapping from k^*/k^{*m} to Hom (G, E_m). This mapping is a homomorphism since, taking notation as before, $f_a(\sigma)f_b(\sigma) = \sigma(x)x^{-1}\sigma(y)y^{-1} = \sigma(xy)(xy)^{-1} = f_{ab}(\sigma)$ for all $\sigma \in G$. (Here, xy is an mth root of ab since x and y are mth roots of a and b respectively.) The homomorphism is injective on k^*/k^{*n}, for if $f_a(G) = \mathbf{1}$, then $1 = f_a(\sigma) = \sigma(x)x^{-1}$ for $\sigma \in G$, and $x \in k^*$ so that $a = x^m \in k^{*m}$. Finally, we show that the homomorphism is surjective. Thus, let $f \in$ Hom (G, E_m). Then the kernel of f is of finite index in G so that f is continuous from G to ${}_nk_{\text{Abel}}$. By 3.7.13, there consequently exists $x \in {}_nk_{\text{Abel}}$ such that $f(\sigma) = \sigma(x)x^{-1}$ for $\sigma \in G$. Now $f(\sigma) \in E_m$, so that $1 = (f(\sigma))^m = \sigma(x^m)x^{-m}$ for $\sigma \in G$. Thus, $x^m \in k^*$ and $f = f_a$ where $a = x^m$. Thus, the homomorphism is surjective. □

3.10.3 Theorem (Artin–Schreier–Witt). The additive groups Hom $(G, \mathbb{Z}_{p^e})$ and $W_ek/(\pi - 1)(W_ek)$ are canonically isomorphic, where $(\pi - 1)(W_ek)$ is the subgroup $\{x^\pi - x \mid x \in W_ek\}$ of the additive group of W_ek.

Proof. Recall that the prime field of k is denoted π_p and that the prime subring of W_ek is $\pi_{p^e} = W_e\pi_p$ and is isomorphic to $\mathbb{Z}_{p^e}$. For $a \in W_ek$, we define $f_a \in$ Hom (G, π_{p^e}) as follows. Let x be an element of $W_e({}_nk_{\text{Abel}})$ such that $x^\pi - x = a$ (see 3.9.13). For $\sigma \in G$, let $f_a(\sigma) = \sigma(x) - x$. Since any two x_1, x_2 such that $x_1{}^\pi - x_1 = a = x_2{}^\pi - x_2$ satisfy $(x_2 - x_1)^\pi = x_2 - x_1$, we have $x_2 - x_1 \in W_e(\pi_p)$. Thus, $\sigma(x_2 - x_1) = x_2 - x_1$ and $\sigma(x_2) - x_2 = \sigma(x_1) - x_1$. It follows that $f_a(\sigma)$ is independent of the choice of x. Since $x^\pi - x = a \in W_ek$, we have $\pi(f_a(\sigma)) = \pi(\sigma(x) - x) = \sigma(x^\pi) - x^\pi = \sigma(x + a) - (x + a) = \sigma(x) - x = f_a(\sigma)$, so that $f_a(\sigma) \in W_e\pi_p = \pi_{p^e}$. Since $f_a(\sigma) \in \pi_{p^e} = W_e(\pi_p)$, we have $f_a \in$ Hom (G, π_{p^e}), for $f_a(\sigma\tau) = (\sigma\tau)(x) - x = \sigma(\tau(x)) - x = \sigma(x + f_a(\tau)) - x = \sigma(x) - x + f_a(\tau) = f_a(\sigma) + f_a(\tau)$ for $\sigma, \tau \in G$. Finally, $f_a = f_b$ if the cosets $a + (\pi - 1)W_ek = b + (\pi - 1)W_ek$. For let $b = a + (\pi - 1)c$ with $c \in W_ek$ and let $y \in W_e({}_nk_{\text{Abel}})$ satisfy $y^\pi - y = b$. Then $(\pi - 1)y = a + (\pi - 1)c = (\pi - 1)x + (\pi - 1)c$, so that

$$(\pi - 1)(x + c - y) = 0$$

and $x + c - y \in W_e\pi_p \subset W_e k$. Consequently, $x - y \in W_e k$ and $f_b(\sigma) = \sigma(y) - y = \sigma(x) - x = f_a(\sigma)$ for all $\sigma \in G$, so that $f_b = f_a$. We now have shown that $a \mapsto f_a$ induces a mapping from $W_e k/(\pi - 1)W_e k$ to Hom (G, π_e). This mapping is a homomorphism, since, taking notation as before, $f_a(\sigma) + f_b(\sigma) = \sigma(x) - x + \sigma(y) - y = \sigma(x + y) - (x + y) = f_{a+b}(\sigma)$ for all $\sigma \in G$. (Here, $(x + y)^\pi - (x + y) = a + b$ since $x^\pi - x = a$ and $y^\pi - y = b$). The homomorphism is injective on $W_e k/(\pi - 1)W_e k$, for if $f_a(G) = \{0\}$, then $0 = f_a(\sigma) = \sigma(x) - x$ for $\sigma \in G$ and $x \in W_e k$, so that $a = (\pi - 1)(x) \in (\pi - 1)W_e k$. Finally, we show that the homomorphism is surjective. Thus, let $f \in$ Hom (G, π_{p^e}). Then Kernel f is a subgroup of G of finite index, so that f is continuous from G to $W_e({}_n k_{\text{Abel}})$ (see 0.4.3). By 3.9.11, there exists $x \in W_e({}_n k_{\text{Abel}})$ such that $f(\sigma) = \sigma(x) - x$ for $\sigma \in G$. Now $f(\sigma) \in \pi_{p^e}$, so that $0 = (\pi - 1)f(\sigma) = \sigma((\pi - 1)(x)) - (\pi - 1)(x)$ for $\sigma \in G$. Thus, $(\pi - 1)(x) \in W_e k$ and $f = f_a$ where $a = (\pi - 1)(x)$. We have now shown that the homomorphism is also surjective. ▯

3.11 Solvable extensions

The objective of the present section is to describe the finite dimensional solvable extensions in terms of extensions of the form $k(x)/k$ where $x^m \in k$ for some m or $x^p - x \in k$. Throughout the section, K is a finite dimensional extension of a field k of exponent characteristic p, k_{Alg} is an algebraic closure of k containing K and K^{norm} is the normal closure of K in k_{Alg} over k.

We begin with an elementary property of solvable extensions.

3.11.1 Proposition. Let K/k be a Galois extension and let k' be a subfield of K containing k such that k'/k is Galois. Then K/k is solvable if and only if K/k' and k'/k are solvable.

Proof. Let $G = \text{Aut}_k K$ and $G' = \text{Aut}_{k'} K$. Then G' is a normal subgroup of G so that G is solvable if and only if G' and G/G' are solvable (see E.0.76). Thus, K/k is solvable if and only if K/k' and k'/k are solvable, for the extensions K/k, K/k', k'/k are Galois with Galois groups G, G', G/G' respectively. ▯

The following definition is taken from the theory of equations (see 3.12).

3.11.2 Definition. The extension K/k is *solvable by radicals* if there exists a tower $k = K_0 \subset \cdots \subset K_r$ of subfields of k_{Alg} such that $K \subset K_r$ and $K_i = K_{i-1}(x_i)$ for some $x_i \in K_i$ such that $x_i^{m_i} \in K_{i-1}$ for some m_i or $x_i^p - x_i \in K_{i-1}$ $(1 \leq i \leq r)$.

Note that K/k is solvable by radicals if and only if k_{sep}/k is solvable by radicals. For the extension K/k_{sep} is radical and therefore has a tower $k_{\text{sep}} = K_0 \subset \cdots \subset K_r = K$ such that $K_i = K_{i-1}(x_i)$ for some $x_i \in K_i$ such that $x_i^p \in K_{i-1}$ $(1 \leq i \leq r)$ (see 2.2.18). Thus, to determine all extensions K/k which are solvable by radicals, it suffices to determine all separable extensions K/k which are solvable by radicals. This is done in the following theorem.

3.11.3 Theorem. Let K/k be a finite dimensional separable extension. Then K/k is solvable if and only if K/k is solvable by radicals.

Proof. Let K/k be solvable and let $G = \mathrm{Aut}_k K^{\mathrm{norm}}$. Let the order of G be $n = mp^e$ where m and p are relatively prime. Taking x to be a primitive mth root of unity in k_{Alg}, the extension $B = k(x)$ is cyclic (see 3.8.4). Let $K' = K^{\mathrm{norm}}B$ and $G' = \mathrm{Aut}_k K'$. Then K'/k is Galois and solvable (see 2.4.2, 2.4.9). Now G' has a tower $G' = G'_0 \supset \cdots \supset G'_r = \mathbf{1}$ such that G'_i is a normal subgroup of G'_{i-1} and G'_{i-1}/G'_i is cyclic of prime order dividing n $(1 \le i \le r)$ (see 0.2). Letting $K'_i = K'^{G'_i}$, we have a tower $k = K'_0 \subset \cdots \subset K'_r = K'$ such that the extensions K'_i/K'_{i-1} are cyclic of prime degree dividing n $(1 \le i \le r-1)$. Since the K'_{i-1} contain the group E_n of nth roots of unity of k_{Alg}, we have $K'_i = K'_{i-1}(x_i)$ where $x_i^{m_i} \in K'_{i-1}$ for some m_i or $x_i^p - x_i \in K'_{i-1}$ $(1 \le i \le r)$ (see 3.9.5, 3.9.7). Thus, K/k is solvable by radicals.

Suppose, conversely, that K/k is solvable by radicals, and take a tower $k = K_0 \subset \cdots \subset K_r$ such that $K \subset K_r$ and $K_i = K_{i-1}(x_i)$ where $x_i^{m_i} \in K'_{i-1}$ for some m_i or $x_i^p - x_i \in K'_{i-1}$ $(1 \le i \le r)$. Since K/k is finite dimensional, the set $\{\sigma(K) \mid \sigma \in \mathrm{Aut}_k k_{\mathrm{Alg}}\}$ of conjugates of K over k is finite (see 1.4.10). Let K' be the composite $K' = K_1 \cdots K_s k'$ where $K_1, \ldots, K_s$ are the distinct conjugates of K over k and $k' = k(x)$ where x is a primitive mth root of unity and m is defined by $n = mp^e$ where $m = m_1 \cdots m_r$ and m and p are relatively prime. The extensions $K_1 \cdots K_s/k$ and k'/k are normal and separable, hence Galois. Thus, K'/k' is Galois. Moreover, K' has a tower $k' = K'_0 \subset \cdots \subset K'_t = K'$ where $K'_j = K'_{j-1}(y_j)$ and $y_j^n \in K'_{j-1}$ or $y_j^p - y_j \in K'_{j-1}$ $(1 \le j \le t)$, for we can take the y_j's to be conjugates of the x_i's such that for each j, $y_j = \sigma(x_i)$ and $K'_{j-1} \supset \sigma(K_{i-1})$ for some i and σ. Letting $G'_j = \mathrm{Aut}\, K'/K'_j$, we have a tower $\mathrm{Aut}_{k'} K' = G'_0 \supset G'_1 \supset \cdots \supset G'_t = \mathbf{1}$. But the extensions K'_j/K'_{j-1} are cyclic since K'_{j-1} contains E_n $(1 \le j \le t)$ (see 3.9.6 and 3.9.7). Thus, G'_j is a normal subgroup of G'_{j-1} and G'_{j-1}/G'_j is cyclic $(1 \le j \le t)$. The group $\mathrm{Aut}_{k'} K'$ is therefore solvable. Thus, K'/k' is solvable and one sees from the following diagram that K^{norm}/k is solvable, so that K/k is solvable:

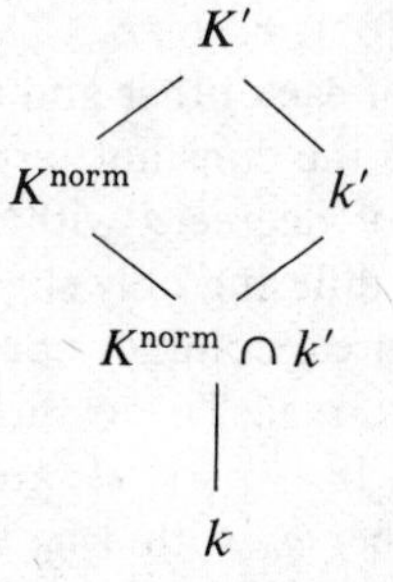

More specifically, $K^{\mathrm{norm}}/(K^{\mathrm{norm}} \cap k')$ is solvable since K'/k' is solvable (see 2.4.4). And $(K^{\mathrm{norm}} \cap k')/k$ is cyclic since k'/k is cyclic (see 3.8.4). But then K^{norm}/k is solvable, by 3.11.1, and K/k is solvable (see 2.4.8). $\square$

3.12 Theory of equations

Let k be a field of characteristic 0 or characteristic $p > 1$. Let $f(x)$ be an irreducible polynomial with coefficients in k, $k(f)$ a splitting field for $f(x)$ over k, $G(f)$ the automorphism group $\mathrm{Aut}_k\, k(f)$ and $R(f) = \{s_1, \ldots, s_m\}$ the set of roots of $f(x)$ in $k(f)$. We say that $f(x)$ is *solvable by radicals* if there exists a tower $k = k_0 \subset \cdots \subset K_r$ of subfields of k_{Alg} and elements $x_i \in K_i$ $(1 \leq i \leq r)$ such that $k(f) \subset K_r$ and, for each i with $1 \leq i \leq r$, $K_i = K_{i-1}(x_i)$ where where $x_i^{m_i} \in K_{i-1}$ for some m_i or $x_i^p - x_i \in K_{i-1}$. If k is of characteristic 0, then $f(x)$ is solvable by radicals when it is possible to eventually reach a full set $R(f)$ of roots of $f(x)$ "by successively adjoining m_ith roots x_i of previously constructed elements to k, $k(x_1)$, $k(x_1)(x_2)$, etc." It is clear from 3.11 that the equation $f(x) = 0$ is solvable by radicals if and only if $G(f)$ is solvable.

Since $R(f)$ is $G(f)$-stable, we may regard $R(f)$ as a $G(f)$-space. The corresponding homomorphism ρ from $G(f)$ to the group $\mathbf{S}(R(f))$ of permutations of $R(f)$ (bijective mappings from $R(f)$ to $R(f)$) is given by $\rho(\sigma) = \sigma|_{R(G)}$ $(\sigma \in G)$, and ρ is injective since $k(f) = k(R(f))$.

The group $\mathbf{S}_n = \mathbf{S}(\{1, \ldots, n\})$ of permutations of $\{1, \ldots, n\}$, called the *symmetric group on n letters*, is solvable if $n \leq 4$ and is not solvable if $n > 4$ (see E.0.84, E.0.85). It follows that if $f(x)$ has degree at most 4, then the equation $f(x) = 0$ is solvable by radicals. For $G(f)$ is then isomorphic to a subgroup of the solvable group $\mathbf{S}_4$.

We next show that for any prime number q, there exists an irreducible polynomial $f(x)$ of degree q with coefficients in the field $\mathbb{Q}$ of rational numbers such that $\mathbf{S}_q$ is isomorphic to $G(f)$. Since any finite group is isomorphic to a subgroup of $\mathbf{S}_q$ for a sufficiently large prime number q (see 0.3.5), it follows that for any finite group G, there exists an irreducible polynomial $f(x)$ with coefficients in $\mathbb{Q}$ such that G is isomorphic to a subgroup of the Galois group $G(f)$ of the Galois extension $\mathbb{Q}(f)/\mathbb{Q}$. In particular, it follows that equations $f(x) = 0$ of degree greater than 4 are solvable by radicals only under very special circumstances.

Thus, let q be a prime number. Let r_1 be a nonreal complex number such that $r_1 + \bar{r}_1$ and $r_1\bar{r}_1$ are integers, and let $r_2 = \bar{r}_1$. Let $r_3, \ldots, r_q$ be distinct nonzero integers. Then $f_0(x) = \prod_1^q (x - r_i)$ is a polynomial with integer coefficients having q distinct roots $r_1, r_2, r_3, \ldots, r_q$, the first two of which are nonreal complex conjugates of each other and the remaining $q - 2$ of which are real. By slightly changing the constant term of $f_0(x)$, we now obtain an irreducible polynomial $f(x)$ of degree q with rational coefficients having q distinct roots $s_1, s_2, s_3, \ldots, s_q$ differing only slightly from the $r_1, r_2, r_3, \ldots, r_q$, the first two of which are nonreal complex conjugates of each other and the remaining $q - 2$ of which are real. To do this, choose $\epsilon > 0$ such that the neighborhoods $N_i = \{z \in \mathbb{C} \mid |z - r_i| < \epsilon\}$ are pairwise disjoint for $1 \leq i \leq q$. Choose a prime number q_0 with $1/q_0$ sufficiently small that $f(x) = f_0(x) + 1/q_0$ has roots $s_1, s_2, s_3, \ldots, s_q$ such that $s_i \in N_i$ for $1 \leq i \leq q$. The roots $s_1, s_2, s_3, \ldots, s_q$ are distinct, and $f(x)$ has only q roots since its degree is q. Moreover, the nonreal roots of $f(x)$ occur in conjugate pairs. Thus, s_1 is nonreal, $s_2 = \bar{s}_1$, and $s_3, \ldots, s_q$ are real. Finally, $f(x)$ is irreducible over $\mathbb{Q}$.

For $f(x) = (q_0 f(x) + 1)/q_0$ and the polynomial $q_0 f(x) + 1$ is irreducible by Eisentein's criteria (see E.0.50).

Finally, we show that $\mathbf{S}_q$ is isomorphic to $G(f)$, where $f(x)$ is the polynomial just constructed. What we must show is that the injective homomorphism $\rho: G(f) \to \mathbf{S}(R(f))$ is surjective. Since $f(x)$ is irreducible over $\mathbb{Q}$ and s_1 is a root of $f(x)$, we have $\mathbb{Q}(s_1):\mathbb{Q} = q$. Thus, q divides the order of $G(f)$. Since q is prime, it follows that $G(f)$ has an element σ of order q (see 0.3.1). Since $\rho(\sigma)$ has order q and is a permutation of the q elements of $R(f)$, $\rho(\sigma)$ must be a q-cycle (see 0.3) and the distinct elements of $R(f)$ are $s_1, \sigma(s_1), \sigma^2(s_1), \ldots, \sigma^{q-1}(s_1)$. Since complex conjugation preserves $k(f)$, G must also have a transposition τ such that $\tau(s_1) = s_2$, $\tau(s_2) = s_1$ and $\tau(s_i) = s_i$ for $3 \le i \le q$. But then the entire symmetric group $\mathbf{S}(R(f))$ on $R(f) = \{s_1, s_2, s_3, \ldots, s_q\}$ is generated by $\rho(\sigma)$ and $\rho(\tau)$, by 0.3.5. It follows that ρ is surjective from $G(f)$ to $\mathbf{S}(R(f))$, so that ρ is an isomorphism from $G(f)$ to the symmetric group $\mathbf{S}(R(f))$ on q elements.

We have now proved the following.

3.12.1 Theorem (Galois). An irreducible polynomial $f(x)$ is solvable by radicals if and only if the Galois group $G(f)$ of a splitting field for $f(x)$ is solvable.

3.12.2 Theorem. For any prime number q, there exists an irreducible polynomial $f(x) \in \mathbb{Q}[x]$ such that the Galois group $G(f)$ of a splitting field for $f(x)$ is isomorphic to the symmetric group $\mathbf{S}_q$ on q elements.

3.13.2 Corollary. For any finite group G, there exists an irreducible polynomial $f(x) \in \mathbb{Q}[x]$ such that G is isomorphic to a subgroup of the Galois group $G(f)$ of a splitting field for $f(x)$.

E.3 Exercises to Chapter 3

E.3.1. Let $\sigma_1, \ldots, \sigma_n$ be distinct homomorphisms from a field K to a field L. Using the L-independence of the σ_j, show that there exist $a_1, \ldots, a_n \in K$ such that the $n \times n$ matrix $(\sigma_j(a_i))$ is nonsingular.

E.3.2. Let G be a subgroup of Aut K, k a subfield of K^G, V a vector space over k. Show that there is a G-product on the K-space $V_K = K \otimes_k V$ such that $V_K{}^G \supset 1 \otimes V$. Show that $1 \otimes V$ is a k-form of V_K.

E.3.3. Let K/k be a finite dimensional extension and k' a subfield of K containing k. Let σ' be a homomorphism from the normal closure K^{norm} of K over k to k_{Alg} (algebraic closure of k containing K^{norm}) and let τ be a homomorphism from K to k_{Alg} such that $\tau|_{k'} = \sigma'|_{k'}$. Show that $\tau = \sigma'\sigma$ for some homomorphism σ from K to k_{Alg}. (Hint: Pass from σ', τ to elements of $\text{Aut}_k K^{\text{norm}}$).

E.3.4 (Symmetric Functions). Let $k[X_1, \ldots, X_n]$ be the polynomial ring in n variables over k and let G be the symmetric group on $\{X_1, \ldots, X_n\}$. Regard G as a subgroup of $\text{Aut}_k K$ where K is the field of quotients $k(X_1, \ldots, X_n)$ of $k[X_1, \ldots, X_n]$. Show that the set $K^G = \{f \in K \mid \sigma(f) = f \text{ for all } \sigma \in G\}$ is

$K^G = k(f_1, \ldots, f_n)$ where $f_1, \ldots, f_n$ are the elementary symmetric polynomials in $X_1, \ldots, X_n$.

E.3.5 (Kronecker). Let $f(X), g(X)$ be irreducible elements of $k[X]$ and let K be an extension field of k containing roots x, y of $f(X)$ and $g(X)$ respectively. Let $f(X) = f_1(X) \cdots f_m(X)$ and $g(X) = g_1(X) \cdots g_n(X)$ be the decompositions of $f(X)$ and $g(X)$ as product of irreducible elements of $k(y)$ and $k(x)$ respectively. Show that $m = n$ and, upon rearranging the indices, $\operatorname{Deg} f(x) \operatorname{Deg} g_i(x) = \operatorname{Deg} g(x) \operatorname{Deg} f_i(x)$ for all i.

E.3.6. Let K be the splitting field over $\mathbb{Q}$ of $(X^2 - 2)(x^2 - 3)$.

(a) Describe a normal basis for K over $\mathbb{Q}$.

(b) Compute $N_{K/\mathbb{Q}}(x)$ and $Tr_{K/k}(x)$ for $x = \sqrt{2}, x = \sqrt{3}, x = \sqrt{2} + \sqrt{3}$, $x = \sqrt{2}\sqrt{3}$.

E.3.7. Let $z \in \mathbb{C}$. Describe $N_{\mathbb{C}/\mathbb{R}} z$ and $Tr_{\mathbb{C}/\mathbb{R}} z$.

E.3.8. Let K be the splitting field of $X^2 + 2X + 2$ over $\mathbb{Q}$ and let $G = \{\epsilon, \sigma\}$ be the Galois group of $K/\mathbb{Q}$, ϵ being the identity of G. (See E.1.10.)

(a) Show that $f\colon G \to K^*$, defined by $f(\epsilon) = 1$ and $f(\sigma) = -1$ is a cocycle.

(b) Find $a \in K$ such that f is the coboundary determined by a.

(c) Determine $N_{K/\mathbb{Q}}(a)$ and $Tr_{K/\mathbb{Q}}(a)$.

E.3.9. Show that $\sigma \cdot x = f(\sigma)^{-1}\sigma(x)$ defines a G-product on K as vector space over K, for any field K, any group G of automorphisms of K and any continuous cocycle f from G to K.

E.3.10. Let K/k be a finite dimensional Galois extension with Galois group G. Let $x \in K$ and define $f(\sigma) = \sigma(x^{-1})x$ for $\sigma \in G$. Let $\sigma \cdot v$ be defined for $v = (x_1, \ldots, x_n) \in K^n$ by $\sigma(v) = (\sigma \cdot x_1, \ldots, \sigma \cdot x_n)$ and $\sigma \cdot x_i = f(\sigma)^{-1}\sigma(x_i)$. Show that $\sigma \cdot v$ is a G-product on K^n and determine $(K^n)^G$.

E.3.11 (Regular Representation). Let K/k be a finite dimensional extension. For $x \in K$, let $L_x\colon K \to K$ be defined by $L_x(y) = xy$ for $y \in K$.

(a) Show that $L\colon K \to \operatorname{End}_k K$ is an injective k-linear ring homomorphism from K to the ring $\operatorname{End}_k K$ of k-linear endomorphisms of K.

(b) Show that the minimum polynomial $f(X)$ of x over k is the same as the minimum polynomial of L_x as linear transformation of K as vector space over k.

(c) Letting the characteristic polynomial of L_x be $a_0 + a_1X + \cdots + a_{n-1}X^{n-1} + X^n$, show that $Tr_{K/k}(x) = -a_{n-1}$ and $N_{K/k}(x) = (-1)^n a_0$. (Hint: The roots of the characteristic polynomial are the conjugates of x).

E.3.12 (Trace Form). Let k/k be a finite dimensional extension. Show that the function $Tr_{K/k}(xy)$ $(x, y \in K)$ is a bilinear form on K as vector space over k. This bilinear form is called the *trace form* of K/k, and is nondegenerate if K/k is separable. (See E.3.13.)

E.3.13 (Different and Discriminant). Let K/k be a finite dimensional separable extension, choose $x \in K$ such that $K = k(x)$ and let $f(X)$ be the minimum polynomial of x over k. The *discriminant* $\mathrm{Disc}_x\, K/k$ of K/k with respect to x is the discriminant of $f(X)$. (See E.1.27.) The *different* $\mathrm{Diff}_x\, K/k$ of K/k with respect to x is $f'(x)$.

(a) Show that $\mathrm{Disc}_x\, K/k = (-1)^{\binom{n}{2}} N_{K/k}\,(\mathrm{Diff}_x\, K/k)$.
(b) Show that $\mathrm{Disc}_x\, K/k = \mathrm{Det}\, M$ where $n = K{:}k$ and M is the $n \times n$ matrix whose (i, j)th coefficient is $Tr_{K/k}(x^{i-1}x^{j-1})$ for $1 \le i, j \le n$.
(c) Show that the trace form of K/k is nondegenerate. (Hint: Use part (b).)

E.3.14 (Cyclotomic Polynomials). Let $x_1, \ldots, x_{\varphi(n)}$ be the primitive nth roots of unity in $\mathbb{C}$ and let $f_n(X) = \prod_1^{\varphi(n)} (X - x_i)$. Show that

(a) $f_n(X) \in \mathbb{Q}[X]$;
(b) $X^n - 1 = \prod_{d|n} f_d(X)$;
(c) $f_n(X) = \dfrac{X^n - 1}{\prod_{\substack{d|n \\ d \neq n}} f_d(X)}$

E.3.15 (Cyclotomic Polynomials). Using the recursion described in the preceding exercise, calculate $f_1(X), f_2(X), f_3(X), f_4(X), f_5(X), f_6(X), f_{10}(X)$.

E.3.16 (Cyclotomic Polynomials). Let p be a prime number. Show that $f_p(X) = X^{p-1} + X^{p-2} + \cdots + X + 1$. Show that $f_p(X)$ is irreducible over $\mathbb{Q}$.

E.3.17 (Cyclotomic Polynomials). Show that $f_n(X)$ has integer coefficients. (Hint: $X^n - 1 = \prod_{d|n} f_d(X) = f_n(X)g(X)$ where $g(X) \in \mathbb{Z}[X]$ by induction. Write $X^n - 1 = m(X)g(X) + r(X)$, where $\mathrm{Deg}\, r(X) < \mathrm{Deg}\, g(X)$ and $m(X), r(X) \in \mathbb{Z}[X]$, and show that $f_n(X) = m(X)$).

E.3.18 (Cyclotomic Polynomials). Show that $f_n(X)$ is irreducible over $\mathbb{Q}$ by writing $f_n(X) = g(X)h(X)$ where $g(X), h(X) \in \mathbb{Z}[X]$ and $g(X)$ is monic and irreducible (see 0.1.11 and E.0.39) and showing that $f_n(X) = g(X)$ as follows:

(a) for any prime p not dividing n, the polynomials $\bar{g}(X), \bar{h}(X) \in \mathbb{Z}_p[X]$ obtained from $g(X), h(X)$ by reducing coefficients modulo p are relatively prime;
(b) for any root x of $g(X)$ and for any element $a(X) \in \mathbb{Z}[X]$ vanishing at x^p, $g(X)$ divides $a(X^p)$;
(c) notation as above, $\bar{g}(X)$ divides $\bar{a}(X)^p$;
(d) therefore, $h(X)$ does not vanish at x^p;
(e) $g(X)$ vanishes at x^p for every root x of $g(X)$ and every prime p not dividing n;
(f) therefore $g(X) = f_n(X)$.

E.3.19 (Structure of $\mathbb{Z}_n^*$). Show that $\mathbb{Z}_n^*$ is isomorphic to the direct product of the $\mathbb{Z}_{p_i^{e_i}}^*$ where $n = \prod_i p_i^{e_i}$ and $\mathbb{Z}_n^*$ is the multiplicative group of units of $\mathbb{Z}_n$. For p a prime, show that the order of $\mathbb{Z}_{p^e}^*$ is $p^{e-1}(p - 1)$. For p odd, show that

(a) $\mathbb{Z}_{p^e}^* = HN$ (direct product) where H and N are subgroups of orders p^{e-1} and $p - 1$ respectively;

(b) there exists a positive integer s such that the elements $s^i + \mathbb{Z}p^e$ ($1 \leq i \leq p - 1$) form a subgroup of $\mathbb{Z}_{p^e}^*$ of order $p - 1$ (choose r such that $r + \mathbb{Z}p, \ldots, r^{p-1} + \mathbb{Z}p$ are distinct and let $s = r^{p^{e-1}}$);
(c) N is cyclic (use part (b));
(d) show that if $0 < u < p^e$ and $u^p \equiv_{p^e} 1$ then $u = 1 + vp^{e-1}$ where $0 \leq v < p$ (show that $u \equiv_p 1$ and $u = 1 + vp^f + wp^{f+1}$ for some $1 \leq f \leq e - 1$, $0 \leq v < p$, $0 \leq w$; then show that $u^p \equiv_{p^{f+2}} 1 + vp^{f+1}$, using the Binomial Theorem, and that f must be $e - 1$);
(e) show that H contains at most $p - 1$ elements of order p (use part (d));
(f) show that H is cyclic (use part (e) and the Decomposition Theorem for Abelian groups);
(g) show that $\mathbb{Z}_{p^e}^*$ is cyclic (use parts (c) and (f)).

For $p = 2$, show that $\mathbb{Z}_{2^e}^*$ is cyclic for $e = 1, 2$ and show for $e \geq 3$ that

(h) $\mathbb{Z}_{p^e}^*$ has three distinct elements x of order 2 (consider the integers $-1, 1 + 2^{e-1}, -1 + 2^{e-1}$);
(i) $\mathbb{Z}_{2^e}^*$ has order 2^{e-1} and is not cyclic (use part (h));
(j) for any odd integer r, $r^{2^{e-2}} \equiv_{2^e} 1$ (use part (i));
(k) $5^{2^{e-3}} \not\equiv_{2^e} 1$ (investigate the congruences of the form $5^{2^{e-3}} \equiv_{2^f} 1$ for $e = 3, 4, \ldots$, showing for each e that the maximal f is $f = e - 1$);
(l) $\mathbb{Z}_{2^e}^*$ is direct product of a cyclic subgroup of order 2^{e-2} and a subgroup of order 2 (use parts (i) and (k)).

Summarize the above, by describing the structure of $\mathbb{Z}_n^*$.

E.3.20 (Duality for Abelian Groups). Let G be an Abelian group of exponent dividing n and let $\chi_n(G)$ be the *Character Group* Hom $(G, \mathbb{Z}_n)$ of homomorphisms from G to $\mathbb{Z}_n$ as group with addition. Let $H^\perp = \{f \in \chi_n(G) \mid f(H) = \{0\}\}$ for $H \subset G$ and $S^\perp = \{\sigma \in G \mid f(\sigma) = 0 \text{ for } f \in S\}$ for $S \subset \chi_n(G)$. Show that

(a) $\chi_n(G)$ together with the multiplication $f_1 f_2$ defined by $(f_1 f_2)(\sigma) = f_1(\sigma) + f_2(\sigma)$ is a group of exponent dividing n;
(b) for any subgroup H of G, $H^\perp$ is a subgroup of $\chi_n(G)$ and $H^\perp$ is mapped isomorphically to $\chi_n(G/H)$, an element f of $H^\perp$ being mapped to the function f' on G/H defined by $f'(\sigma H) = f(\sigma)$;
(c) if H is a subgroup of G of finite index, then $H = H^{\perp\perp}$. (Hint: Use the Basis Theorem for G/H.)

Show that the condition in (c) that H be of finite index can be dropped by showing that $\chi_n(G)^\perp = \mathbf{1}$ as below and applying that to G/H:

(d) if I is a subgroup of G and the exponents of G and I are equal, then $G = IN$ (direct) for some subgroup N (show first in the case $G{:}I < \infty$, then consider "independent families" of finite subgroups $\{N_i\}$ such that $I \cap N = \mathbf{1}$ where N is generated by the family $\{N_i\}$, show that there exists a maximal family and maximal N and show that $G = IN$);

(e) every element g of G is contained in a finite subgroup I such that $G = IN$ (direct) for some subgroup N;

(f) $\chi_n(G)^\perp = \mathbf{1}$.

Show that for any finite subgroup S of $\chi_n(G)$, then $S = S^{\perp\perp}$. Show that if $n = rs$ where r and s are relatively prime, then $\chi_n(G)$ is isomorphic to $\text{Hom}(G, \mathbb{Z}_r) \times \text{Hom}(G, \mathbb{Z}_s)$ (direct).

E.3.21. Let K/k be a finite dimensional Galois extension and let $f(X)$ be an irreducible element of $k[X]$. Show that the irreducible factors of $f(X)$ in $K[X]$ are all of the same degree.

E.3.22. Find a field K having subfields k_1, k_2 such that K/k_1 is finite dimensional Galois for $i = 1, 2$ and $K:k_1 \cap k_2 = \infty$.

E.3.23. How many subfields does the field of p^n elements have? Describe them.

E.3.24. Let q be a prime number and let K/k be a Galois extension of degree q^n $(n \geq 1)$. Show that

(a) K has a subfield K_G containing k such that K_G/k is a Galois extension of degree q^{n-1};

(b) K has precisely one subfield K_q containing k such that $K_q:k = q$, and the extension K_q/k is Galois.

E.3.25. Let K/k be an Abelian Galois extension of degree 6. Show that

(a) K has precisely four subfields containing k;

(b) $K = K_2K_3$ (internal tensor product over k) where K_2 and K_3 are subfields of K containing k of degrees 2 and 3 over k respectively.

E.3.26. What is the discriminant of $X^m - c$? What is the resultant of $X^m - c$ and $X^n - d$?

E.3.27. What is the discriminant of $X^p - X - c$? What is the resultant of $X^p - X - c$ and $X^p - x - d$? (Here, p is the characteristic.)

E.3.28. Let $f(X) = X^3 - 3X + 1 \in k[X]$. Show that $k(f)/k$ is cyclic.

E.3.29. Show for $x, y \in \mathbb{Z}$ that $x \equiv_{p^j} y$ implies $x^p \equiv_{p^{j+1}} y^p$ for $j \geq 1$.

E.3.30. Show that for $f(X) = X^4 + cX^2 + d$ in $\mathbb{Q}[X]$, $\mathbb{Q}(f)/\mathbb{Q}$ is solvable.

E.3.31. Describe an element $f(X)$ of $\mathbb{Q}[X]$ of lowest degree such that the extension $\mathbb{Q}(f)/\mathbb{Q}$ is not solvable.

E.3.32. Determine what groups (up to isomorphism) occur as Galois groups of polynomials of degree 1, 2 and 3, and of polynomials of the form $X^4 + cX^2 + d$.

E.3.33. Suppose that the characteristic of k is not 2 and that $f(X)$ is a monic irreducible element of $k[X]$ such that the quadratic roots of the discriminant of $f(X)$ are in k. Show that $G(f)$ consists of even permutations.

E.3.34. Find the discriminant of $f(X) = X^3 + 2X + 1$ over $\mathbb{Q}$. Then determine $G(f)$.

E.3.35. Describe polynomials $f(X) \in \mathbb{Q}[X]$ whose Galois groups are isomorphic to:

(a) $\mathbb{Z}_4$

(b) $\mathbb{Z}_3$

(c) $\mathbb{Z}_{12}$
(d) the alternating group A_4.

E.3.36. Determine whether the equation $X^5 + 5X - 10 = 0$ is solvable by radials over $\mathbb{Q}$.

E.3.37. Determine the Galois group of $X^3 - 2$ over $\mathbb{Q}$.

E.3.38. Determine the Galois group of $(X^2 - 5)(X^2 - 7)$ over $\mathbb{Q}$.

E.3.39. Prove that not every algebraic subgroup of the automorphism group Aut K of automorphisms of the field $K = (\mathbb{Z}_p)_{\text{Alg}}$ is algebraic. (Hint: Consider the Frobenius homomorphism.)

E.3.40. Describe an algebraic subgroup G of the automorphism group Aut K of $K = (\mathbb{Z}_p)_{\text{Alg}}$ such that $k' \mapsto G \cap \text{Aut}_{k'} K$ is not a bijection from the set $\mathbf{F}_k$ of subfields k' of K containing k such that K/k' is Galois to the set of closed subgroups of G.

E.3.41. Let G and H be finite groups, let d be the greatest common denominator of $|G|$, $|H|$ and suppose that H is Abelian. Show that $H^1(G, H)$ has exponent dividing d. (Hint: Let $f \in H^1(G, H)$. Show that if $x = \prod_{g \in G} f(g)$, then $f(g)^{|G|} = xg(x)^{-1}$. Show that $f^{|G|}$ and $f^{|H|}$ and therefore f^d are cohomologous to the identity.)

E.3.42. Let G be a finite group of automorphisms of K and let Hom (G, K^+), Hom (G, K^*) be the group of homomorphisms from G to K^+, K^* respectively. Let A be the subgroup $\{x \in K^* \mid x^{\text{Exp}G} \in K^G \text{ and } g(x)^{-1}x \in K^G \text{ for all } g \in G\}$ of K^* and let $k^* = (K^G)^*$. Prove the following.

(a) Hom $(G, K^+) = Z^1(G, K^+)$ and Hom $(G, K^*) = Z^1(G, K^*)$.
(b) Hom (G, K^*) is isomorphic to A/k^*.
(c) G is isomorphic to Hom (G, K^*) if and only if K^* contains a primitive Exp Gth root of unity.
(d) G is isomorphic to A/k^* if and only if K^* contains a primitive Exp Gth root of unity.

(Here, Exp G is the exponent of G).

E.3.43 (Cartier). Let G be a finite group of automorphisms of K. Show that $H^1(G, GL_nK) = \mathbf{1}$. (Hint: Let $f \in Z^1(G, GL_nK)$ and let $b: K^n \to K^n$ be defined by $b(x) = \sum_{g \in G} f(g)g(x)$ $(x \in K^n)$. Show that $\{b(x) \mid x \in K^n\}$ spans K^n by showing that for $u \in K^{n*}$ (dual space of K^n) and $u(b(ax)) = 0$ for all $a \in K$, $x \in K^n$, $0 = \sum_{g \in G} g(a)u(f(g)g(x))$ for all a, x, hence $0 = u(f(g)g(x))$ for all x, hence $u = 0$. Then show that $b: K^n \to K^n$ is surjective.)

4. Algebraic function fields

In this chapter, we study the structure of field extensions K of k of the form $K = k(x_1, \ldots, x_n)$. Extensions K of k of this form are called finitely generated extensions or algebraic function fields. We use the latter term here, since the main theorems of this chapter are most naturally interpreted as theorems about fields of functions associated with varieties.

In 4.1, we relate algebraic function fields to affine varieties, in order to give the reader the proper geometric orientation for the chapter. In 4.2, we briefly discuss algebraic function fields of transcendency degree 1. In 4.3, we relate algebraic function fields K/k and their derivation algebras $\mathrm{Der}_k K$. We introduce the notion of separability for an arbitrary extension K/k. And we introduce the notion of p-basis for a field extension K/k and use it to show that an algebraic function field K/k of transcendence degree d is separable if and only if it is separably generated, and if and only if d is the dimension of $\mathrm{Der}_k K$ over K.

4.1 Algebraic function fields and their geometrical interpretation

In this section, we briefly relate algebraic function fields to affine varieties and to surfaces in affine space. This account is for motivational purposes and is not used later in the book.

4.1.1 Definition. An *algebraic function field* over k is a field extension K of k of the form $K = k(x_1, \ldots, x_n)$.

We now introduce the notion of affine variety. Our definition is more general than the classical definition of affine variety in that the classical affine varieties are the separable affine varieties in the sense of 4.1.2 and 4.3.19.

4.1.2 Definition. An *affine variety* over k is the set $V = \mathrm{Spec}_{\bar{k}}\, k[x_1, \ldots, x_n]$ of k-linear ring homomorphisms from a subring $k[x_1, \ldots, x_n]$ of a field extension L of k into the algebraic closure $\bar{k}$ of k. The *algebraic function field* of an affine variety $V = \mathrm{Spec}_{\bar{k}}\, k[x_1, \ldots, x_n]$ over k is the algebraic function field $K = k(x_1, \ldots, x_n)$ over k, and the *ring of regular functions* of V is the subring $k[x_1, \ldots, x_n]$ of K. The *dimension* of an affine variety V over k is the transcendence degree of its algebraic function field over k.

Let V be an affine variety over k with algebraic function field K and ring of regular functions $k[x] = k[x_1, \ldots, x_n]$. (Here we let $x = \{x_1, \ldots, x_n\}$ for notational convenience.) Each element $f \in k[x]$ determines a function $\hat{f}\colon V \to \bar{k}$ defined by $\hat{f}(v) = v(f)$ for $v \in V$. By a fundamental theorem of algebraic geometry, the Hilbert Nullstellensatz, the set $V = \text{Spec}_{\bar{k}}\, k[x]$ separates the points of $k[x]$ (see E.4.1). It follows that the mapping $f \mapsto \hat{f}$ on $k[x]$ is injective. Thus, we can regard $k[x]$ as a ring and k-space of functions from V to $\bar{k}$. The algebraic function field K of V consists of the quotients f/g ($f \in k[x]$, $g \in k[x] - \{0\}$). Since f/g can be regarded via $\hat{f}$ and $\hat{g}$ as a function on $V_g = \{v \in V \mid \hat{g}(v) \neq 0\}$, the elements of K can be regarded as "functions" from V to $\bar{k}$ which are not everywhere defined. Thus, the terminology "field of algebraic functions of V" and "ring of regular functions of V" is justified.

We imbed V into affine n-space $\bar{k}^n$ (n-fold Cartesian product of $\bar{k}$) by the injective mapping from V to $\bar{k}^n$ sending $v \in V$ to $(\hat{x}_1(v), \ldots, \hat{x}_n(v)) \in \bar{k}^n$. The regular functions $x_1, \ldots, x_n$ may be thought of as a *set of coordinate functions* on V. Let $X = \{X_1, \ldots, X_n\}$ and let $k[X]$ be the ring of polynomials over k in the n algebraically independent elements X_i. Let $I_k(V)$ be the *ideal of relations* $I_k(V) = \{f \in k[X] \mid f(x_1, \ldots, x_n) = 0\}$ in $k[X]$, so that $I_k(V)$ is the kernel of the homomorphism from $k[X]$ to $k[x]$ sending each $f(X_1, \ldots, x_n) \in k[X]$ to $f(x_1, \ldots, x_n) \in k[x]$. Then $k[X]/I_k(V)$ is isomorphic to $k[x]$. Since $k[x]$ is an integral domain, $I_k(V)$ is a prime ideal.

4.1.3 Proposition. The image $\bar{V} = \{(\hat{x}_1(v), \ldots, \hat{x}_n(v)) \mid v \in V\}$ of V in $\bar{k}^n$ is the *surface* $\bar{V} = \{(v_1, \ldots, v_n) \in \bar{k}^n \mid f(v_1, \ldots, v_n) = 0 \text{ for all } f \in I_k(V)\}$ of zeros in $\bar{k}^n$ of the prime ideal $I_k(V)$.

Proof. Suppose first that $(v_1, \ldots, v_n) \in \bar{k}^n$ and $f(v_1, \ldots, v_n) = 0$ for all $f \in I_k(V)$. Then the k-linear homomorphism v from $k[X]$ to $\bar{k}$ such that $v(X_i) = v_i$ for $1 \leq i \leq n$ vanishes on the ideal $I_k(V)$, since $v(f(X_1, \ldots, X_n)) = f(v(X_1), \ldots, v(X_n)) = f(v_1, \ldots, v_n) = 0$ for $f \in I_k(V)$. It follows that there is a k-linear homomorphism from $k[x]$ to $\bar{k}$ mapping x_i to v_i for $1 \leq i \leq n$, so that $(v_1, \ldots, v_n) \in \bar{V}$. Suppose, conversely, that $(v_1, \ldots, v_n) \in \bar{V}$. Since there is a homomorphism from $k[x]$ to $\bar{k}$ mapping x_i to v_i for $1 \leq i \leq n$, there exists a homomorphism v from $k[X]$ to $\bar{k}$ which vanishes on $I_k(V)$ and maps X_i to v_i for $1 \leq i \leq n$. But then

$$f(v_1, \ldots, v_n) = f(v(X_1), \ldots, v(X_n)) = v(f(X_1, \ldots, X_n)) = 0$$

for $f \in I_k(V)$. Thus,

$$\bar{V} = \{(v_1, \ldots, v_n) \in \bar{k}^n \mid f(v_1, \ldots, v_n) = 0 \text{ for all } I_k(V)\}. \qquad \square$$

4.1.4 Proposition. Let $X = \{X_1, \ldots, X_n\}$ and let $k[X]$ be the ring of polynomials in the n algebraically independent elements X_i. Let I be a prime ideal of $K[X]$. Then there exists an affine variety V with ring of regular functions $k[x_1, \ldots, x_n]$ such that $\bar{V} = \{(\hat{x}_1(v), \ldots, \hat{x}_n(v)) \mid v \in V\}$ is the surface $\{(v_1, \ldots, v_n) \in \bar{k}^n \mid f(v_1, \ldots, v_n) = 0 \text{ for all } f \in I\}$ of zeros in $\bar{k}^n$ of I and $I = I_k(V)$.

Proof. Note that $k[X]/I$ is an integral domain consisting of all polynomial expressions in $X_1 + I, \ldots, X_n + I$ over k. Let $x_1 = X_1 + I, \ldots, x_n = X_n + I$. Then $k[X]/I$ is the subring $k[x_1, \ldots, x_n]$ of the field of quotients

$k(x_1, \ldots, x_n)$ of $k[X]/I$. Let V be the affine variety with ring of regular functions $k(x_1, \ldots, x_n]$ and algebraic function field $k(x_1, \ldots, x_n)$. Since $K[X]/I = k[x_1, \ldots, x_n]$, the elements of V can be regarded as the k-linear homomorphisms v from $k[X]$ to $\bar{k}$ which vanish on I, so that we may write

$$v(f(X_1, \ldots, X_n)) = f(v(X_1), \ldots, v(X_n)) = f(v(x_1), \ldots, v(x_n))$$
$$= f(\hat{x}_1(v), \ldots, \hat{x}_n(v))$$

for $v \in V$. It follows that for $f \in I$, $f(v_1, \ldots, v_n) = 0$ for all $(v_1, \ldots, v_n) \in \overline{V} = \{(\hat{x}_1(v), \ldots, \hat{x}_n(v)) \mid v \in V\}$. Suppose, conversely, that $(v_1, \ldots, v_n) \in \bar{k}^n$ and $f(v_1, \ldots, v_n) = 0$ for all $f \in I$. Let $v\colon k[X] \to \bar{k}$ be the k-linear homomorphism such that $v(X_i) = v_i$. Then for $f(X_1, \ldots, X_n) \in I$, $v(f(X_1, \ldots, X_n)) = f(v_1, \ldots, v_n) = 0$. Thus, $v \in V$ and $(v_1, \ldots, v_n) \in \overline{V}$. We have now shown that $\overline{V}$ is the surface of zeros in $\bar{k}^n$ of I. Furthermore, we have

$$I = \{f \in k[X] \mid f(x_1, \ldots, x_n) = 0\} = I_k(V). \qquad \square$$

The above two propositions show that the affine varieties V over k with rings of regular functions $k[x_1, \ldots, x_n]$ having n generators $x_1, \ldots, x_n$ correspond exactly to the surfaces of zeros in $\bar{k}^n$ of prime ideals of the polynomial ring $k[X]$ in n algebraically independent elements $X_1, \ldots, X_n$. Since it can be shown that every surface $\{(v_1, \ldots, v_n) \in \bar{k}^n \mid f_i(v_1, \ldots, v_n) = 0 \text{ for } 1 \leq i \leq r\}$ of zeros in $\bar{k}^n$ of elements $f_1, \ldots, f_r \in k[X]$ is a finite union of surfaces of zeros of prime ideals of $k[X]$, the affine varieties over k are the basic objects of study of affine geometry (see E.4.2).

The algebraic function fields of two affine varieties $V = \operatorname{Spec}_{\bar{k}} k[x_1, \ldots, x_m]$ and $W = \operatorname{Spec}_{\bar{k}} k[y_1, \ldots, y_n]$ over k are k-isomorphic if and only if the associated surfaces $\overline{V}$ in $\bar{k}^m$ and $\overline{W}$ in $\bar{k}^n$ are birationally equivalent in the sense of E.4.4.

4.2 Algebraic function fields of transcendency degree 1

We now prove an important theorem due to Lüroth on algebraic function fields K/k of transcendency degree 1. Letting x be a transcendental element of K, $K/k(x)$ is algebraic. Lüroth's Theorem states that every subfield k' of $k(x)$ properly containing k is of the form $k' = k(y)$ for some transcendental element y of K.

We begin with a few comments on the elements y of $k(x) - k$ where x is transcendental over k. Writing $y = u(x)/v(x)$ where $u(x)$ and $v(x)$ have no nonconstant common factor in $k[x]$ and letting $P(X) = yv(X) - u(X)$ in the polynomial ring $k(x)[X]$ in an indeterminant X over the field $k(x)$, we have $P(x) = 0$. The polynomial $P(X)$ is nonzero, for if $P(X) = 0$ and $u(X) = \sum_0^r u_i X^i$, $v(X) = \sum_0^s v_j X^j$, then upon taking a nonzero v_j we would have $0 = yv_j - u_j$ and $y \in k$, a contradiction. It follows that x is algebraic over $k(y)$. Consequently, y is transcendental over k, for otherwise x would be algebraic over k. Suppose that $P(X)$ has a factorization $P(X) = Q(X)R(X)$ with $Q(X), R(X) \in k[y][X]$. Since the degree of $P(X)$ in y is 1, we can take the degree of $Q(X)$ in y to be 0 and the degree of $R(X)$ in y to be 1. Since y is transcendental over k and $u(X)$, $v(X)$ have no nonconstant common factor in $k[X]$, it is impossible to write $yv(X) - u(X) = Q(X)R(X)$ where $Q(X)$ is a nonconstant element of $k[X]$ and $R(X) \in k[y][X]$ (see E.0.24). Thus, $Q(X) \in k$ and the only factorizations $P(X) = Q(X)R(X)$ of $P(X)$ in $k[y][X]$ are trivial. It follows from 0.1.11 that $P(X)$ is an irreducible element of $k(y)[X]$. Since $P(x) = 0$, we have $k(x):k(y) = m$ where m is the degree of $P(X)$ in X.

4.2.1 ***Theorem*** (Lüroth). Let x be transcendental over k and let k' be a subfield of $k(x)$ properly containing k. Then $k' = k(y)$ for some transcendental element y over k.

Proof. We have seen that x is algebraic over $k(y)$ and y is transcendental over k for any $y \in k' - k$. Consequently, x is algebraic over k'. Let $f(X)$ be the minimum polynomial of x over k', where $f(X) \in k'[X] \subset k(x)[X]$. Let $f(X) = cf^*(X)$ where $c \in k(x)$ and where $f^*(X)$ is a primitive element of $k[x][X]$ (see 0.1.8). Let $f(X) = \sum_0^n a_i(x)X^i$ and $f^*(X) = \sum_0^n b_i(x)X^i$. Choose i such that $a_i(x) \notin k$ and set $y = a_i(x)$. We claim that $k' = k(y)$. Write $y = u(x)/v(x)$ where $u(x)$ and $v(x)$ have no nonconstant factor in $k(x)$. Let m be the degree of $f^*(X)$ in x (the maximum of the degrees of the $b_i(x)$) and note that $\operatorname{Deg} u(x) \leq m$ and $\operatorname{Deg} v(x) \leq m$. Let $P(X) = yv(X) - u(X)$ so that $P(X)$ is nonzero and $P(x) = 0$ as in the discussion preceding this theorem. It follows that $P(X) = Q(X)f(X)$ for some $Q(X) \in k'[X]$. Now $P(X) = aP^*(X)$, $Q(X) = bQ^*(X)$ and $f(X) = cf^*(X)$ where $a = v(x)$, b is a suitable element of $k(x)$, c is as before and where $P^*(X)$, $Q^*(X)$, $f^*(X)$ are primitive elements of $k[x][X]$ (see E.1.21). Then $P^*(X) = dQ^*(X)f^*(X)$ with $d \in k$, by 0.1.9. Since the degree of the left hand side in x is at most m (recall that $\operatorname{Deg} u(x) \leq m$ and $\operatorname{Deg} v(x) \leq m$) and the degree of $f^*(X)$ in x is m, the degree of $P^*(X)$ in x is m and the degree of $Q^*(X)$ in x is 0. Since $Q^*(X)$ is a factor of $P(X)$ contained in $k[x]$, $Q^*(X) \in k$ by the discussion preceding this theorem. It follows that $d^*f^*(X) = P^*(X) = u(x)v(X) - v(x)u(X)$ with $d^* \in k$. By the symmetry in x and X, the degree m of $f^*(X)$ in x coincides with the degree n of $f^*(X)$ in X. It is now clear that $m = n$ and the degree of $P^*(X)$ in X is n. By the discussion preceding this theorem, $k(x):k(y)$ is the degree of $P(X)$ in X. Thus, $k(x):k(y) = n = k'(x):k$. Since $k(x) = k'(x) \supset k' \supset k(y)$, it follows that $k' = k(y)$. □

4.3 Separably generated algebraic function fields

The derivation algebra $\operatorname{Der}_k K$ of an algebraic function field K/k (see 4.3.1) is related to the spaces of tangents at the various points of an affine variety V with algebraic function field K over k (see E.4.5). The classical affine varieties are those affine varieties V such that the dimension of V coincides with the dimension of $\operatorname{Der}_k K$ over K. The purpose of this section is to study the class of algebraic function fields corresponding to the classical affine varieties. Throughout the section, p denotes the exponent characteristic of K (see 1.1.4).

We begin by discussing derivations. For this, let L be a field, K a subfield of L and k a subfield of K.

4.3.1 Definition. A *derivation* from K to L over k is a k-linear mapping D from K to L such that $D(xy) = D(x)y + xD(y)$ for $x, y \in K$. The set of derivations from K to L over k is denoted $\mathrm{Der}_k(K, L)$. We denote $\mathrm{Der}_k(K, K)$ by $\mathrm{Der}_k K$. The elements of $\mathrm{Der}_k K$ are called *derivations* of K/k and $\mathrm{Der}_k K$ is the *derivation algebra* of K/k.

For $D, E \in \mathrm{Der}_k(K, L)$ and $a \in K$, we define $D + E$ and aD as mappings from K to L by $(D + E)(b) = D(b) + E(b)$ and $(aD)(b) = a(D(b))$ for $b \in K$. One then easily shows that $D + E$ and aD are elements of $\mathrm{Der}_k(K, L)$ and that $\mathrm{Der}_k(K, L)$ together with the addition $D + E$ and scalar multiplication aD is a vector space over K. The k-linearity of an element $D \in \mathrm{Der}_k(K, L)$ easily leads to the equation $D(a) = 0$ for $a \in k$ (see E.4.6).

4.3.2 Proposition. Let k' be a subfield of K containing k and suppose that K/k' is a finite dimensional separable extension. Then each derivation D' from k' to L over k has a unique extension to a derivation D from K to L over k.

Proof. $K = k'(s)$ for some $s \in K$, by 2.2.14. Letting $f(X) = \sum_0^n a_i X^i$ be the minimum polynomial of s over k', $f'(s) \neq 0$, by 2.1.2. Suppose that D is a derivation from K to L over k and let $D|_{k'} = D'$. Then we have

$$\begin{aligned} 0 = D(f(s)) &= \sum_0^n D(a_i s^i) = \sum_0^n D(a_i)s^i + \sum_0^n a_i i s^{i-1} D(s) \\ &= f^{D'}(s) + f'(s)D(s) \end{aligned}$$

where $f^{D'}(X) = \sum_0^n D'(a_i)X^i$ and $f'(X) = \sum_0^n a_i i X^i$. Thus, $D(s) = -f^{D'}(s)/f'(s)$ is determined completely by D' and D is the only derivation from K to L extending D. Conversely, let D' be a derivation from k' to L over k. Define $D(s) = -f^{D'}(s)/f'(s)$ and then define

$$D\left(\sum_0^{n-1} b_i s^i\right) = \sum_0^{n-1} D(b_i)s^i + \sum_0^{n-1} b_i i s^{i-1} D(s)$$

for any $b_0, \ldots, b_{n-1} \in k'$. Then D is a k-linear mapping from K to L and $D|_{k'} = D'$. By the choice of $D(s)$, one can show that in fact

$$D\left(\sum_0^{m} b_i s^i\right) = \sum_0^{m} D(b_i)s^i + \sum_0^{m} b_i i s^{i-1} D(s)$$

for any integer $m \geq 0$ and $b_0, \ldots, b_m \in k'$. It then follows easily that D is a derivation from K to L over k. □

4.3.3 Proposition. Let K/k be finite dimensional. Then K/k is separable if and only if $\mathrm{Der}_k K = \{0\}$.

Proof. Suppose first that K/k is separable and $D \in \operatorname{Der}_k K$. Then $D(a) = 0$ for all $a \in k$. Thus, D and the zero derivation 0 of K/k extend the zero derivation $0'$ of k/k. By 4.3.2, we therefore have $D = 0$. Thus, $\operatorname{Der}_k K = \{0\}$. Suppose next that K/k is not separable. Let k' be a maximal proper subfield of K containing k_{sep}. Since K/k_{sep} is radical (see 2.2.13), K/k' is radical. Letting $u \in K - k'$, we have $K \supset k'(u) \supsetneqq k'(u^p) \supset k'$ (see 2.2.5). Thus, by the maximality of k', we have $K = k'(u)$ and $u^p \in k'$. Thus, $1, u, \ldots, u^{p-1}$ is a basis for K over k'. Let D be the k'-linear mapping from K to K such that $D(u^i) = iu^{i-1}$ for $0 \leq i \leq p - 1$. Using the ensuing fact that $D(s) = 0$ for $a \in k'$, one shows easily that $D(u^i) = iu^{i-1}$ for all $i \geq 0$ and therefore that D is a nonzero derivation of K/k. Thus, K/k is not separable only if $\operatorname{Der}_k K \neq \{0\}$. $\square$

In order to study the derivations of algebraic function fields which are not finite dimensional, we now determine $\operatorname{Der}_k K$ in the important case where $K = k(x_1, \ldots, x_d)$ and the $x_1, \ldots, x_d$ are algebraically independent. For this, let $\partial/\partial x_j$ be the derivation of $k(x_1, \ldots, x_d)$ defined by

$$\frac{\partial}{\partial x_j} f = \frac{1}{x_j} \sum e_j a_{e_1 \cdots e_d} x_1^{e_1} \cdots x_d^{e_d}$$

for

$$f = \sum a_{e_1 \cdots e_d} x_1^{e_1} \cdots x_d^{e_d} \in k[x_1, \ldots, x_d]$$

and by

$$\frac{\partial}{\partial x_j} \frac{f}{g} = \frac{1}{g} \frac{\partial}{\partial x_j} f - \frac{f}{g^2} \frac{\partial}{\partial x_j} g \qquad \text{for } f, g \in k[x_1, \ldots, x_d].$$

(We leave it to the reader to show that $\partial/\partial x_j$ is well defined and is a derivation.) Note that $\partial/\partial x_j(x_i) = \delta_{ij}$ for $1 \leq i, j \leq d$. It follows that the $\partial/\partial x_1, \ldots, \partial/\partial x_d$ are linearly independent over k. For any $D \in \operatorname{Der}_k K$, D and $\sum_1^d D(x_j)(\partial/\partial x_j)$ are equal at each x_i $(1 \leq i \leq d)$, so that $D = \sum_1^d D(x_j)\, \partial/\partial x_j$. Thus, $\partial/\partial x_j, \ldots, \partial/\partial x_d$ is a basis for $\operatorname{Der}_k K$ and $\operatorname{Der}_k K : K = d$.

4.3.4 Proposition. Let K/k be an algebraic function field. Then $\operatorname{Der}_k K = \{0\}$ if and only if K/k is finite dimensional and separable.

Proof. If K/k is finite dimensional and separable, then $\operatorname{Der}_k K = \{0\}$ by 4.3.3. Suppose, conversely, that $\operatorname{Der}_k K = \{0\}$. We assert first that K/k has transcendency degree 0. Assume, to the contrary, that $x_1, \ldots, x_d$ $(d \geq 1)$ is a transcendency basis for K/k and let $k' = k(x_1, \ldots, x_d)$. If K/k' is not separable, then $\operatorname{Der}_{k'} K \neq \{0\}$, by 4.3.3, so that $\operatorname{Der}_k K \neq \{0\}$, a contradiction. Thus, K/k' is separable. But then the derivation $\partial/\partial x_1$ of k'/k extends to a derivation of K/k, by 4.3.2, so that $\operatorname{Der}_k K \neq \{0\}$, a contradiction. Thus, K/k has transcendency degree 0. But then K/k is finite dimensional, and K/k is separable by 4.3.3. $\square$

If K/k is a field extension with $p > 1$, then each $D \in \operatorname{Der}_k K$ vanishes on $K^p = \{x^p \mid x \in K\}$, since $D(x^p) = px^{p-1}D(x) = 0$ for $x \in K$. Thus, each

$D \in \operatorname{Der}_k K$ vanishes on $k(K^p)$. The key to the relationship between $\operatorname{Der}_k K$ and K/k is therefore the extension $K/k(K^p)$.

4.3.5 Definition. A *p-basis* for an extension K/k is a subset S of K such that $K = k(K^p)(S)$ and $K \neq k(K^p)(T)$ for $T \subsetneq S$.

For algebraic function fields K/k, it is obvious that K/k has a p-basis and that every p-basis of K/k is finite. It can in fact be shown that every extension K/k has a p-basis (see E.4.7).

4.3.6 Proposition. Let S be a p-basis for an extension K/k. Then for each $s \in S$, there exists $D_s \in \operatorname{Der}_k K$ such that $D_s(t) = \delta_{st}$ for all $t \in S$.

Proof. Let $s \in S$. Then $K = k'(s)$, $K \neq k'$ and $s^p \in k'$, where $k' = k(K^p)(S - \{s\})$. Thus, there exists a derivation D_s of K/k' such that $D_s(s^i) = is^{i-1}$ for all i, as in the proof of 4.3.3. Clearly, $D_s(s) = 1$. And $D_s(t) = 0$ for $t \in S - \{s\}$, since $S - \{s\} \subset k'$. □

4.3.7 Corollary. Let K/k be an algebraic function field and let $p > 1$. Then for any p-basis S of K/k, the number of elements $|S|$ of S is the dimension $\operatorname{Der}_k K{:}K$ of $\operatorname{Der}_k K$ over K.

Proof. The set $\{D_s \mid s \in S\}$ is a basis for $\operatorname{Der}_k K$ over K. For if $D \in \operatorname{Der}_k K$, D and $\sum_{s \in S} D(s)D_s$ are equal at each $t \in S$ and vanish on $k(K^p)$, so that they are equal on $K = k(K^p)(S)$. And if $\sum_{s \in S} c_s D_s = 0$ ($c_s \in K$ for $s \in S$), then $0 = \sum_{s \in S} c_s D_s(t) = c_t$ for all $t \in S$. □

4.3.8 Proposition. Let K/k be an algebraic function field with $p > 1$ and let S be a p-basis for K/k. Then $K/k(S)$ is finite dimensional and separable.

Proof. It suffices, by 4.3.4, to show that $\operatorname{Der}_{k(S)} K = \{0\}$. Thus, let $D \in \operatorname{Der}_{k(S)} K$. Since D vanishes at each $s \in S$ and on $k(K^p)$, D vanishes on $K = k(K^p)(S)$ and $D = 0$. □

4.3.9 Definition. An algebraic function field K/k is *separably generated* if K/k has a transcendency basis $y_1, \ldots, y_d$ such that $K/k(y_1, \ldots, y_d)$ is separable. Such a transcendency basis is called a *separating transcendency basis* for K/k.

Every algebraic function field K/k of exponent characteristic $p = 1$ is separably generated, since every transcendency basis for such an extension K/k is a separating transcendency basis for K/k.

4.3.10 Theorem. Let K/k be an algebraic function field and let the transcendency degree of K/k be d. Then

1. $\operatorname{Der}_k K{:}K \geq d$;
2. K/k is separably generated if and only if $\operatorname{Der}_k K = d$;
3. if K/k is separably generated and $p > 1$, then the p-bases for K/k are the separating transcendency bases for K/k.

Proof. Suppose first that K/k is separably generated and let $y_1, \ldots, y_d$ be a separating transcendency basis for K/k. By 4.3.2, each of the derivations $\partial/\partial y_i$ of $k(y_1, \ldots, y_d)/k$ can be extended to a derivation D_i of K/k $(1 \leq i \leq d)$. Since $D_j(y_i) = \delta_{ij}$ $(1 \leq i, j \leq d)$, the $D_1, \ldots, D_d$ are linearly independent over K. If $D \in \operatorname{Der}_k K$, then D and $\sum_1^d D(y_j)D_j$ are equal at each y_i $(1 \leq i \leq d)$, hence are equal on $k(y_1, \ldots, y_d)$. Since $K/k(y_1, \ldots, y_d)$ is separable, it follows that $D = \sum_1^d D(y_j)D_j$ for $D \in \operatorname{Der}_k K$, so that $D_1, \ldots, D_d$ is a basis for $\operatorname{Der}_k K$ and $\operatorname{Der}_k K{:}K = d$. Suppose conversely that $\operatorname{Der}_k K{:}K = d$. We assert that K/k is separably generated. If $p = 1$, this is true. Thus, assume that $p > 1$ and let S be a p-basis for K/k. Then $K/k(S)$ is finite dimensional and separable, by 4.3.8, and $|S| = \operatorname{Der}_k K{:}K = d$, by 4.3.7. It follows that S is a separating transcendency basis for K/k and therefore that K/k is separably generated. We have now proved (2).

For (3), suppose that K/k is separably generated and $p > 1$. If S is a p-basis for K/k, then S is a separating transcendency basis for K/k as in the proof of (2) above. Suppose conversely that $S = \{y_1, \ldots, y_d\}$ is a separating transcendency basis for K/k. Then $K/k(S)$ is a finite dimensional separable extension, so that $K = k(S)(K^p)$, by 2.2.5. Thus, $K = k(K^p)(S)$ and S contains a p-basis T for K/k. But then T is also a separating transcendency basis for K/k, as noted above. Thus, $S = T$ and S is a p-basis for K/k.

We finally prove (1). If $p = 1$, then $K \mid k$ is separably generated and $\operatorname{Der}_k K{:}K = d$ by (2). Let $p > 1$ and let S be a p-basis for K/k. Then $K/k(S)$ is a finite dimensional separable extension, by 4.3.8, and consequently $|S| \geq d$. Thus, $\operatorname{Der}_k K{:}K \geq d$, by 4.3.7. □

4.3.11 Theorem (MacLane). Let $K = k(x_1, \ldots, x_n)$ be a separably generated algebraic function field. Then a separating transcendency basis $x_{i_1}, \ldots, x_{i_d}$ can be selected from the given set $x_1, \ldots, x_n$ of generators of K/k.

Proof. If $p = 1$, this is obvious. If $p > 1$, then $x_1, \ldots, x_n$ contains a p-basis $S = \{x_{i_1}, \ldots, x_{i_d}\}$ for K/k which, by 4.3.10, is a separating transcendency basis for K/k. □

4.3.12 Definition. Let L be an extension field of k and let k_1 and k_2 be subrings of K containing k. The ring of linear combinations over k of the products xy $(x \in k_1, y \in k_2)$ is denoted k_1k_2. Elements $x_1, \ldots, x_n$ of k_1 are said to be *k_2-linearly independent* if for $y_1, \ldots, y_n \in k_2$, $\sum_1^n x_iy_i = 0$ implies $0 = y_1 = \cdots = y_n$. If every set $x_1, \ldots, x_n$ of k-linearly independent elements of k_1 are k_2-linearly independent, then k_1 and k_2 are said to be *linearly disjoint* over k.

We now relate the linear disjointness of subrings k_1 and k_2 of L containing k to tensor products and give some equivalent conditions for linear disjointness. The reader is referred to Appendixes T and A for the details. For k_1 and k_2 to be linearly disjoint over k, it is necessary and sufficient that $k_1 \otimes_k k_2$ and k_1k_2 be isomorphic under the k-linear mapping from $k_1 \otimes_k k_2$ to k_1k_2 mapping $x \otimes y$ to xy for $x \in k_1$ and $y \in k_2$. Consequently, k_1 and k_2 are linearly disjoint over k if and only if k_2 and k_1 are linearly disjoint over k. Furthermore, k_1 and k_2 are linearly disjoint over k if and only if k_1 has a basis x_a $(a \in A)$ such that for $y_a \in k_2$ $(a \in A)$, $\sum_{a \in A} x_a y_a = 0$ implies that $y_a = 0$ for all $a \in A$.

4.3.13 ***Definition.*** An extension K/k is *separable* if for any algebraically closed field L containing K, K and $k^{p^{-1}} = \{x \in L \mid x^p \in k\}$ are linearly disjoint over k. (If $p = 1$, every extension K/k is separable.)

The above definition of separability of an extension K/k coincides with our earlier definition of separability for algebraic extensions, as we now show. Suppose that K/k is an algebraic extension. We assert that K/k is separable in the sense of 2.2.6 if and only if for any set $x_1, \ldots, x_n$ of k-linearly independent elements of K, the elements $x_1{}^p, \ldots, x_n{}^p$ are k-linearly independent. The equivalence of the two definitions of separability is thus a consequence of 4.3.14 below. Suppose first that K/k is not separable and choose $x \in K$ such that x is not separable over k. Then $k(x) \supsetneqq k(x^p)$, by 2.2.5. Letting $d = k(x)\!:\!k$, $1, x, \ldots, x^{d-1}$ is a basis for $k(x)/k$. But the d elements $1^p, x^p, \ldots, (x^{d-1})^p$ of $k(x^p)$ are not k-linearly independent since $k(x^p)\!:\!k < d$. This proves one direction of our assertion. For the other, suppose that K/k is separable and let $x_1, \ldots, x_n$ be k-linearly independent elements of K. The extension $k(x_1, \ldots, x_n)/k$ is a finite dimensional separable extension, so that there exists x such that $k(x_1, \ldots, x_n) = k(x)$, by 2.2.14. Since $k(x) = k(x^p)$ (see 2.2.5), $\{x^i \mid 0 \le i \le d - 1\}$ and $\{(x^p)^i \mid 0 \le i \le d - 1\}$ are bases for $k(x) = k(x^p)$ over k, where $d = k(x)\!:\!k$. Thus, $k(x)$ has a basis $y_1, \ldots, y_d$ over k such that $y_1{}^p, \ldots, y_d{}^p$ is a basis for $k(x)$ over k. Let $z_1, \ldots, z_d$ be any basis for $k(x)$ over k. Upon writing the $y_i{}^p$ in terms of the $z_j{}^p$, we see that $z_1{}^p, \ldots, z_d{}^p$ spans $k(x)$ over k and therefore is a basis for $k(x)$ over k. In particular, if we extend $x_1, \ldots, x_n$ to a basis $x_1, \ldots, x_n, z_{n+1}, \ldots, z_d$ for $k(x)$ over k, $x_1{}^p, \ldots, x_n{}^p, z_{n+1}^p, \ldots, z_d{}^p$ is a basis for $k(x)$ over k and the $x_1{}^p, \ldots, x_n{}^p$ are k-linearly independent, as asserted.

4.3.14 ***Proposition.*** An extension K/k is separable if and only if for any set $x_1, \ldots, x_n$ of k-linearly independent elements of K, the elements $x_1{}^p, \ldots x_n{}^p$ are k-linearly independent.

Proof. Let L be an algebraically closed field containing K. Suppose first that $x_1, \ldots, x_n$ are k-linearly independent elements of K. Suppose that $\sum_1^n c_i(x_i)^p = 0$ where the $c_1, \ldots, c_n$ are elements of k. Choose $y_i \in k^{p^{-1}}$ such that $y_i{}^p = c_i$ $(1 \le i \le n)$. Then

$$0 = \sum_1^n y_i{}^p x_i{}^p = \left(\sum_1^n y_i x_i\right)^p,$$

so that $\sum_1^n x_i y_i = 0$. But then $0 = y_1 = \cdots y_n$, so that $0 = c_1 = \cdots = c_n$, and the elements $x_1{}^p, \ldots, x_n{}^p$ are k-linearly independent. Suppose next that K and $k^{p^{-1}}$ are not linearly disjoint over k and choose k-linearly independent elements $x_1, \ldots, x_n$ and elements $y_1, \ldots, y_n$ (not all zero) of $k^{p^{-1}}$ such that $\sum_1^n x_i y_i = 0$. Then

$$0 = \left(\sum_1^n x_i y_i\right)^p = \sum_1^n x_i^p y_i^p = \sum_1^n c_i x_i^p$$

where $c_1 = y_1^p, \ldots, c_n = y_n^p$ (not all zero) are elements of k. Thus, $x_1^p, \ldots, x_n^p$ are not k-linearly independent. □

4.3.15 Proposition. Let K/k be purely transcendental. Then K/k is separable.

Proof. Since any finite subset of K is contained in $k(x_1, \ldots, x_d)$ where $x_1, \ldots, x_d$ are suitable algebraically independent elements of K over k, it suffices to show that such a $k(x_1, \ldots, x_d)/k$ is separable. Thus, let L be an algebraically closed field containing $k(x_1, \ldots, x_d)$. Obviously, $k(x_1, \ldots, x_d)$ and $k^{p^{-1}}$ are linearly disjoint over k if and only if $k[x_1, \ldots, x_d]$ and $k^{p^{-1}}$ are linearly disjoint over k. The set of monomials $M = x_1^{e_1} \cdots x_d^{e_d} \mid e_i \geq 0$ for $1 \leq i \leq d\}$ form a k-basis for $k[x_1, \ldots, x_d]$ over k. Since the set $M^p = \{x^p \mid x \in M\}$ is a subset of M, M^p is a k-linearly independent set. It follows as in the proof of 4.3.14 that M is a $k^{p^{-1}}$-linearly independent set. Since M is a basis for $k[x_1, \ldots, x_d]$ over k, it follows that $k[x_1, \ldots, x_d]$ and $k^{p^{-1}}$ are linearly disjoint over k. Thus, $k(x_1, \ldots, x_d)/k$ is separable. □

4.3.16 Theorem. Let K/k' and k'/k be separable extensions. Then K/k is separable.

Proof. Let L be an algebraically closed field containing K. Then $k^{p^{-1}}$ and k' are linearly disjoint over k and $k'^{p^{-1}}$ and K are linearly disjoint over k'. We assert that $k^{p^{-1}}$ and K are linearly disjoint over k. For this, let $x_1, \ldots, x_n$ be k-linearly independent elements of $k^{p^{-1}}$. Since $k^{p^{-1}}$ and k' are linearly disjoint over k, the $x_1, \ldots, x_n$ are k'-linearly independent elements of $k'^{p^{-1}}$. Since $k'^{p^{-1}}$ and K are linearly disjoint, the $x_1, \ldots, x_n$ are K-linearly independent. Thus, $k^{p^{-1}}$ and K are linearly disjoint over k and K/k is separable. □

4.3.17 Theorem. Let K/k be a separable extension of exponent characteristic $p > 1$ and let S be a p-basis for $K \mid k$. Then the set S is algebraically independent over k.

Proof. Suppose that S is not algebraically independent and let $s_1, \ldots, s_n$ be a minimal set of distinct elements of S which are not algebraically independent over k. Choose a finite set of monomials $s_e = s_1^{e_1} \cdots s_n^{e_n}$ $(e \in A)$ such that

1. $\{s_e \mid e \in A\}$ is a linearly dependent set over k;
2. A has as few elements as possible;
3. $\sum_{e \in A} (e_1 + \cdots + e_n)$ is as small as possible.

Let $\sum_{e \in A} c_e s_e = \sum c_e s_1^{e_1} \cdots s_n^{e_n} = 0$ be a nontrivial dependence relation among the s_e $(e \in A)$ over k and note that $c_e \neq 0$ for all $e \in A$, by (2). Choose derivations $D_i = D_{s_i} \in \mathrm{Der}_k K$ $(1 \leq i \leq n)$ such that $D_i(s_j) = \delta_{ij}$ for $1 \leq i,$

$j \leq n$ (see 4.3.6). Applying D_i to the above dependence relation, we get the equation

$$\sum_{e \in A} e_i c_e s_1^{e_1} \cdots s_{i-1}^{e_{i-1}} s_i^{e_i-1} s_{i+1}^{e_{i+1}} \cdots s_n^{e_n} = 0.$$

By the minimality assumptions (1), (2), (3), the monomials appearing in this equation are k-linearly independent, so that $e_i c_e = 0$ for all $e \in A$. This is of course true for $1 \leq i \leq n$. Since $c_e \neq 0$ for all $e \in A$, it follows that the exponent e_i is divisible by p for all $e \in A$ and $1 \leq i \leq n$. But then the dependence relation $\sum_{e \in A} c_e s_e = 0$ says that the pth powers $(s_1^{e_1/p} \cdots s_n^{e_n/p})^p$ $(e \in A)$ are linearly dependent over k. Since K/k is separable, it follows that the monomials $s_1^{e_1/p}, \ldots, s_n^{e_n/p}$ $(e \in A)$ are linearly dependent over k (see 4.3.14). But this contradicts the minimality assumption (3). Thus, S is algebraically independent over k. □

4.3.18   Theorem (MacLane).   Let K/k be an algebraic function field. Then K/k is separable if and only if K/k is separably generated.

Proof.   There is nothing to prove if $p = 1$. Thus, let $p > 1$. If K/k is separably generated and $y_1, \ldots, y_d$ is a separating transcendency basis for K/k, then $K/k(y_1, \ldots, y_d)$ is separable and $k(y_1, \ldots, y_d)/k$ is separable, by 4.3.15, so that K/k is separable, by 4.3.16. Suppose, conversely, that K/k is separable and let S be a p-basis for K/k. Then $K/k(S)$ is finite dimensional and separable, by 4.3.8, and S is algebraically independent over k, by 4.3.17. Thus, S is a separating transcendency basis for K/k and K/k is separably generated. □

4.3.19   Definition.   An affine variety V is *separable* if its algebraic function field is separable.

4.3.20   Definition.   A field k is *perfect* if every extension K of k is separable.

Any field of exponent characteristic 1 is perfect. The perfect fields are those fields k such that $k = k^p$ (see E.2.22).

4.3.21   Theorem (Schmidt).   Let K/k be an algebraic function field and let k be perfect. Then K/k is separably generated.

Proof.   Since K/k is separable, K/k is separably generated by 4.3.18. □

E.4   Exercises to Chapter 4

E.4.1 (Hilbert Nullstellensatz).   Let $X = \{X_1, \cdots, X_n\}$ and let $k[X]$ be the polynomial ring in the algebraically independent elements $X_1, \cdots, X_n$. For I an ideal of $k[X]$, the *Hilbert Nullstellensatz* states that for any $f \in k[X]$ such that $v(f) = 0$ for every k-linear ring homomorphism $v : k[X] \to \bar{k} = k_{\text{Alg}}$ which vanishes on I, some power f^m of f is contained in I. Using this, prove the following.

   (a) Let V be a variety with ring of regular functions $k[X]$. Then the mapping $f \mapsto \hat{f}$ is injective on $k[X]$, $\hat{f} : V \to \bar{k}$ being defined by $\hat{f}(v) = v(f)$ for $v \in V = \text{Spec}_k k[X]$.

(b) Let I be an ideal of $k[X]$ and let $C(I)$ be the surface $\{(v_1, \ldots, v_n) \in \bar{k}^n \mid f(v_1, \ldots, v_n) = 0 \text{ for } f \in I\}$ of zeros of I in $\bar{k}^n$. If $f \in k[X]$ and $f(v_1, \ldots, v_n) = 0$ for all $(v_1, \ldots, v_n) \in C(I)$, then $f^m \in I$ for some m.

(c) If I is a prime ideal of $k[x]$, then $I = \{f \in k[X] \mid f \text{ vanishes on } C(I)\}$.

(d) If I is a maximal ideal of $k[X]$ and $k = \bar{k}$, then $C(I)$ consists of a single point $(v_1, \ldots, v_n)$, and $I = \{f_1(X)(X_1 - v_1) + \cdots + f_n(X) \times (X_n - v_n) \mid f_i(X) \in k[X] \text{ for } 1 \leq i \leq n\}$ and $k[X] = k1 + I$ (direct).

E.4.2 (Irreducible Components). Following the above notation, we say that a subset C of $\bar{k}^n$ is *closed* in the *k-Zariski Topology* if $C = C(I)$ for some ideal I of $k[X]$. A closed subset C of $\bar{k}^n$ is *irreducible* if $C = C_1 \cup C_2$ where C_1 and C_2 are closed only if $C = C_1$ or $C = C_2$. Prove the following.

(a) The collection of open subsets of $\bar{k}^n$ is a topology for $\bar{k}^n$ (see E.0.85), U being said to be *open* if $\bar{k}^n - U$ is closed.

(b) The closed subsets of $\bar{k}^n$ satisfy the *descending chain condition* that any descending chain $C_1 \supset C_2 \supset C_3 \supset \cdots$ of closed subsets of $\bar{k}^n$ has only finitely many distinct terms. To prove this, assume the *Hilbert Basis Theorem*, which states that the ideals of $k[X]$ satisfy the *ascending chain condition* that any ascending chain $I_1 \subset I_2 \subset I_3 \subset \cdots$ of ideals of $k[X]$ has only finitely many distinct terms.

(c) Show that a closed subset C of $\bar{k}^n$ is irreducible if and only if $C = C(I)$ for a unique prime ideal I of $k[X]$. (For the unicity, refer to the preceding exercise.)

(d) Using the descending chain condition for closed sets, show that for any closed subset C of $\bar{k}^n$, $C = C_1 \cup \cdots \cup C_n$ where the $C_1, \ldots, C_n$ are maximal irreducible closed subsets of C. Show that the C_i are unique up to order of occurrence. The C_i are called the *irreducible components* of C.

E.4.3 (Density). Follow the notation and terminology above and let C be a closed subset of $\bar{k}^n$. A subset U of C is an *open* subset of C if $U = C - D$ for some closed subset D of $\bar{k}^n$. Prove the following.

(a) A subset U of C is open in C if and only if there exist $f_1, \ldots, f_r \in k[X]$ such that $U = \{(v_1, \ldots, v_n) \in C \mid f_i(v_1, \ldots, v_n) \neq 0 \text{ for some } i \text{ with } 1 \leq i \leq r\}$. (Use the descending chain condition for closed sets.)

(b) The collection of open subsets of C is a topology for C.

(c) if C is irreducible and U is a nonempty open subset of C, then U is *dense* in C in the sense that any element $f \in k[X]$ which vanishes on U vanishes on C.

(d) If C is irreducible, any two nonempty open subsets U, V of C intersect.

E.4.4 (Birational Equivalence). Let C and D be irreducible closed subsets of $\bar{k}^m$ and $\bar{k}^n$ respectively. Then C and D are *birationally equivalent* if C and D have nonempty open subsets U and V respectively such that there exists a bijection P from U to V given by

$$P(u_1, \ldots, u_m) = (P_1(u_1, \ldots, u_m), \ldots, P_n(u_1, \ldots, u_m))$$
$$P^{-1}(v_1, \ldots, v_n) = (Q_1(v_1, \ldots, v_n), \ldots, Q_m(v_1, \ldots, v_n))$$

for suitable rational functions P_i, Q_j ($1 \leq i \leq n$, $1 \leq j \leq m$) whose denominators do not vanish at any point of U, V respectively. The *algebraic function field* of C is the field of quotients $k(C)$ of the integral domain $k[X_1, \ldots, X_m]/I$ where I is the prime ideal of $k[X_1, \ldots, X_m]$ such that $C = C(I)$. (See E.4.2.) For $a \in k$, identify a first with the constant polynomial aX^0 of $k[X_1, \ldots, X_m]$, then with the element $a + I$ of the above integral domain, then with the element $a/1$ of $k(C)$, so that $k(C)$ is a field extension of k. Note that $k(C) = k(x_1, \ldots, x_m)$ where $x_i = (X_i + I)/1$ for $1 \leq i \leq m$. Similarly, $k(D) = k(y_1, \ldots, y_n)$ for $y_j = (Y_j + J)/1$ for $1 \leq j \leq n$ where $k[Y_1, \ldots, Y_n]$ is a polynomial ring in n commuting indeterminants Y_j and J is the prime ideal of $k[Y_1, \ldots, Y_n]$ such that $D = C(J)$. Prove the following.

(a) C and D are birationally equivalent if and only if $k(C)$ and $k(D)$ are k-isomorphic. (Hint: Suppose that there is a k-isomorphism α from $k(x_1, \ldots, x_m)$ to $k(y_1, \ldots, y_n)$, and let $\alpha(f) = f'$ for $f \in k(x_1, \ldots, x_m)$. Find a nonzero element $f' \in k[x_1', \ldots, x_m']$ such that $k[x_1', \ldots, x_m', 1/f'] \supset k[y_1, \ldots, y_n]$. Then find a nonzero element $g' \in k[y_1, \ldots, y_n]$ such that $k[x_1', \ldots, x_m, 1/f', 1/g'] = k[y_1, \ldots, y_n, 1/g']$. Choose the open sets U and V above so that the elements of $k[x_1, \ldots, x_m, 1/f, 1/g]$ and $k[y_1, \ldots, y_n, 1/g']$ can be regarded as functions defined throughout U and V respectively).

(b) Using part (*a*), show that affine varieties $V = \operatorname{Spec}_{\bar{k}} k[x_1, \ldots, x_m]$ and $W = \operatorname{Spec}_{\bar{k}} k[y_1, \ldots, x_n]$ have k-isomorphic algebraic function fields if and only if the closed irreducible subsets $\bar{V} = \{(\hat{x}_1(v), \ldots, \hat{x}_m(v)) \mid v \in V\}$ and $\bar{W} = \{(\hat{y}_1(w), \ldots, \hat{y}_n(w)) \mid w \in W\}$ of $\bar{k}^m$ and $\bar{k}^n$ respectively are birationally equivalent.

E.4.5 (Tangent Space). Let V be a variety with ring of regular functions $k[x_1, \ldots, x_n]$ and let $v \in V$. A *tangent* to V at v is a k-linear mapping $T: k[x_1, \ldots, x_n] \to \bar{k}$ such that $T(fg) = T(f)\hat{g}(v) + \hat{f}(v)T(g)$ for $f, g \in k[x_1, \ldots, x_n]$. The set of tangents to V at v is denoted $\operatorname{Tan}_v V$. Prove the following.

(a) $\operatorname{Tan}_v V$ is a $\bar{k}$-vector space of $\bar{k}$-valued functions and the dimension of $\operatorname{Tan}_v V$ over $\bar{k}$ is at most n.

(b) The mapping $T \mapsto (T(x_1), \ldots, T(x_n))$ is a $\bar{k}$-linear embedding of $\operatorname{Tan}_v V$ in $\bar{k}^n$ whose image $\operatorname{Tan}_{\bar{v}} \bar{V}$ is the set of *tangents* to the surface $\bar{V} = \{(\hat{x}_1(w), \ldots, \hat{x}_n(w)) \mid w \in V\}$ at $\bar{v} = (\hat{x}_1(v), \ldots, \hat{x}_n(v))$, that is, the set of $(t_1, \ldots, t_n) \in \bar{k}^n$ such that

$$\sum_i \frac{\partial f}{\partial X_i}\bigg|_{\bar{v}} t_i = 0$$

for all $f \in I$ where I is the ideal of functions vanishing on $\bar{V}$.

(c) For $D \in \operatorname{Der}_k k[x_1, \ldots, x_n]$ and $v \in V$, the mapping

$$D_v: k[x_1, \ldots, x_n] \to \bar{k}$$

defined by $D_v(f) = D(f)(v)$ is a tangent to V at v.

E.4.6. Let $D \in \operatorname{Der}_{\bar{k}} K$ and let $L_x: K \to K$, $R_x: K \to K$ be defined by $L_x(y) = xy$, $R_x(y) = yx$ for $y \in K$. Show that $[D, L_x] = L_{D(x)}$ and $[D, R_x] =$

$R_{D(x)}$ for $x \in K$. Show that D commutes with L_a and/or R_a if and only if $D(a) = 0$, for $D \in \operatorname{Der}_k K$, $a \in K$. Show that $D(k) = \{0\}$.

E.4.7 (p-Dependence). Let L/k be a field extension. For $x \in K$ and $S \subset K$, write $x \prec_p S$ for $x \in k(K^p)(S)$. Show that $\prec$ is a dependence relation in the sense of E.1.20. A subset S of K is *p-independent* over k if S is independent in the sense of E.1.20. Show that a subset S of K is a basis for K in the sense of E.1.20 if and only if S is a p-basis for K/k. Conclude that every extension K/k has a p-basis and the cardinalities of any two p-bases of K/k are the same. (See 4.3.5.)

E.4.8 (Leibniz's Rule). For $D \in \operatorname{Der}_k K$ and $x, y \in K$, prove that

(a) $D^n(xy) = \sum \binom{n}{m} D^m(x)D^{n-m}(y)$;

(b) $(D - (a + b)I)^n(xy) = \sum \binom{n}{m}(D - aI)^m(x)(D - bI)^{n-m}(y)$

for $a, b \in k$;

(c) $D^p \in \operatorname{Der}_k K$ for $D \in \operatorname{Der}_k K$ (use part (a));

(d) letting $K_a(D) = \{x \in K \mid (D - aI)^m(x) = 0 \text{ for some } m\}$ for $D \in \operatorname{Der}_k K$ and $a \in k$, show that $K_a(D)K_b(D) \subset K_{a+b}(D)$.

E.4.9. Let A be an algebra over a field k of exponent characteristic p. For $x, y \in A$, let $[x, y] = xy - yx$. Define $ad\, x\colon A \to A$ by $ad\, x(y) = [x, y]$ for $x, y \in A$. Show that

(a) $[x, y] = -[y, x]$ for $x, y \in A$;

(b) $[[x, y], z] + [[y, z], x] + [[z, x], y] = 0$ for $x, y, z \in A$;

(c) $[x, \ldots [x, [x, y]] \ldots]$ (x occurring p-times) is the same as $[x^p, y]$ for $x, y \in A$;

(d) $ad\, [x, y] = [ad\, x, ad\, y]$ for $x, y \in A$, where $[ad\, x, ad\, y] = ad\, x\; ad\, y - ad\, y\; ad\, x$;

(e) $ad\, x^p = (ad\, x)^p$ for $x \in A$.

An additive subgroup B of A such that $[x, y] \in B$ for $x, y \in B$ is called a *Lie ring* in A. Show that $\operatorname{Der}_k K$ is a Lie ring in $\operatorname{End}_k K$.

E.4.10. Show that there are separable extensions K/k which are not separably generated over k.

E.4.11. Let $K = k(X)$ where X is transcendental over k. Let $f = X^3 + X + 1$ and $g = X^{-2} + X^2$, and let $k' = k(f, g)$. Find $y \in k'$ such that $k' = k(y)$.

E.4.12. Let L be a field, G a subgroup of $\operatorname{Aut} L$, K a subfield of L stable under G. Show that K and L^G are linearly disjoint over $K \cap L^G$. (Hint: Review the proof of Dedekind's Lemma.)

E.4.13. Let K be a field, G a subgroup of $\operatorname{Aut} K$. Show that K/K^G is separable. (Hint: Use E.4.12.)

5 Modern Galois Theory

This chapter is devoted to material which, like the Galois Correspondence Theorem, relates the structure of a field extension K/k to that of some associated algebraic structure. In 5.1, we relate the field extension K/k to the ring $\text{End}_k K$ of endomorphisms of K/k. The main theorems here are the Jacobson-Bourbaki Correspondence Theorem (see 5.1.7) and a theorem on descent by rings of endomorphisms (see 5.1.10). In 5.2, we relate K/k to the Lie ring $\text{Der}_k K$ of derivations of K/k. We prove here the Jacobson Differential Correspondence Theorem (see 5.2.6) and a theorem on descent by Lie rings of derivations of K/k (see 5.2.9). We also describe in detail the structure of finite dimensional radical extensions of exponent 1 (see 5.2.12). In 5.3, we relate K/k to the biring $H(K/k)$ of endomorphisms of K/k (see 5.3.2). We prove a Biring Correspondence Theorem (see 5.3.12) and describe the structure of $H(K/k)$ for K/k Galois/radical/normal (see 5.3.20).

5.1 Rings of endomorphisms of K

We begin by considering the ring $\text{End}\, K = \{T: K \to K \mid T(x + y) = T(x) + T(y) \text{ for } x, y \in K\}$ of endomorphisms of a field K. The identity element of End K is denoted by I. We regard End K as a vector space over K with respect to the scalar multiplication xT ($x \in K, T \in \text{End}\, K$) defined by $(xT)(y) = x(T(y))$ for $y \in K$.

5.1.1 Definition. A *K-subring* of End K is a subring $\mathscr{A}$ of End K which is a K-subspace of End K.

Note that the one-dimensional subspace KI of End K containing I is a K-subring of End K contained in every K-subring of End K. For any subfield k of K, the ring $\text{End}_k K$ of k-linear endomorphisms of K is a K-subring of K and $(\text{End}_k K){:}K = K{:}k$ if $K{:}k < \infty$ (see E.5.1).

5.1.2 Definition. The *groundfield* of a K-subring $\mathscr{A}$ of End K is the subfield $K^{\mathscr{A}} = \{y \in K \mid T(xy) = T(x)y \text{ for } T \in \mathscr{A}, x \in K\}$ of K.

Note that if $\mathscr{A}$ is a K-subring of End K, then $\mathscr{A} \subset \text{Hom}_{K^{\mathscr{A}}} K$.

An important example of K-subring of End K is the K-span $K[G]$ of a group G of automorphisms of K. The set G is a basis for $K[G]$ over K (see 3.1). If $\mathscr{A} = K[G]$, then $K^{\mathscr{A}} = K^G$.

Let $\mathscr{A}$ be a K-subring of End K. Let $\hat{x}: \mathscr{A} \mapsto K$ be defined for $x \in K$ by $\hat{x}(T) = T(x)$ for $T \in \mathscr{A}$, and let $\hat{K}_0 = \{\hat{x} \mid x \in K_0\}$ for $K_0 \subset K$. Note that $\hat{K}$ is contained in the dual space $\text{Hom}_K(\mathscr{A}, K)$ of $\mathscr{A}$ and that $\hat{K}$ separates $\mathscr{A}$, that is, $\hat{x}(S) = \hat{x}(T)$ for all $x \in \hat{K}$ only if $S = T$ $(S, T \in \mathscr{A})$.

5.1.3 *Definition.* $\mathscr{A}^{\mathscr{A}} = \{R \in \mathscr{A} \mid T(xR) = T(x)R \text{ for } T \in \mathscr{A}, x \in K\}$.

Note that $\mathscr{A}^{\mathscr{A}}$ is a $K^{\mathscr{A}}$ subspace of $\mathscr{A}$.

5.1.4 *Proposition.* Let K_0 be a subset of K such that $\hat{K}_0$ separates $\mathscr{A}$. Then $\mathscr{A}^{\mathscr{A}} = \{R \in \mathscr{A} \mid R(K_0) \subset K^{\mathscr{A}}\}$.

Proof. Let $R \in \mathscr{A}$. We claim that $T(xR) = T(x)R$ for all $T \in \mathscr{A}$, $x \in K$ if and only if $R(K_0) \subset K^{\mathscr{A}}$, thereby establishing the assertion. From the equations

$$\hat{y}(T(xR)) = T(xR)(y) = T(xR(y)),$$

$$\hat{y}(T(x)R) = (T(x)R)(y) = T(x)R(y)$$

we see that $T(xR) = T(x)R$ for all $T \in \mathscr{A}$, $x \in K$ if and only if $T(xR(y)) = T(x)R(y)$ for all $T \in \mathscr{A}$, $x \in K$ and for all $y \in K_0$. The latter condition is that $R(K_0) \subset K^{\mathscr{A}}$. □

5.1.5 *Corollary.* $\mathscr{A}^{\mathscr{A}} = \{R \in \mathscr{A} \mid R(K) \subset K^{\mathscr{A}}\}$. □

5.1.6 *Theorem.* Let $\mathscr{A}:K < \infty$. Then $\mathscr{A}^{\mathscr{A}}$ is a $K^{\mathscr{A}}$-form of $\mathscr{A}$ and $\mathscr{A}:K = K:K^{\mathscr{A}}$.

Proof. Since $\hat{K}$ is a subset of the dual space $\text{Hom}_K(\mathscr{A}, K)$ and $\hat{K}$ separates $\mathscr{A}$, $\hat{K}$ contains a basis $\hat{K}_0 = \{\hat{x}_1, \ldots, \hat{x}_n\}$ for $\text{Hom}_K(\mathscr{A}, K)$ and $\mathscr{A}$ has a dual basis $R_1, \ldots, R_n$ ($n = \mathscr{A}:K$). Now $R_j(x_i) = \hat{x}_i(R_j) = \delta_{ij}$ for all i, so that $R_j(K_0) \subset K^{\mathscr{A}}$ and $R_j \in \mathscr{A}^{\mathscr{A}}$ ($1 \leq j \leq n$). We claim that $R_1, \ldots, R_n$ is a $K^{\mathscr{A}}$-basis for $\mathscr{A}^{\mathscr{A}}$. Thus, let $R \in \mathscr{A}^{\mathscr{A}}$ and take the unique $y_1, \ldots, y_n \in K$ such that $R = \sum_1^n y_j R_j$. Then $y_i = \sum_1^n y_j R_j(x_i) = R(x_i) \in K^{\mathscr{A}}$. Thus, $R_1, \ldots, R_n$ is a basis for $\mathscr{A}^{\mathscr{A}}$ over $K^{\mathscr{A}}$ and $\mathscr{A}^{\mathscr{A}}$ is a $K^{\mathscr{A}}$-form of $\mathscr{A}$. To show that $\mathscr{A}:K = K:K^{\mathscr{A}}$, it suffices to show that $x_1, \ldots, x_n$ is a basis for K over $K^{\mathscr{A}}$. If $x = \sum_1^n x_i y_i$ where $y_1 \cdots y_n \in K^{\mathscr{A}}$, then $R_j(x) = \sum_1^n R_j(x_i) y_i = y_j$. Since $R_1(x), \ldots, R_n(x) \in K^{\mathscr{A}}$ (see 5.1.5), what we must therefore show is that $x = \sum_1^n x_i R_i(x)$. Since I is a linear combination of $R_1, \ldots, R_n$ over K, it suffices to show that $R_j(x)$ and $R_j(\sum_1^n x_i R_i(x)) = \sum_1^n R_j(x_i) R_i(x)$ are equal for $1 \leq j \leq n$. But this is clear from the orthogonality $R_j(x_i) = \delta_{ij}$ ($1 \leq i, j \leq n$). □

5.1.7 *Theorem* (Jacobson-Bourbaki). Let $\mathscr{A}:K < \infty$. Then $\mathscr{A} = \text{End}_{K^{\mathscr{A}}} K$.

Proof. Clearly $\mathscr{A} \subset \text{End}_{K^{\mathscr{A}}} K$. But $(\text{End}_{K^{\mathscr{A}}} K):K = K:K^{\mathscr{A}} = \mathscr{A}:K$ by the preceding theorem. Thus, $\mathscr{A} = \text{End}_{K^{\mathscr{A}}} K$. □

The above theorem shows that $k \mapsto \text{End}_k K$ defines a bijective inclusion reversing correspondence between the set of subfields k of K of finite codimension $K:k$ and the set of K-subrings $\mathscr{A}$ of End K of finite dimension $\mathscr{A}:K$. This correspondence is called the *Jacobson-Bourbaki Correspondence*.

We now briefly describe a process of descent by K-subrings of End K which is analogous to the Galois descent of 3.2. For this, let $\mathscr{A}$ be a K-subring of End K, V a vector space over K.

5.1.8 Definition. An *$\mathscr{A}$-product* on V is a mapping $\mathscr{A} \times V \to V$, denoted $(T, v) \mapsto T(v)$, such that

1. $I(v) = v$ for $v \in V$;
2. $S(T(v)) = (ST)(v)$ for $S, T \in \mathscr{A}, v \in V$;
3. $(S + T)(v) = S(v) + T(v)$ for $S, T \in \mathscr{A}, v \in V$;
4. $T(v + v') = T(v) + T(v')$ for $T \in \mathscr{A}, v, v' \in V$;
5. $(xT)(v) = x(T(v))$ for $x \in K, T \in \mathscr{A}, v \in V$.

5.1.9 Definition. The *ground subspace* of a vector space V over K with $\mathscr{A}$-product is the $K^{\mathscr{A}}$-subspace $V^{\mathscr{A}} = \{v \in V \mid T(xv) = T(x)v \text{ for } T \in \mathscr{A}, x \in K\}$ of V.

Note that $V^{\mathscr{A}}$ *is* a $K^{\mathscr{A}}$-subspace, being closed under addition and satisfying $T(x(yv)) = T((xy)v) = T(xy)v = (T(x)y)v = T(x)(yv)$ for $T \in \mathscr{A}, x \in K$, $y \in K^{\mathscr{A}}, v \in V^{\mathscr{A}}$.

The multiplication mapping $\mathscr{A} \times \mathscr{A} \to \mathscr{A}$ of the ring $\mathscr{A}$ is an $\mathscr{A}$-product on $\mathscr{A}$, and the corresponding ground subspace of $\mathscr{A}$ is the already familiar $\mathscr{A}^{\mathscr{A}} = \{R \in \mathscr{A} \mid T(xR) = T(x)R \text{ for } T \in \mathscr{A}, x \in K\}$.

For $V = K^n$, the mapping $\mathscr{A} \times V \to V$ defined by $T(x_1, \ldots, x_n) = (T(x_1), \ldots, T(x_n))$ for $T \in \mathscr{A}, (x_1, \ldots, x_n) \in V$ is an $\mathscr{A}$-product on V, and the corresponding ground subspace is $\mathscr{A}^{\mathscr{A}} = (K^{\mathscr{A}})^n$.

More generally, if $V = K \otimes_k W$ where k is a subfield of $K^{\mathscr{A}}$ and W is a vector space over k, the product $\mathscr{A} \times V \to V$ such that

$$T\left(\sum_1^n x_i \otimes w_i\right) = \sum_1^n T(x_i) \otimes w_i \quad \text{for } T \in \mathscr{A},$$

$\sum_1^n x_i \otimes w_i \in V$ is an $\mathscr{A}$-product on V. If $k = K^{\mathscr{A}}$, the corresponding ground subspace is $V^{\mathscr{A}} = 1 \otimes_k W$.

Finally, if G is a group of automorphisms of K and $\mathscr{A} = K[G]$, then a G-product $G \times V \to V$, denoted $(\sigma, v) \mapsto \sigma(v)$, determines an $\mathscr{A}$-product $\mathscr{A} \times V \to V$ on V, defined by $(\sum_\sigma x_\sigma \sigma)(v) = \sum_\sigma x_\sigma \sigma(v)$ for $\sum_\sigma x_\sigma \sigma \in \mathscr{A}, v \in V$. In showing this, only the verification of condition 2 of 5.1.3 presents a problem, and it suffices to consider the case $S = \sigma$ and $T = x\tau$ where $x \in K$, σ, $\tau \in G$. Then $ST = \sigma(x)(\sigma\tau)$ since $(ST)(y) = S(T(y)) = \sigma((x\tau)(y)) = \sigma(x\tau(y)) = (\sigma(x)(\sigma\tau))(y)$ for $y \in K$. But then we have $S(T(v)) = \sigma((x\tau)(v)) = \sigma(x(\tau(v))) = \sigma(x)(\sigma(\tau(v))) = \sigma(x)((\sigma\tau)(v)) = (\sigma(x)(\sigma\tau))(v) = (ST)(v)$ for $v \in V$. Thus, condition 2 is satisfied. The corresponding ground subspace of V is $V^{\mathscr{A}} = V^G$.

It is a consequence of 5.1.10 below that if $\mathscr{A} = K[G]$ where G is a finite group of automorphisms of K and if $\mathscr{A} \times V \to V$ is an $\mathscr{A}$-product on a K-vector space V, then the restriction $G \times V \to V$ is a G-product on V.

We know from the preceding section that $\mathscr{A}^{\mathscr{A}}$ is a $K^{\mathscr{A}}$-form of $\mathscr{A}^{\mathscr{A}}$ if $\mathscr{A} : k < \infty$. This is also true for any vector space V over K with $\mathscr{A}$-product, as we now show.

5.1.10 Theorem (Jacobson). If $\mathscr{A}^{\mathscr{A}}$ is a $K^{\mathscr{A}}$-form of $\mathscr{A}$, then $V^{\mathscr{A}}$ is a $K^{\mathscr{A}}$-form of V and $V^{\mathscr{A}} = \mathscr{A}^{\mathscr{A}}(V)$ where $\mathscr{A}^{\mathscr{A}}(V)$ is the additive subgroup of

V generated by $\{R(v) \mid R \in \mathscr{A}^{\mathscr{A}}, v \in V\}$. In particular $V^{\mathscr{A}} = \mathscr{A}^{\mathscr{A}}(V)$ and $V^{\mathscr{A}}$ is a $K^{\mathscr{A}}$-form of V if $\mathscr{A}:K < \infty$.

Proof. Let $R \in \mathscr{A}^{\mathscr{A}}$ and $v \in V$. Then $T(xR) = T(x)R$ and consequently $T(xR(v)) = (T(xR))(v) = (T(x)R)(v) = T(x)R(v)$ for $T \in \mathscr{A}, x \in K$. Thus, $R(v) \in V^{\mathscr{A}}$ for $R \in \mathscr{A}^{\mathscr{A}}, v \in V$ and $\mathscr{A}^{\mathscr{A}}(V) \subset V^{\mathscr{A}}$. Assume now that $\mathscr{A}^{\mathscr{A}}$ is a $K^{\mathscr{A}}$-form of $\mathscr{A}$. Then $I = \sum_1^n x_i R_i$ for suitable $x_i \in K$, $R_i \in \mathscr{A}^{\mathscr{A}}$, and we have $v = I(v) = \sum_1^n x_i R_i(v)$ for $v \in V$. Thus, $\mathscr{A}^{\mathscr{A}}(V)$ spans V over K. Since $\mathscr{A}^{\mathscr{A}}(V) \subset V^{\mathscr{A}}$, $V^{\mathscr{A}}$ spans V over K. Now let $\{v_\alpha \mid \alpha \in J\}$ be a $K^{\mathscr{A}}$-basis for $V^{\mathscr{A}}$. Since $V^{\mathscr{A}}$ spans V over K, $\{v_\alpha \mid \alpha \in J\}$ spans V over K. Suppose that $\sum_\alpha y_\alpha v_\alpha = 0$, the x_α being elements of K. Then

$$0 = R_i\left(\sum_\alpha y_\alpha v_\alpha\right) = \sum_\alpha R_i(y_\alpha)v_\alpha$$

and $R_i(y_\alpha) \in \mathscr{A}^{\mathscr{A}}(K) \subset K^{\mathscr{A}}$ $(1 \leq i \leq n)$. By the $K^{\mathscr{A}}$-independence of the v_α, $R_i(y_\alpha) = 0$ $(1 \leq i \leq n)$. But then $y_\alpha = I(y_\alpha) = \sum_1^n x_i R_i(y_\alpha) = 0$ for all α. Thus, the v_α are K-independent, so that $\{v_\alpha \mid \alpha \in J\}$ is a basis for V over K and $V^{\mathscr{A}}$ is a $K^{\mathscr{A}}$-form of V. Since $\mathscr{A}^{\mathscr{A}}(V) \subset V^{\mathscr{A}}$ and $\mathscr{A}^{\mathscr{A}}(V)$ is a $K^{\mathscr{A}}$-space which spans V over K, it follows that $\mathscr{A}^{\mathscr{A}}(V) = V^{\mathscr{A}}$. □

5.2 Lie rings of derivations of K

We now consider the Lie ring $\operatorname{Der} K = \{D \in \operatorname{End} K \mid D(xy) = D(x)y + xD(y)\}$ of derivations of a field K (see E.4.9). Note that $\operatorname{Der} K$ is a K-subspace of $\operatorname{End} K$. We let $\operatorname{Der}_k K = \operatorname{Der} K \cap \operatorname{End}_k K$. Since $\operatorname{Der}_k K = \{0\}$ for finite dimensional separable extensions (see 4.3.3) and since finite dimensional field extensions of characteristic 0 are separable, we assume in this chapter that K has nonzero characteristic p.

Letting $[D, E] = DE - ED$ for $D, E \in \operatorname{End} K$, we have $D, E \in \operatorname{Der} K \Rightarrow [D, E] \in \operatorname{Der} K$. Furthermore, $D \in \operatorname{Der} K \Rightarrow D^p \in \operatorname{Der} K$ (see E.4.6, E.4.8, E.4.9).

5.2.1 **Definition.** A *K-sub Lie ring* of $\operatorname{Der} K$ is a K-subspace $\mathscr{D}$ of $\operatorname{Der} K$ such that $[D, E] \in \mathscr{D}$ and $D^p \in \mathscr{D}$ for $D, E \in \mathscr{D}$.

Note that for any subfield k of K, $\operatorname{Der}_k K$ is a K-sub Lie ring of $\operatorname{Der} K$.

5.2.2 **Definition.** The *groundfield* of a K-sub Lie ring $\mathscr{D}$ of $\operatorname{Der} K$ is the subfield $K^{\mathscr{D}} = \{y \in K \mid D(xy) = D(x)y \text{ for all } D \in \mathscr{D}, x \in K\}$.

For any K-sub Lie ring $\mathscr{D}$ of $\operatorname{Der} K$, $K^{\mathscr{D}}$ is the subfield $\{y \in K \mid D(y) = 0$ for $D \in \mathscr{D}\}$.

Let $K^{p^e} = \{x^{p^e} \mid x \in K_p\}$ for any positive integer e.

5.2.3 **Proposition.** $K^p \subset K^{\mathscr{D}}$ for any K-sub Lie ring $\mathscr{D}$ of $\operatorname{Der} K$.

Proof. Let $x \in K$ and $D \in \mathscr{D}$. Then $D(x^p) = px^{p-1}D(x) = 0$. □

5.2.4 **Proposition.** Let k be a subfield of K such that $K^p \subset K$. Then $K^{\mathscr{D}} = k$ for $\mathscr{D} = \operatorname{Der}_k K$.

Proof. Let $x \in K - k$ and let k' be a maximal subfield of K not containing x. (Such a k' exists, by Zorn's Lemma.) Suppose that $K \supsetneqq k'(x)$ and take $y \in K - k'(x)$. Then $x \in k'(y)$, by the maximality of k', so that $k' \subsetneqq k'(x) \subset k'(y)$. Now $x^p \in k'$ and $y^p \in k'$, so that $k'(x):k = p$ and $k'(y):k = p$. Thus, $k'(x) = k'(y)$ and $y \in k'(x)$, a contradiction. It follows that $K = k'(x)$. Let T be the k'-linear transformation on K such that $T(x^i) = ix^i$ for $0 \leq i \leq p - 1$. Since $x^p \in k'$, one shows easily that $T(x^i) = ix^i$ for all i, so that $T \in \operatorname{Der}_k K$. Consequently, $x \notin K^{\operatorname{Der}_k K}$ for $x \in K - k$ and $K^{\operatorname{Der}_k K} = k$. □

Let $\mathscr{D}$ be a K-sub Lie ring of Der K. Let $\hat{x}: \mathscr{D} \to K$ be defined for $x \in K$ by $\hat{x}(D) = D(x)$ for $D \in \mathscr{D}$, and let $\hat{K}_0 = \{x \mid x \in K_0\}$ for $K_0 \subset K$. Then $\hat{K}$ is a subset of the dual space Hom $(\mathscr{D}, K)$ of $\mathscr{D}$ and $\hat{K}$ separates $\mathscr{D}$. Let π be the prime field of K.

5.2.5 Theorem. Let $\mathscr{D}:K < \infty$. Then $\mathscr{D}$ has a π-form $\mathscr{T}_\pi$ consisting of pairwise commuting π-diagonalizable derivations. The dimensions $\mathscr{D}:K$ and $K:K^{\mathscr{D}}$ are related by $K:K^{\mathscr{D}} = p^{\mathscr{D}:K}$.

Proof. Since $\hat{K}$ separates $\mathscr{D}$, $\hat{K}$ contains a basis $\hat{x}_1, \ldots, \hat{x}^n$ for $\operatorname{Hom}_K(\mathscr{D}, K)$. Let $T_1, \ldots, T_n$ be the basis for $\mathscr{D}$ defined by the equations $\hat{x}_j(T_i) = \delta_{ij}x_j$ $(1 \leq i, j \leq n)$. We then have $T_i(x_j) = \delta_{ij}x_j$ for $1 \leq i, j \leq n$ and therefore

$$[T_i, T_j](x_r) = T_i(T_j(x_r)) - T_j(T_i(x_r)) = 0$$
$$T_i^p(x_r) = \delta_{ir}x_r = T_i(x_r)$$

for $1 \leq i, j, r \leq n$. Since the $\hat{x}_r$ $(1 \leq r \leq n)$ separate $\mathscr{D}$ and since $[T_i, T_j]$, 0, T_i^p, T_i are elements of $\mathscr{D}$, we have $[T_i, T_j] = 0$ and $T_i^p = T_i$ for $1 \leq i, j \leq n$. Since T_i satisfies the separable polynomial $X^p - X$ and since the roots of $X^p - X = \prod_{a \in \pi} (X - a)$ lie in π, T_i is π-diagonalizable for $1 \leq i \leq n$ (see E.5.6). Let $\mathscr{T}_\pi$ be the π-span of $T_1, \ldots, T_n$. Clearly, $\mathscr{T}_\pi$ is a π-form of $\mathscr{D}$ consisting of pairwise commuting π-diagonalizable derivations. Letting $K_\alpha = \{x \in K \mid T(x) = \alpha(T)x \text{ for } T \in \mathscr{T}_\pi\}$ for $\alpha \in \mathbf{R} = \operatorname{Hom}_\pi(\mathscr{T}_\pi, \pi)$, we have $K_0 = K^{\mathscr{D}}$, $K = \sum_{\alpha \in \mathbf{R}} \oplus K_\alpha$ and $K_\alpha K_\beta \subset K_{\alpha+\beta}$ for $\alpha, \beta \in \mathbf{R}$ (see E.5.6). Taking $x_\beta \in K_\beta - \{0\}$, we have $\sum_{\alpha \in \mathbf{R}} \oplus K_\alpha = Kx_\beta = \sum_{\alpha \in \mathbf{R}} K_\alpha x_\beta$. Since $K_\alpha x_\beta \subset K_{\alpha+\beta}$ for all $\alpha \in \mathbf{R}$, it follows that $K_\alpha x_\beta = K_{\alpha+\beta}$ for all α. In particular, $K_\alpha = K^{\mathscr{D}}x_\alpha$ for $\alpha \in \mathbf{R}_0$ where $\mathbf{R}_0 = \{\beta \in \mathbf{R} \mid K_\beta \neq \{0\}\}$, so that $K = \sum_{\alpha \in \mathbf{R}_0} \oplus K^{\mathscr{D}}x_\alpha$. We have seen that the finite set $\mathbf{R}_0$ is closed under addition. It follows that $\mathbf{R}_0$ is a π-subspace of $\mathbf{R} = \operatorname{Hom}_\pi(\mathscr{T}_\pi, \pi)$. Since $\mathbf{R}_0$ separates $\mathscr{T}_\pi$, it follows that $\mathbf{R}_0 = \mathbf{R}$, so that $K = \sum_{\alpha \in \mathbf{R}} \oplus K^{\mathscr{D}}x_\alpha$ and $K:K^{\mathscr{D}} = |\mathbf{R}|$ (number of elements of $\mathbf{R}$). But $|\mathbf{R}| = p^{\mathscr{T}_\pi:\pi}$ and $\mathscr{T}_\pi:\pi = \mathscr{D}:K$. Thus $K:K^{\mathscr{D}} = p^{\mathscr{D}:K}$. □

5.2.6 Theorem (Jacobson). Let $\mathscr{D}:K < \infty$. Then $\mathscr{D} = \operatorname{Der}_{K^{\mathscr{D}}} K$.

Proof. We have $\mathscr{D} \subset \operatorname{Der}_{K^{\mathscr{D}}} K$. But $p^{\mathscr{D}:K} = K:K^{\mathscr{D}} = K^{(\operatorname{Der}_{K^{\mathscr{D}}} K)} = p^{(\operatorname{Der}_{K^{\mathscr{D}}} K)}$ and $\mathscr{D} = \operatorname{Der}_{K^{\mathscr{D}}} K$. □

The above theorem shows that $k \mapsto \operatorname{Der}_k K$ defines a bijective inclusion reversing correspondence between the set of subfields of K such that $K:k < \infty$ and $K^p \subset k$ and the set of K-sub Lie rings $\mathscr{D}$ of Der K of finite dimension $\mathscr{D}:K$ (see 5.2.3 and 5.2.4). This correspondence is called the *Jacobson Differential Correspondence*.

We next discuss a process of descent by K-sub Lie rings of Der K. For this, let $\mathscr{D}$ be a K-sub Lie ring of Der K, V a nonzero vector space over K.

5.2.7 *Definition.* A *$\mathscr{D}$-product* on V is a mapping $\mathscr{D} \times V \to V$, denoted $(D, v) \mapsto D(v)$, such that

1. $D(E(v)) = [D, E](v) + E(D(v))$ for $D, E \in \mathscr{D}, v \in V$;
2. $(D + E)(v) = D(v) + E(v)$ for $D, E \in \mathscr{D}, v \in V$;
3. $D(v + w) = D(v) + D(w)$ for $D \in \mathscr{D}, v, w \in V$;
4. $D(xv) = D(x)v + xD(v)$ for $D \in \mathscr{D}, x \in K, v \in V$;
5. $(xD)(v) = x(D(v))$ for $x \in K, D \in \mathscr{D}, v \in V$;
6. $(D_L)^p = (D^p)_L$ for $D \in \mathscr{D}$, where $D_L \colon V \to V$ is defined by $D_L(v) = D(v)$ for $D \in \mathscr{D}, v \in V$.

5.2.8 *Definition.* The *groundsubspace* of a vector space V with $\mathscr{D}$-product is the $K^{\mathscr{D}}$-subspace $V^{\mathscr{D}} = \{v \in V \mid D(xv) = D(x)v$ for all $x \in K, D \in \mathscr{D}\}$ of V.

The Lie ring product $\mathscr{D} \times \mathscr{D} \to \mathscr{D}$ fails to be a $\mathscr{D}$-product on $\mathscr{D}$ because condition 5 is not satisfied. All of the other conditions, however, are satisfied.

For $V = K^n$, the mapping $\mathscr{D} \times V \to V$ defined by $D(x_1, \ldots, x_n) = (D(x_1), \ldots, D(x_n))$ for $D \in \mathscr{D}$, $(x_1, \ldots, x_n) \in V$ is a $\mathscr{D}$-product on V with ground subspace $V^{\mathscr{D}} = (K^{\mathscr{D}})^n$.

If $V = K \otimes_k W$ where k is a subfield of $K^{\mathscr{D}}$ and W is a vector space over k, then the product $\mathscr{D} \times V \to V$ such that $D(\sum_1^n x_i \otimes w_i) = \sum_1^n D(x_i) \otimes w_i$ for $D \in \mathscr{D}$ and $\sum_1^n x_i \otimes w_i \in K \otimes_k W$ is a $\mathscr{D}$-product on V. If $k = K^{\mathscr{D}}$, the corresponding groundsubspace is $V^{\mathscr{D}} = 1 \otimes W$.

5.2.9 *Theorem* (Jacobson). Let $\mathscr{D}\colon K < \infty$ and let $\mathscr{D} \times V \to V$ be a $\mathscr{D}$-product on the nonzero vector space V over K. Then $V^{\mathscr{D}}$ is a $K^{\mathscr{D}}$-form of V. Letting $\mathscr{T}_\pi$ be any π-form of $\mathscr{D}$ consisting of pairwise commuting π-diagonalizable derivations, we have $K = \sum_{\alpha \in \mathbf{R}} \oplus K_\alpha$ and $V = \sum_{\beta \in \mathbf{R}} \oplus V_\beta$ where $\mathbf{R} = \mathrm{Hom}_\pi(\mathscr{T}_\pi, \pi)$, $K_\alpha = \{x \in K \mid T(x) = \alpha(T)x$ for $T \in \mathscr{T}_\pi\}$ and $V_\beta = \{v \in V \mid T(v) = \beta(T)v$ for $T \in \mathscr{T}_\pi\}$ $(\alpha, \beta \in \mathbf{R})$. For all $\alpha, \beta \in \mathbf{R}$, K_α and V_β are nonzero, $K_\alpha K_\beta = K_{\alpha+\beta}$, $K_\alpha V_\beta = V_{\alpha+\beta}$, $K_\alpha = (K^{\mathscr{D}})x_\alpha$ for $\alpha \in K_\alpha - \{0\}$ and $V_\beta = (K^{\mathscr{D}})v_\beta$ for $v_\beta \in V_\beta - \{0\}$.

Proof. Since the elements of $\mathscr{T}_\pi$ are π-diagonalizable and commute pairwise, $K = \sum \oplus K_\alpha$. Furthermore, the π-diagonalizability of an element T of $\mathscr{T}_\pi$ implies that $T^p = T$, hence that $(T_L)^p = T_L$ (see condition 6 of 5.2.7). Thus, T_L is a π-linear transformation of V at which the polynomial $X^p - X = \prod_{a \in \pi} (X - a)$ vanishes. Since $X^p - X$ is separable and its roots lie in π, T_L is π-diagonalizable on V (see E.5.6). Since the elements T of $\mathscr{T}_\pi$ commute pairwise, the elements T_L of $\mathscr{T}_{\pi L} = \{T_L \mid T \in \mathscr{T}_\pi\}$ commute pairwise. It follows that $V = \sum_{\beta \in \mathbf{R}} \oplus V_\beta$.

Letting $T \in \mathscr{T}_\pi$, $x \in K_\alpha$ and $y \in K_\beta$ $(\alpha, \beta \in \mathbf{R})$, we have $T(xy) = T(x)y + xT(y) = \alpha(T)xy + \beta(T)xy = (\alpha + \beta)(T)(xy)$. Similarly, $T(xv) = T(x)v +$

$x(T(v)) = (\alpha + \beta)(T)(xv)$ for $v \in V_\beta$. Thus, $K_\alpha K_\beta \subset K_{\alpha+\beta}$ and $K_\alpha V_\beta \subset V_{\alpha+\beta}$. Take $x_\alpha \in K_\alpha - \{0\}$ and note that

$$\sum_{\beta \in \mathbf{R}} \oplus K_\beta = K = x_\alpha K = \sum_{\beta \in \mathbf{R}} \oplus x_\alpha K_\beta \subset \sum_{\beta \in \mathbf{R}} \oplus K_{\alpha+\beta} = K.$$

It follows that $x_\alpha K_\beta = K_{\alpha+\beta}$ for all $\alpha, \beta \in R_0 = \{\alpha \in \mathbf{R} \mid K_\alpha \neq \{0\}\}$. Consequently, $K_\alpha = K_0 x_\alpha$ and $K_\alpha K_\beta = K_{\alpha+\beta}$ for $\alpha, \beta \in \mathbf{R}_0$. Since $\mathbf{R}_0$ is a nonempty subset of the finite additive group $\mathbf{R}$ and $\mathbf{R}_0$ is closed under the operation $+$, $\mathbf{R}_0$ is an additive subgroup of $\mathbf{R}$, hence a π-subspace of $\mathbf{R}$. Since $\mathbf{R}_0$ separates points, it follows that $\mathbf{R}_0 = \operatorname{Hom}_\pi(\mathscr{T}_\pi, \pi) = \mathbf{R}$. Thus, $K_\alpha \neq \{0\}$ for $\alpha \in \mathbf{R}$.

Next, take $x_\alpha \in K_\alpha - \{0\}$ and note that

$$\sum_{\beta \in \mathbf{R}} \oplus V_\beta = V = x_\alpha V = \sum_{\beta \in \mathbf{R}} \oplus x_\alpha V_\beta \subset \sum_{\beta \in \mathbf{R}} \oplus V_{\alpha+\beta}.$$

It follows that $x_\alpha V_\beta = V_{\alpha+\beta}$ and $K_\alpha V_\beta = V_{\alpha+\beta}$ for all $\alpha, \beta \in \mathbf{R}$. Since V is nonzero, V_β is nonzero for some $\beta \in \mathbf{R}$. Thus, $V_{\alpha+\beta} = x_\alpha V_\beta$ is nonzero for all $\alpha \in \mathbf{R}$. Since $\alpha + \mathbf{R} = \mathbf{R}$, it follows that $V_\gamma \neq \{0\}$ for all $\gamma \in \mathbf{R}$.

We now show that V_0 is a K_0-form of V. We see from the foregoing discussion that $V_0 \neq \{0\}$ and $V = \sum_{\alpha \in \mathbf{R}} \oplus V_\alpha = \sum_{\alpha \in \mathbf{R}} \oplus K_\alpha V_0$, so that V is the K-span KV_0 of V_0. We claim that, moreover, a K_0-independent subset of the K_0-space V_0 is a K-independent subset of V, thereby establishing that V_0 is a K_0-form of V. Suppose not and let $v_1, \ldots, v_m$ be a minimal K_0-independent subset of V_0 which is K-dependent. Choose elements $x_1, \ldots, x_m$ (not all zero) of K such that $\sum_{i=1}^m x_i v_i = 0$. We may take $x_m = 1$ with no loss of generality. Then

$$0 = T\left(\sum_1^m x_i v_i\right) = \sum_1^{m-1} T(x_i) v_i \qquad \text{for } T \in \mathscr{T}_\pi,$$

since $v_i \in V_0$ for all i and $T(x_m) = T(1) = 0$. Since $v_1, \ldots, v_{m-1}$ are K-independent, we have $T(x_i) = 0$ for $T \in \mathscr{T}_\pi$. Thus, $x_i \in K_0$ for all i. But then $v_1, \ldots, v_m$ are K_0-dependent, a contradiction. It follows that V_0 is a K_0-form of V.

Finally, $\mathscr{D} = K\mathscr{T}_\pi$, so that $K_0 = K^{\mathscr{D}}$ and $V_0 = V^{\mathscr{D}}$. Thus, $V^{\mathscr{D}}$ is a $K^{\mathscr{D}}$-form of V. □

Let K/k be a radical extension.

5.2.10 Definition. The *exponent* of K/k is e if $x^{p^e} \in k$ for all $x \in K$ and e is minimal with respect to this property.

5.2.11 Definition. A *p-basis* for K/k is a minimal subset S of K such that $K = k(K^p)$ where $K^p = \{x^p \mid x \in K\}$ (see 4.3.5).

We have seen that $K^{\mathscr{D}} = k$ for $\mathscr{D} = \operatorname{Der}_k K$ if and only if K/k is a radical extension of exponent 1. Throughout the remainder of this section, we assume K/k is a finite dimensional radical extension of exponent 1. We furthermore fix a p-basis $\{x_1, \ldots, x_n\}$ for K/k.

Taking k_i to be a maximal subfield of K containing k and $\{x_1, \ldots, x_n\} - \{x_i\}$ but not containing x_i, we obtain as in the proof of 5.2.4 a derivation T_i of

K/k_i such that $T_i(x_i) = x_i$. Since $T_i(k_i) = \{0\}$, we have $T_i(x_j) = 0$ for $i \neq j$. Since $\{x_1, \ldots, x_n\}$ generates K over k, $\{x_1, \ldots, x_n\}$ separates $\mathrm{Der}_k\, K$. Since a derivation D of $\mathscr{D}$ coincides with $\sum_1^n D(x_j)x_j^{-1}T_j$ at the x_i $(1 \leq i \leq n)$, $D = \sum_1^n D(x_j)x_j^{-1}T_j$. Thus, $T_1, \ldots, T_n$ is a K-basis for $\mathrm{Der}_k\, K$. The field K is the k-span of $\{x_1^{d_1} \cdots x_n^{d_n} \mid 0 \leq d_i \leq p-1 \text{ for all } i\}$, since the latter is a subfield of K containing k and $\{x_1, \ldots, x_n\}$. Since $K{:}k = p^{\mathscr{D}:K} = p^n$, the set $\{x_1^{d_1} \cdots x_n^{d_n} \mid 0 \leq d_i \leq p-1 \text{ for all } i\}$ is a k-basis for K. Since $T_i(x_1^{d_1} \cdots x_n^{d_n}) = d_i x_1^{d_1} \cdots x_n^{d_n}$, the π-span $\mathscr{T}_\pi$ of $T_1, \ldots, T_n$ is a π-form of $\mathrm{Der}_k\, K$ consisting of pairwise commuting π-diagonalizable derivations of K/k. The decomposition $K = \sum_{\alpha \in \mathbf{R}} \oplus K_\alpha$ of K with respect to $\mathscr{T}_\pi$ described in the proof of 5.2.9 is given by $K_\alpha = kx_1^{d_1} \cdots x_n^{d_n}$ where α is the π-linear function on $\mathscr{T}_\pi$ such that $\alpha(T_i) = d_i$ $(1 \leq i \leq n)$. The K-span $\mathscr{A}$ of $\{T_1^{d_1} \cdots T_n^{d_n} \mid 0 \leq d_i \leq p-1$ for all $i\}$ is a K-subring of End K, as one sees by applying the easily verified equations

$$(xS)(yT) = xS(y)T + (xy)(ST) \qquad (x \in K,\ S, T \in \mathscr{T}_\pi)$$

$$T_i^p = T_i \qquad (1 \leq i \leq n).$$

Since $K^{\mathscr{A}} = K^{\mathscr{D}} = k$, we have $\mathscr{A} = \mathrm{End}_k\, K$ under the Jacobson-Bourbaki Correspondence. We have now proved the following theorem.

5.2.12 Theorem (Jacobson). Let K/k be a finite dimensional radical extension of exponent 1. Let $x_1, \ldots, x_n$ be a p-basis for K/k. Then K has k-basis $\{x_1^{d_1} \cdots x_n^{d_n} \mid 0 \leq d_i \leq p-1 \text{ for all } i\}$ and $\mathrm{End}_k\, K$ has K-basis $\{T_1^{d_1} \cdots T_n^{d_n} \mid 0 \leq d_i \leq p-1 \text{ for all } i\}$ where $T_1, \ldots, T_n$ is a K-basis for $\mathrm{Der}_k\, K$ such that $T_i(x_j) = \delta_{ij}x_j$ for $1 \leq i, j \leq n$. The π-span $\mathscr{T}_\pi$ of $T_1, \ldots, T_n$ is a π-form of $\mathrm{Der}_k\, K$ consisting of pairwise commuting π-diagonalizable derivations and the decomposition $K = \sum_{\alpha \in \mathbf{R}} \oplus K_\alpha$ of K with respect to $\mathscr{T}_\pi$ described in the proof of 5.2.9 is given by $K_\alpha = kx_1^{d_1} \cdots x_n^{d_n}$ where α is the π-linear function on $\mathscr{T}_\pi$ such that $\alpha(T_i) = d_i$ $(1 \leq i \leq n)$. □

5.3 Birings of endomorphisms of K

Finally, we describe for any field K a unique maximal K-subring $H(K)$ of End K having the property that for each $x \in H(K)$, there exist ${}_1x, x_1, {}_2x, x_2, \ldots, {}_nx, x_n$ in $H(K)$ such that $x(ab) = \sum_i {}_ix(a)x_i(b)$ for all $a, b \in K$. The elements x of $H(K)$ can be regarded informally as generalized homomorphisms from K to K. This $H(K)$ is called the *K-biring* of K. The finite dimensional K-subbirings H of $H(K)$ (see 5.3.1) are of the form $H = \mathrm{End}_{K^H}\, K$ and therefore correspond bijectively to the subfields of K of finite codimension. For any finite dimensional subbiring H of $H(K)$, K/K^H is Galois/radical/normal if and only if H is co-Galois/coradical/conormal (see 5.3.20). If K/K^H is normal with decomposition $K = K_{\mathrm{Gal}}K_{\mathrm{rad}}$ (internal tensor product of K^H-algebras) where K_{Gal}/K^H and K_{rad}/K^H are Galois and radical extensions of K^H respectively (see 2.3.16), then H has a decomposition $H = H_{\mathrm{Gal}}H_{\mathrm{rad}}$ (internal tensor product of K^H-algebras) where H_{Gal} and H_{rad} are co-Galois and coradical subbirings of $H = H(K/K^H)$. The component H_{Gal} of H is the

K_{Gal}-span of the Galois group $G(K/K^H)$ and the component K_{Gal} of K is studied using H_{Gal} as described in Chapter 3. The components H_{rad} of H and K_{rad} of K are the objects of study of Chapter 6.

Throughout the section, K is a field, k is a subfield of K, π is the prime field of K and p is the exponent characteristic of K.

5.3.1 Definition. A *coclosed* subset of End K/End$_k$ K is a subset H of End K/End$_k$ K such that for each $x \in H$, there exist ${}_1x, x_1, \ldots, {}_nx, x_n$ in H such that $x(ab) = \sum_i {}_ix(a)x_i(b)$ for all $a, b \in K$. A *K-subbiring* of End K/End$_k$ K is a coclosed K-subring of End K/End$_k$ K.

5.3.2 Definition. $H(K)/H(K/k)$ is the union of all coclosed subsets of End K/End (K/k).

Note that $H(K) = H(K/\pi)$.

5.3.3 Proposition. For any two coclosed subsets H_1, H_2 of End K, the set $H_1H_2 = \{xy \mid x \in H_1, y \in H_2\}$ is a coclosed subset of End K.

Proof. Observe that if $x(ab) = \sum_i {}_ix(a)x_i(b)$ and $y(ab) = \sum_j {}_jy(a)y_j(b)$ for all $a, b \in K$, then $(xy)(ab) = \sum_{i,j} {}_ix_jy(a)x_iy_j(b)$ for all $a, b \in K$. □

Since unions and K-spans of coclosed subsets of End K are coclosed, and since $\{I\}$ (I being the identity mapping from K to K) is coclosed, the following corollary is an immediate consequence of 5.3.3.

5.3.4 Corollary. $H(K)$ and $H(K/k)$ are K-subbirings of End K and End$_k$ K respectively. □

5.3.5 Lemma. Let ${}_ix, x_i, {}_jy, y_j \in H(K/k)$ for $1 \le i \le m$, $1 \le j \le n$. Then

$$\sum_i {}_ix(a)x_i(b) = \sum_j {}_jy(a)y_j(b) \qquad \text{for all } a, b \in K$$

if and only if

$$\sum_i {}_ix \otimes_K x_i = \sum_j {}_jy \otimes_K y_j.$$

Proof. Fix $a, b \in K$. Then $u(a)v(b)$ is K-linear in u and v ($u, v \in H(K/k)$), so that there exists a K-linear mapping $H(K/k) \otimes_K H(K/k) \to H(K/k)$ such that $u \otimes_K v \mapsto u(a)v(b)$ for all $u, v \in H(K/k)$. Thus, $\sum_i {}_ix \otimes_K x_i = \sum_j {}_jy \otimes_K y_j$ implies that $\sum_i {}_ix(a)x_i(b) = \sum_j {}_jy(a)y_j(b)$ for all $a, b \in K$. Conversely, suppose that $\sum_i {}_ix(a)x_i(b) = \sum_j {}_jy(a)y_j(b)$ for all $a, b \in K$. Let e_r be a K-basis for $H(K/k)$ and $x_i = \sum_r x_{ir}e_r$, $y_j = \sum_r y_{jr}e_r$ for all i, j. Then $\sum_r \sum_i {}_ix(a)x_{ir}e_r = \sum_r \sum_j {}_jy(a)y_{jr}e_r$, so that $\sum_i {}_ix(a)x_{ir} = \sum_j {}_jy(a)y_{jr}$ for all r and all $a \in K$. Thus, $\sum_i x_{ir}({}_ix) = \sum_j y_{jr}({}_jy)$ for all r, whence

$$\sum_r \sum_i x_{ir}({}_ix \otimes_K e_r) = \sum_r \sum_j y_{jr}({}_jy \otimes_K e_r)$$

and $\sum_i {}_ix \otimes_K x_i = \sum_j {}_jy \otimes_K y_j$. □

5.3.6 Definition. $\Delta: H(K/k) \to H(K/k) \otimes_K H(K/k)$ is the mapping such that for $x \in H(K/k)$, $\Delta x = \sum_i {}_ix \otimes_K x_i$ if and only if the ${}_ix, x_i$ are elements of $H(K/k)$ such that $x(ab) = \sum_i {}_ix(a)x_i(b)$. And $\varepsilon: H(K/k) \to K$ is the mapping defined by $\varepsilon(x) = x(1)$ for $x \in H(K/k)$.

5.3.7 Proposition.

$$x(abc) = \sum_{i,r} {}_ix(a)_r(x_i)(b)(x_i)_r(c) = \sum_{i,r} {}_r({}_ix)(a)({}_ix)_r(b)x_i(c)$$

and

$$x(a) = \sum_i \varepsilon({}_ix)x_i(a) = \sum_i \varepsilon(x_i){}_ix(a) \qquad \text{for } a, b, c \in K.$$

Proof. $x(abc) = x(a(bc)) = \sum_{i,r} {}_ix(a)_r(x_i)(b)(x_i)_r(c)$ and $x(a) = x(1a) = \sum_i {}_ix(1)x_i(a) = \sum_i \varepsilon({}_ix)x_i(a)$ for $a, b, c \in K$. These observations establish two of the above assertions. The other two are similarly established using $x(abc) = x((ab)c)$ and $x(a) = x(a1)$ for $a, b, c \in K$. □

5.3.8 Proposition. Δ and ε are K-linear mappings such that

1. $\sum_{i,r} {}_ix \otimes_r (x_i) \otimes (x_i)_r = \sum_{i,r} {}_r({}_ix) \otimes ({}_ix)_r \otimes x_i$ for $x \in H(K/k)$ (Δ is *coassociative*);
2. $x = \sum_i \varepsilon({}_ix)x_i = \sum_i \varepsilon(x_i){}_ix$ for $x \in H(K/k)$ (ε is a *coidentity*);
3. $\Delta(I) = I \otimes I$ and $\Delta(xy) = \sum_{i,j} {}_ix_jy \otimes x_iy_j$ for $x, y \in H(K/k)$ (Δ *preserves products*);
4. $\varepsilon(I) = I$ and $\varepsilon(xy) = \varepsilon(x)\varepsilon(y)$ for $x, y \in H(K/k)$ and $\varepsilon(y) \in k$ (ε *preserves products*);
5. $x(ay) = \sum_i {}_ix(a)x_iy$ for $x, y \in H(K/k)$ and $a \in K$.

Proof. It is clear that Δ and ε are K-linear. That Δ is coassociative and ε is a coidentity follows from 5.3.7 (see E.5.3). That Δ preserves products was shown while proving 5.3.3, it being obvious that $\Delta(I) = I \otimes I$. And we have $\varepsilon(I) = I(1) = 1$ and $\varepsilon(xy) = (xy)(1) = x(y(1)) = x(1)y(1) = \varepsilon(x)\varepsilon(y)$ for $x, y \in H(K/k)$ and $\varepsilon(y) = y(1) \in k$. Finally, we observed while proving 5.3.3 that $x(ay) = \sum_i {}_ix(a)x_iy$ for $x, y \in H(K/k)$ and $a \in K$, as one sees from the equations $x(ay(b)) = \sum_i {}_ix(a)x_i(y(b))$ ($b \in K$). □

Properties (1) and (2) of 5.3.8 say that $H(K/k)$ together with Δ and ε is a K-coalgebra (see C.1). Properties (1) through (4) of 5.3.8 say that $H(K/k)$ together with its k-algebra structure and K-coalgebra structure is a K/k-bialgebra (see B.1). Property (5) of 5.3.8 says that left translation in $H(K/k)$ is *semilinear*.

5.3.9 Definition. A *subbiring* of $H(K)$ is a subring H of $H(K)$ such that for each $x \in H$, there exist ${}_ix\, x_i \in H$ such that $\Delta x = \sum_i {}_ix \otimes_K x_i$, and such that $K_H = \varepsilon(H)$ is a subfield of K, $H \supset K_HI$ and a K_H-basis for H is a K-basis for the K-span KH of H. A *subbiring*/*k-subbiring* of $H(K/k)$ is a subbiring H of $H(K)$ contained in $H(K/k)$ such that $K_H \supset k/K_H = k$. For any subfield k' of K, a *k'-subbiring* of $H(K)$ is a subbiring H of $H(K)$ such that $K_H = k'$.

Letting H be a subbiring of $H(K)$, note that there is a unique mapping $\Delta_H: H \to H \otimes_{K_H} H$ such that $\Delta_H(x) = \sum_i {}_ix \otimes_{K_H} x_i$ if and only if the ${}_ix$, x_i are elements of H such that $x(ab) = \sum_i {}_ix(a)x_i(b)$ for all $a, b \in K$ (see E.5.4). Letting $\varepsilon_H: H \to K_H$ be defined by $\varepsilon_H(x) = \varepsilon(x)$ for $x \in H$, we have K_H-linear mappings Δ_H, ε_H satisfying properties (1), (2), (3), (5) of 5.3.8 with K being replaced by K_H. If H is a subbiring of $H(K/k)$, then property (4) of 5.3.8 is also satisfied and H together with Δ_H, ε_H is a K_H/k-bialgebra in the sense of B.1.

We now prove a correspondence theorem for K-subbirings of End K, which we refer to as the *Biring Correspondence Theorem*. For this, let K be a field and let

$$\mathbf{F} = \{k \mid k \text{ is a subfield of } K \text{ and } K{:}k < \infty\}.$$

$$\mathbf{S} = \{H \mid H \text{ is a } K\text{-subbiring of End } K \text{ and } H{:}K < \infty\}.$$

5.3.10 Theorem. For $k \in \mathbf{F}$, $H(K/k) = \operatorname{End}_k K$.

Proof. Let $A = \operatorname{Hom}_K(\operatorname{End}_k K, K)$ be the K-dual space of $\operatorname{End}_k K$. For $a \in K$, let $\hat{a}$ be the element of A defined by $\hat{a}(x) = x(a)$ for $x \in \operatorname{End}_k K$. Letting $e_1, \ldots, e_n$ be a k-basis for K and choosing $x_1, \ldots, x_n \in \operatorname{End}_k K$ such that $x_i(e_j) = \delta_{ij}$ for $1 \leq i, j \leq n$, we have a K-basis $x_1, \ldots, x_n$ for $\operatorname{End}_k K$ and a dual K-basis $\hat{e}_1, \ldots, \hat{e}_n$ for A. Clearly $\hat{e}_1, \ldots, \hat{e}_n$ is a k-basis for the k-subspace $\hat{K} = \{\hat{a} \mid a \in K\}$ of A and $A = K\hat{K}$ (K-span of $\hat{K}$). It follows that A has a unique commutative K-algebra product $\pi: A \otimes_K A \to A$ such that $\pi(\hat{a}, \hat{b}) = \widehat{ab}$ for all $\hat{a}, \hat{b} \in \hat{K}$ (see A.2). Since $\operatorname{End}_k K$ and A are finite dimensional and dual over K, π induces $\pi^*: \operatorname{End}_k K \to \operatorname{End}_k K \otimes_K \operatorname{End}_k K$ such that $\pi^*(x) = \sum_i {}_ix \otimes_K x_i$ if and only if $(\pi(f, g))(x) = \sum_i f({}_ix)g(x_i)$ for all $f, g \in A$ (see C.1). Thus, $\pi^*(x) = \sum_i {}_ix \otimes_K x_i$ if and only if $\widehat{ab}(x) = \sum_i \hat{a}({}_ix)\hat{b}(x_i)$ for all $a, b \in K$, hence if and only if $x(ab) = \sum_i {}_ix(a)x_i(b)$ for all $a, b \in K$. Thus, $H(K/k) = \operatorname{End}_k K$. ▯

5.3.11 Definition. For $H \subset \operatorname{End} K$, K^H is the subfield $K^H = \{a \in K \mid x(ab) = ax(b) \text{ for all } x \in H \text{ and } b \in K\}$ of K.

We now have the following version of the Jacobson-Bourbaki Correspondence Theorem, which we refer to as the *Biring Correspondence Theorem*.

5.3.12 Theorem. $\mathbf{F}$ is mapped bijectively to $\mathbf{S}$ by $k \mapsto H(K/k)$.

Proof. In view of 5.3.10, this follows from 5.1.7. ▯

Finally, we describe the structure of $H(K/k)$ for K/k Galois/radical/normal. For this, let H be a subbiring of End K, let K_H, Δ_H, ε_H be as described at the end of 5.3.9 and let $K_H{}^H = K_H \cap K^H$.

5.3.13 Definition. An element g of H is *grouplike* if $\Delta_H g = g \otimes g$ and $\varepsilon_H(g) = 1$. The set of grouplike elements of H is denoted $G(H)$.

Note that $G(H)$ is a semigroup with identity I consisting of K^H-linear field homomorphisms from K to K, so that $G(H)$ is a K-independent set by the Dedekind Independence Theorem (see 3.1.3).

5.3.14 Definition. H is *co Galois* if $G(H)$ is a group and $H = K_H G(H)$ (K_H-span of $G(H)$).

Note that the product in $K_H G(H)$ is semidirect in the sense that $(ag)(bh) = ag(b)(gh)$ for $a, b \in K_H$, $g, h \in G(H)$.

5.3.15 *Definition.* For $S \subset H$, $S^0 = \{x \in S \mid \varepsilon_H(x) = 0\}$.

5.3.16 *Definition.* An element $x \in H$ is *primitive* if $\varepsilon_H(x) = 0$ and $\Delta_H x = x \otimes I + I \otimes x$. The set of primitive elements of H is denoted $D(H)$. For $x \in H$ and $S \subset H$, x is *primitive modulo* S if $\varepsilon_H(x) = 0$ and $\Delta_H x = x \otimes I + I \otimes x + \sum_3^\infty {}_ix \otimes x_i$ where the ${}_ix$, x_i are in S^0 for $i \geq 3$.

Note that $D(H)$ is a K-sub Lie ring of Der K for H a K-subbiring of End K.

5.3.17 *Definition.* A *filtration* for H is a sequence H_i of K_H-subspaces of H such that $H_0 = K_H I$, $H_i \subset H_{i+1}$ and each element x of H_{i+1}^0 is primitive modulo H_i for all i.

5.3.18 *Definition.* H is *coradical* if H has a filtration.

H is coradical if and only if KH is coradical (see E.7.1).

5.3.19 *Definition.* H is *conormal* if H has a co Galois subbiring H_{Gal} and a coradical subbiring H_{rad} such that $H = H_{\text{Gal}} H_{\text{rad}}$ (internal tensor product of $K_H{}^H$-algebras) (see A.2).

5.3.20 *Theorem.* Let $K:k < \infty$. Then K/k is Galois/radical if and only if $H(K/k)$ is co Galois/coradical.

Proof. Let $H = H(K/k)$ and note that since $H:k < \infty$, $G(H) = \operatorname{Aut}_k K$. If K/k is Galois, then $H = KG(H)$ since $KG(H)$ is a K-subbiring of End K and $K^H = k = K^{G(H)} = K^{KG(H)}$ (see E.5.2 and 5.3.12). If, conversely, $H = KG(H)$, then $k = K^H = K^{G(H)}$ and K/k is Galois. Thus, K/k is Galois if and only if $H(K/k)$ is co Galois.

Suppose next that K/k is radical. As in the proof of 5.3.10, the dual K-algebra $A = H^*$ of H as K-coalgebra is $A = K\hat{K}$ (K-span of $\hat{K}$) where $\hat{K}$ is a k-subalgebra of A isomorphic to K under a k-linear mapping $a \mapsto \hat{a}$ from K to $\hat{K}$. Let $M = \{\sum_1^n a_i\hat{b}_i \mid n \geq 1, a_i, b_i \in K \text{ for } 1 \leq i \leq n \text{ and } \sum_1^n a_i b_i = 0\}$. Then M is an ideal of A and $A = K1_A + M$ where 1_A is the identity element of A. We claim that M consists of nilpotent elements, whence A is a split local K-algebra (see A.3). Thus, let $\sum_1^n a_i\hat{b}_i \in M$ and choose e such that $a_i{}^{p^e} \in k$ for $1 \leq i \leq n$. Then

$$\left(\sum_1^n a_i\hat{b}_i\right)^{p^e} = \sum_1^n a_i{}^{p^e}\hat{b}_i{}^{p^e} = \widehat{\sum_1^n a_i{}^{p^e}b_i{}^{p^e}} = \widehat{\left(\sum_1^n a_i b_i\right)^{p^e}} = \hat{0},$$

so that $\sum_1^n a_i\hat{b}_i$ is nilpotent. Thus, $A = H^*$ is a split local K-algebra, and H is coradical by C.4.

Suppose conversely that H is coradical and let H_i be a filtration of H. Then $H = \bigcup_0^n H_i$ for some n, since $K:k < \infty$. Note that $K = K^{H_0}$. We claim that $(K^{H_i})^p \subset K^{H_{i+1}}$ for all i, so that $K^{p^n} \subset K^{H_n} = K^H = k$. Thus, let

$x \in H_{i+1}^0$, so that $\Delta x = x \otimes I + I \otimes x + \sum_3^\infty {}_i x \otimes x_i$ where the ${}_i x$, x_i are in H_i^0 for $i \geq 3$. For $a \in K^{H_i}$ and $b \in K$, we then have $x(ab) = x(a)b + ax(b)$. Iterating, we have $x(a^n) = na^{n-1}x(a)$. In particular, $x(a^p) = 0$ for all $a \in K^{H_i}$, $x \in H_{i+1}^0$. Thus, $(K^{H_i})^p \subset K^{H_{i+1}}$ for all i and $K^{p^n} \subset K^{H_n} = k$. Thus, K/k is radical for $H(K/k)$ coradical. □

5.3.21 Theorem. Let $K:k < \infty$. Then K/k is normal if and only if $H(K/k)$ is conormal. If K/k is normal, then $K = K_{\text{Gal}}K_{\text{rad}}$ (internal tensor product of k-algebras) and $H(K/k) = H_{\text{Gal}}H_{\text{rad}}$ (internal tensor product of k-algebras) where

1. K_{Gal}/k is Galois and H_{Gal} is a co Galois subbiring of $H(K/k)$ stabilizing K_{Gal} such that $x \mapsto x|_{K_{\text{Gal}}}$ is an isomorphism from H_{Gal} to $H(K_{\text{Gal}}/k)$;
2. K_{rad}/k is radical and H_{rad} is a coradical subbiring of $H(K/k)$ stabilizing K_{rad} such that $y \mapsto y|_{K_{\text{rad}}}$ is an isomorphism from H_{rad} to $H(K_{\text{rad}}/k)$.

Proof. Suppose first that K/k is normal, so that $K = K_{\text{Gal}}K_{\text{rad}}$ (internal tensor product of k-algebras). Let H be the k-span of the set $\{x \otimes y \mid x \in H(K_{\text{Gal}}/k)$ and $y \in H(K_{\text{rad}}/k)\}$ where $x \otimes y$ denotes the element of $\text{End}_k K$ such that $(x \otimes y)(a \otimes b) = x(a)y(b)$ for $a \in K_{\text{Gal}}$, $b \in K_{\text{rad}}$. Then H is a K-subring of $\text{End}_k K$. For H is clearly a subring of $\text{End}_k K$ and, for $a, b \in K$ with $a \in K_{\text{Gal}}$, $b \in K_{\text{rad}}$ and $x \otimes y \in H$ with $x \in H(K_{\text{Gal}}/k)$, $y \in H(K_{\text{rad}}/k)$, we have $(ab)(x \otimes y) = (ax) \otimes (by) \in H$, so that H is a K-subspace of $\text{End}_k K$. If $x \in H(K_{\text{Gal}}/k)$ and $\Delta x = \sum_i {}_i x \otimes_{K_{\text{Gal}}} x_i$, and if $y \in H(K_{\text{rad}}/k)$ and $\Delta y = \sum_j {}_j y \otimes_{K_{\text{rad}}} y_j$, then it is easily verified that $\Delta(x \otimes y) = \sum_{i,j} ({}_i x \otimes_j y) \otimes_K (x_i \otimes y_i)$. It follows that H is a K-subbiring of $\text{End}_k K$. Since $K^{H(K/k)} = k = K^H$, we have $H(K/k) = H$, by 5.3.12, so that $H(K/k) = H_{\text{Gal}}H_{\text{rad}}$ (internal tensor product of k-algebras) where $H_{\text{Gal}} = \{x \otimes I \mid x \in H(K_{\text{Gal}}/k)\}$ and $H_{\text{rad}} = \{I \otimes y \mid y \in H(K_{\text{rad}}/k)\}$. It is clear from 5.3.20 that H_{Gal} is a co Galois subbiring of $H(K/k)$ and that H_{rad} is a coradical subbiring of $H(K/k)$.

Suppose, conversely, that $H = H(K/k)$ is conormal, so that $H = H_{\text{Gal}}H_{\text{rad}}$ (internal tensor product of k-algebras). Let $K_{\text{Gal}} = K^{H_{\text{rad}}}$ and $K_{\text{rad}} = K^{H_{\text{Gal}}}$, and note that $KH_{\text{Gal}} = H(K/K_{\text{rad}})$ and $KH_{\text{rad}} = H(K/K_{\text{Gal}})$ by 5.3.12 (see E.5.5). Then K/K_{rad} is radical and K/K_{rad} is Galois by 5.3.20 (see E.5.5), and $K/K_{\text{Gal}}K_{\text{rad}}$ is Galois and radical. Thus, $K = K_{\text{Gal}}K_{\text{rad}}$. Next, note that K_{Gal} and K_{rad} are stable under H_{rad} and H_{Gal} respectively, since the elements of H_{Gal} commute with the elements of H_{rad} (see A.2). It follows that $(K_{\text{Gal}})^{H_{\text{Gal}}} \subset K^{H_{\text{Gal}}} \cap K^{H_{\text{rad}}} = K^H = k$ and $(K_{\text{rad}})^{H_{\text{rad}}} \subset K^H = k$. Thus, K_{Gal}/k is Galois and K_{rad}/k is radical (see E.5.5). Thus, K is normal over k. □

E.5 Exercises to Chapter 5

E.5.1. Prove that for any field extension K/k, $\text{End}_k K$ is a K-subring of K. If K/k is finite dimensional, show that $(\text{End}_k K):K = K:k$.

E.5.2. Verify that if G is a group of automorphisms of K, then the K-span $K[G]$ of G is a K-subring of End K such that $K^{K[G]} = K^G$.

E.5.3. Prove the coassociativity of Δ and coidentity of ε in the proof of 5.3.8 using 5.3.7 and 5.3.5.

E.5.4. Let K/k be a field extension, k' a subfield of K containing k. Let C be a k_1-subspace of $H(K/k)$ such that a k'-basis for C is a K-basis for KC. Suppose furthermore that $\varepsilon(C) = k'$ and that for each $x \in C$, there exist ${}_ix, x_i \in C$ such that $\Delta x = \sum_i {}_ix \otimes_K x_i$. Show that there exist mappings $\Delta_C: C \to C \otimes_{k'} C$, $\varepsilon_C: C \to k'$ with respect to which C is a k'-coalgebra such that $\Delta_C x = \sum_i {}_ix \otimes_{k'} x_i$ with $x, {}_ix, x_i \in C$ if and only if $x(ab) = \sum_i {}_ix(a)x_i(b)$, for all $a, b \in K$ and $\varepsilon_C(x)$ for $x \in C$. Show that KC is a K-subcoalgebra of $H(K/k)$.

E.5.5. Let H be a subbiring of End K. Show that KH is a subbiring of End K and

(a) H is coradical if and only if KH is coradical;
(b) KH is co Galois if H is co Galois;
(c) KH is conormal if H is conormal.

E.5.6. Let D be a nonzero derivation of a field K of characteristic $p > 0$ such that $D = D^p$. Let π be the prime field of K.

(a) Show that $K = \sum_{i \in \pi} K_i$ (direct sum of additive groups) where $K_i = \{x \in K \mid D(x) = ix\}$ for $i \in \pi$.
(b) Show that $K^D = K_0$ is a subfield of K and all the K_i $(i \in \pi)$ are K^D-subspaces of K.
(c) Show that the K^D-span K_iK_j of K_i and K_j is K_{i+j} for $i, j \in \pi$.
(d) Show that $K = K^D(x)$ and the K_i $(i \in \pi)$ are the K^Dx^i $(i = 0, 1, \ldots, p-1)$ for any $x \in K_1 - \{0\}$.

E.5.7. Let K/k be a (possibly infinite dimensional) field extension, let $D_1, \ldots, D_n \in \mathrm{Der}_k K$ and let $x_1, \ldots, x_n$ be nonzero elements of K such that $D_i(x_j) = \delta_{ij}x_j$ for $1 \le i, j \le n$. Show that there is a linear combination D of $D_1, \ldots, D_n$ over K^p such that $K:K^D \ge p^n$ by proving the following.

(a) The restrictions $t_1, \ldots, t_n$ of $D_1, \ldots, D_n$ to $L = K^p(x_1, \ldots, x_n)$ are derivations of the finite dimensional extension L of $l = K^p$, $x_1, \ldots, x_n$ is a p-basis for L/l and $L:l = p^n$.
(b) The l-span T of $t_1, \ldots, t_n$ is a diagonalizable l-subspace of $\mathrm{Der}_l L$ and contains an element t such that $L^t = l$.
(c) The l-span S of $t, t^p, t^{p^2}, \ldots$ is T itself. (Hint: Use 5.2.5 and 5.2.6 and observe that $LS = LT = \mathrm{Der}_l L$).
(d) $K:K^D \ge p^n$ for any linear combination D of $D_1, \ldots, D_n$ over K^p such that $D|_L = t$. (Hint: The K^p-span of $D, D^p, D^{p^2}, \ldots$ contains elements $D_1', \ldots, D_n'$ of $\mathrm{Der}_k K$ such that $D_i'|_L = t_i$, by part (c). The equations $D_i'(x_j) = \delta_{ij}x_j$ imply that the $x_1, \ldots, x_n$ are p-independent over K^D.)

E.5.8 (Gerstenhaber). Let K be a field, $\mathscr{D}$ a finite dimensional K-subspace of Der K such that $D^p \in \mathscr{D}$ for all $D \in \mathscr{D}$. Let $K^{\mathscr{D}}$ be the subfield of constants $K^{\mathscr{D}} = \{x \in K \mid D(x) = 0 \text{ for } D \in \mathscr{D}\}$. Show that $K:K^{\mathscr{D}} = p^{\mathscr{D}:K}$ and $\mathscr{D} = \mathrm{Der}_{K^{\mathscr{D}}} K$ by proving the following.

(a) For $D \in \mathscr{D}$ and $E = D^p - D$, $K^E = K^D(x)$ and $x^p \in K^D$ for some $x \in K^E$, where K^D denotes the subfield $\{y \in K \mid D(y) = 0\}$. (Use the earlier exercise E.5.6.)

(b) For each $D \in \mathscr{D}$, there exists a sequence $D_1, D_2, \ldots, D_n$ of elements of $\mathscr{D}$ and a sequence $x_1, \ldots, x_n$ of nonzero elements of K such that $K^{D_{i+1}} = K^{D_i}(x_i)$ for $1 \leq i \leq n - 1$, and $D_n{}^p - D_n = 0$. (Hint: Choose x_1 such that $D(x_1) \neq 0$ and replace D by a multiple D_1 such that $D_1(x_1) = x_1$. Then choose x_2 such that $(D_1{}^p - D_1)(x_2) \neq 0$ and let D_2 be a multiple of $D_1{}^p - D_1$ such that $D_2(x_2) = x_2$, noting that $D_2(x_1) = 0$ so that D_1 and D_2 are linearly independent over K. Continue this process, showing that it must terminate with $D_n{}^p - D_n = 0$ for some $n \leq \mathscr{D}:K$ such that $K:K^D \leq p^n$.)

(c) Show that $K:K^D \leq p^{\mathscr{D}:K}$ for $D \in \mathscr{D}$. (Use part (*b*).)

(d) Show that for any element $D \in \mathscr{D}$ such that $K:K^D$ is maximal, $K:K^D = K:K^{\mathscr{D}}$. (Hint: Consider the sequences $x_1, \ldots, x_n$ and $D_1, \ldots, D_n$ constructed in part (*b*). Suppose that $K^D \supsetneqq K^{\mathscr{D}}$, so that there exists $y \in K^D$ and $E \in \mathscr{D}$ such that $E(y) \neq 0$. Since $D_i(x_j) = 0$ for $j < i$ and $D_i(x_i) = x_i$, E can be chosen such that $E(x_i) = 0$ for $1 \leq i \leq n$ and $E(y) = y$. Show, using E.5.7, that for some linear combination F of the D_i $(1 \leq i \leq n)$ and E over K, $K:K^F > K:K^D$, a contradiction.)

(e) Show that $K:K^{\mathscr{D}} \leq p^{\mathscr{D}:K}$.

(f) Show that $\mathscr{D} = \operatorname{Der}_{K^{\mathscr{D}}} K$ and $K^{\mathscr{D}} = p^{\mathscr{D}:K}$. (Hint: Compare the dimensions of K over $K^{\mathscr{D}}$ and $\operatorname{Der}_{K^{\mathscr{D}}} K$ over K.)

E.5.9 (Gerstenhaber and Zaromp). Let $D_1, \ldots, D_n$ be commuting derivations of a radical extension K/k which are linearly independent over k. Prove the following.

(a) $D_1, \ldots, D_n$ are linearly independent over K.

(b) $K:k \geq n$.

(c) In order that $K:k = n$, it is necessary and sufficient that $D_i{}^p$ be in the k-span of $D_1, \ldots, D_n$ for $1 \leq i \leq n$.

E.5.10 (Hochschild). Show that for $D \in \operatorname{Der} K$ and $a \in K$, $(aD)^p = bD^p + cD$ for suitable $b, c \in K$.

E.5.11. Let H be a k'subbiring of $H(K/k)$ and let k' be a subfield of K containing k such that k' is stable under H and $k'^H = k$. Show that if H is coradical/co Galois/conormal, then k'/k is radical/Galois/normal.

6 Tori and the structure of radical extensions

In this chapter, we enter into a detailed study of the biring $H(K/k)$ of a field extension K/k in terms of its toral subbirings (see 6.2.3). In 6.1, we define the notion of torus and develop basic properties of tori. In 6.2, we discuss the diagonalizable toral subbirings of $H(K/k)$ and show that they correspond to certain tensor product decompositions of K. In 6.3, we study the coradical toral subbirings of $H(K/k)$. These are the toral subbirings of $H(K/k)$ for which the corresponding extension K/K^T is radical (see 6.3.2), and we accordingly restrict ourselves to the case of radical extensions K/k. In this case, a toral subbiring T can be studied inductively by considering a filtration $T = \bigcup_0^\infty T_i$ (see 5.3.17) and T can be studied by comparing T with a diagonalizable toral subbiring $\overline{T}$ of an extension $\overline{K}/\overline{k}$ obtained from K/k by ascent to the separable closure of k (see 6.3.4). Using these methods for studying a toral subbiring T of $H(K/k)$, we prove that the centralizer $H(K/k)^T$ of T in $H(K/k)$ is a K^T-subbiring of $H(K/k)$ and $H(K/k) = KH(K/k)^T$ (see 6.3.11), so that the entire structure of $H(K/k)$ is determined by the structure of $H(K/k)^T$ and its action on K. In 6.4, we discuss the use of toral subbirings in studying the structure of radical field extensions and generalize the theorem of Jacobson on the finitude of the dimension of an extension $K/K^{\mathscr{D}}$ where $\mathscr{D}$ is a finite dimensional K-sub Lie ring of Der K.

6.1 Tori

Let k be a field of exponent characteristic p and prime field π, and let V be a finite dimensional vector space over k.

6.1.1 Definition. A linear transformation t of V is *semisimple* if the minimum polynomial of t is the product of distinct irreducible separable polynomials over k.

6.1.2 Definition. A *splitting field* of a set T of linear transformations of V is a splitting field of the set of minimum polynomials of the elements of T.

6.1.3 Definition. If L/k is an extension field of k and t a linear transformation of the k-vector space V, then t_L is the linear transformation of the L-vector space $V_L = L \otimes_k V$ such that $t_L(a \otimes_k v) = a \otimes_k t(v)$ for $a \in L$, $v \in V$.

6.1.4 Definition. A linear transformation t on V is *diagonalizable* if V has a basis with respect to which the matrix of t is diagonal. A set T of linear transformations on V is *diagonalizable* if V has a basis with respect to which the matrix of t is diagonal for every $t \in T$.

6.1.5 Proposition. Let t be a linear transformation of V and let L/k be a splitting field of t. Then t is semisimple if and only if L/k is Galois and t_L is diagonalizable on V_L.

Proof. If t is semisimple, then L/k is Galois (see 2.3.12) and the minimum polynomial of t_L is a product of distinct linear factors $X - a_i$, whence t_L is diagonalizable on V_L. Conversely, suppose that L/k is Galois and t_L diagonalizable on V_L. Then the minimum polynomial of t is a product of distinct irreducible separable polynomials over k and t is semisimple. □

6.1.6 Proposition. Let $p > 1$ and let t be a linear transformation of V. Then t^{p^e} is semisimple for some e.

Proof. Let L/k be a splitting field of t and let the eigenvalues of t_L on V_L be $a_1, \ldots, a_n$. Choose e such that $(t_L)^{p^e}$ is diagonalizable on V_L (see E.6.1) and such that $a_1{}^{p^e}, \ldots, a_n{}^{p^e}$ are separable over k (see 2.2.4). Letting $L' = k(a_1{}^{p^e}, \ldots, a_n{}^{p^e})$, the splitting field of t^{p^e} is the Galois extension L'/k and $(t^{p^e})_{L'}$ is diagonalizable on $V_{L'}$. Thus, t^{p^e} is semisimple. □

6.1.7 Definition. A *k-torus* on V is a set T of pairwise commuting semisimple linear transformations of V such that $as + bt \in T$ and $t^p \in T$ for all $a, b \in k$, $s, t \in T$.

For any set S of pairwise commuting semisimple linear transformations of V, the k-span $\langle S \rangle_p$ of the set $\{s^{p^e} \mid e \geq 0, s \in S\} \bigcup \{I\}$ is a k-torus on V containing S and the identity I (see E.6.2). Obviously, $\langle S \rangle_p$ is diagonalizable on V if and only if S is diagonalizable on V.

6.1.8 Proposition. The splitting field L/k of a k-torus T on V is a finite dimensional Galois extension and the L-span T_L of $\{t_L \mid t \in T\}$ is a diagonalizable L-torus on V_L.

Proof. Let $t_1, \ldots, t_n$ be a basis for T and let L/k be a splitting field for $\{t_1, \ldots, t_n\}$. Then L/k is a finite dimensional Galois extension (see 6.1.5). Since the t_L ($t \in T$) commute pairwise and are diagonalizable on V_L, T_L is a diagonalizable L-torus on V_L (see E.6.2). □

6.1.9 Definition. For any set T of linear transformations of V and for any function $\alpha: T \to k$, $V_\alpha(T) = \{v \in V \mid t(v) = \alpha(t)v \text{ for all } t \in T\}$.

6.1.10 Proposition. Let T be a k-torus on V and let T^* be the k-dual space of T. Then the following statements are equivalent:

1. every element of T is diagonalizable on V;
2. T is diagonalizable on V;
3. $V = \sum_{\alpha \in T^*} V_\alpha(T)$ (direct).

Proof. See E.6.2. □

6.1.11 Definition. Let T be a k-torus on V. Then $T_\pi = \{t \in T \mid t^p = t\}$. Note that T_π is a π-subspace of T.

6.1.12 Proposition. Let $p > 1$ and let T be a k-torus on V. Then $V = \sum_{\alpha \in T_\pi^*} V_\alpha(T_\pi)$ (direct) where T_π^* is the π-dual space of T_π.

Proof. Since each $t \in T_\pi$ satisfies $t^p - t = 0$, the minimum polynomials of the elements of T_π are divisors of the separable polynomial $X^p - X = \prod_{i \in \pi}(X - i)$. Thus, each $t \in T_\pi$ is diagonalizable on V with all eigenvalues in π. It follows that $V = \sum_{\alpha \in T_\pi^*} V_\alpha(T_\pi)$ (direct). □

6.1.13 Theorem. Let $p > 1$. Then a k-torus T on V is diagonalizable on V if and only if T is the k-span of T_π.

Proof. One direction is clear from 6.1.12. For the other, suppose that T is diagonalizable on V, so that $V = \sum_{\alpha \in \mathbf{R}} V_\alpha(T)$ (direct) where $R = \{\alpha \in T^* \mid V_\alpha(T) \neq \{0\}\}$. Since R separates the points of T, R contains a basis $\alpha_1, \ldots, \alpha_n$ for T^*. Let $t_1, \ldots, t_n$ be a dual basis for T. Then $\alpha_j(t_i) = \delta_{ij}$ and $\alpha_j(t_i{}^p) = \alpha_j(t_i)^p = \delta_{ij}$ for all i, j. Thus, $\alpha_j(t_i{}^p) = \alpha_j(t_i)$ for all i, j. Since the α_j separate the points of T, it follows that $t_i{}^p = t_i$ for all i, so that T is the k-span of T_π. □

6.2 Diagonalizable toral k-subbirings of $H(K/k)$

Let K/k be a finite dimensional field extension, and let p and π denote the exponent characteristic and prime subfield of k respectively.

6.2.1 Proposition. Let T be a diagonalizable k-torus on K. Then a k-basis for T is a K-basis for KT (K-span of T).

Proof. We have $K = \sum_{\alpha \in T_\pi^*} K_\alpha(T)$ (direct) where T^* is the k-dual space of T. Since $R = \{\alpha \in T^* \mid K_\alpha(T) \neq \{0\}\}$ separates the points of T, R contains a basis $\alpha_1, \ldots, \alpha_n$ for T^*. Let $t_1, \ldots, t_n$ be a dual basis for T. Taking $x_i \in K_{\alpha_i} - \{0\}$ for $1 \leq i \leq n$, we have $t_i(x_j) = \alpha_j(t_i)x_j = \delta_{ij}x_j$ for $1 \leq i, j \leq n$. It follows easily that the t_i are K-independent, so that $t_1, \ldots, t_n$ is a K-basis for KT. □

The reader should note that in 8.2.1 the assumption that T be diagonalizable cannot be dropped (see E.6.1) except when K/k is radical (see 6.3).

It is convenient to introduce now the following counterpart of definitions 5.3.1 and 5.3.9.

6.2.2 Definition. A *subcoring* of $H(K)$ is a subset C of $H(K)$ containing I such that for each $x \in C$, there exist ${}_ix, x_i \in C$ such that $\Delta x = \sum_i {}_ix \otimes_K x_i$ and such that $K_C = \varepsilon(C)$ is a subfield of K, C is a K_C-subspace of $H(K)$ and a K_C-basis for C is a K-basis for KC (K-span of C). A *subcoring/k-subcoring* of $H(K/k)$ is a subcoring C of $H(K)$ contained in $H(K/k)$ such that $K_C \supset k$/ $K_C = k$. A *k′subcoring* of $H(K)$ is a subcoring C of $H(K)$ such that $K_C = k'$, for any subfield k' of K.

If C is a subcoring of $H(K)$, then $\Delta_C\colon C \to C \otimes_{K_C} C$ and $\varepsilon_C\colon C \to K_C$ are defined by $\Delta_C x = \sum_i {}_ix \otimes_{K_C} x_i$ where the ${}_ix, x_i$ are elements of C such that

$\Delta x = \sum_i {}_ix \otimes_K x_i$ and $\varepsilon_C(x) = \varepsilon(x)$ for $x \in C$ (see E.5.4), and C together with Δ_C and ε_C is a K_C-coalgebra in the sense of C.1.

6.2.3 Definition. A *toral k-subbiring/k-subcoring* of $H(K/k)$ is a k-subbiring/k-subcoring T of $H(K/k)$ which is a k-torus on K.

For a diagonalizable k-torus T on K/k containing I to be a toral k-subcoring of $H(K/k)$, it is necessary and sufficient that $\varepsilon(T) \subset k$ and for each $t \in T$, there exist ${}_it, t_i \in T$ such that $\Delta t = \sum_i {}_it \otimes_K t_i$ (see 6.2.1).

6.2.4 Definition. Let T be a diagonalizable toral k-subcoring of $H(K/k)$. Then $G(T) = \{K_\alpha(T) \mid \alpha \in T^*, K_\alpha(T) \neq \{0\}\}$.

Let T be a diagonalizable k-subcoring on K and let $0 \neq x \in K_\alpha(T) \in G(T)$, $0 \neq y \in K_\beta(T) \in G(T)$. Then

$$t(xy) = \sum_i {}_it(x)t_i(y) = \left(\sum_i \alpha({}_it)\beta(t_i)\right)xy \qquad \text{for } t \in T.$$

Since $xy \neq 0$, $\sum_i \alpha({}_it)\beta(t_i)$ is independent of the particular representation of Δt as $\Delta t = \sum_i {}_it \otimes t_i$, and we denote $\sum_i \alpha({}_it)\beta(t_i) = (\alpha * \beta)(t)$. We then have $K_\alpha(T)K_\beta(T) \subset K_{\alpha*\beta}(T)$ and $K_{\alpha*\beta}(T) \in G(T)$. It follows that $G(T)$ is a splitting of K/k in the sense of the following definition.

6.2.5 Definition. A *splitting* of K/k is a (necessarily finite) collection G of nonzero k-subspaces $U, V \cdots$ of K such that $K = \sum_{U \in G} U$ (direct) and such that for each pair of elements $U, V \in G$, there exists $W \in G$ such that W contains the k-span UV of $\{uv \mid u \in U, v \in V\}$.

6.2.6 Proposition. Let G be a splitting of K/k. Then $UV \in G$ for U, $V \in G$ and G together with the binary composition UV is a group. The identity K^G of G is a subfield of K containing k, the elements of G are one-dimensional K^G-subspaces of K and $K{:}K^G = G{:}\mathbf{1}$ (order of G).

Proof. Let $U, V \in G$. Then there exists precisely one $W \in G$ such that $UV \in W$. We denote W by $W = U \circ V$. Fixing $V \in G$ and $v \in V - \{0\}$, we have $K = Kv = \sum_{U \in G} Uv$ (direct) $\subset \sum_{U \in G} U \circ V$ (direct) where $Uv = \{uv \mid u \in U\}$ for $U \in G$. It follows that $U \circ V \subset Uv \subset UV \subset U \circ V$, so that $U \circ V = Uv = UV$ for all $U \in G$. It also follows that $G = \{UV \mid U \in G\}$ for each $V \in G$. It is clear that the binary composition UV is associative and commutative. Since G is finite and right translations are surjective, they are bijective, so that the cancellation law $UV = U'V \Rightarrow U = U'$ holds in G. The powers $V, V^2 \cdots$ cannot all be distinct and we choose $1 \leq m < n$ such that $V^m = V^n$. Cancelling on the right, we have $V = V^{d+1} = VV^d$ for some $d \geq 1$. Thus, $UV = UVV^d$ for $U \in G$. Cancelling V, we have $U = UV^d$ for $U \in G$, so that $V^d = E$ is an identity in G. Moreover, $V^{d-1}V = E$ so that V has inverse V^{d-1}. Thus, G is a group. Letting K^G be the identity element of G and $u \in K^G - \{0\}$, we have $K^GK^G \subset K^G$ and $K^G = K^Gu$. Since $u \in K^Gu$, we have $1 \in K^G$, whence K^G is a finite dimensional k-subspace of K containing k and closed under multiplication. Thus, K^G is a subfield of K. For $V \in G$, we have $V = K^G V = K^G v$ for $v \in V - \{0\}$, so that $V:K^G = 1$. Since $K = \sum_{V \in G} V$ (direct), it follows that $K:K^G = G:1$. □

Let G be a splitting of K/k and let T be the k-dual space $(kG)^*$ of the group k-algebra kG of G (see A.1). For $s, t \in T$, let st be the element of T such that $(st)(g) = s(g)t(g)$ for $g \in G$. For $t \in T$, let $\Delta_G t$ be the element of $T \otimes_k T$ such that $\Delta_G t = \sum_i {}_i t \otimes_k t_i$ if and only if $t(gh) = \sum_i {}_i t(g) t_i(h)$ for $g, h \in G$ and let $\varepsilon_G(t) = t(e)$ where $e = K^G$ is the identity element of G (see B.1). For $t \in T$, let $\bar{t}$ be the element of $H(K/k)$ such that $\bar{t}(x) = t(U)x$ for $x \in U$ and $U \in G$.

6.2.7 *Definition.* $T(G) = \{\bar{t} \mid t \in (kG)^*\}$.

Note that $\overline{st} = \bar{s}\bar{t}$, $\Delta_G t = \sum_i {}_i t \otimes_k t_i$ implies that $\Delta \bar{t} = \sum {}_i \bar{t} \otimes_K \bar{t}_i$ and $\varepsilon_G(t) = \varepsilon(\bar{t})$ for all $s, t \in (kG)^*$. It follows that $T = T(G)$ is a diagonalizable toral k-subbiring of $H(K/k)$. In fact, letting $\hat{g}: T \to k$ be defined by $\hat{g}(\bar{t}) = t(g)$ for $\bar{t} \in T$, $g \in G$, one sees that $K = \sum_{g \in G} K_{\hat{g}}(T)$ and $K_{\hat{g}}(T) = U$ for $g = U \in G$, since $K_{\hat{g}}(T) = \{x \in K \mid \bar{t}(x) = \hat{g}(\bar{t})x = t(g)x = t(U)x$ for all $\bar{t} \in T\} = U$ for $g = U \in G$. It is clear from this that $G(T(G)) = G$.

6.2.8 *Definition.* For $T \subset \operatorname{End} K$, $\langle T \rangle$ is the subring of End K generated by T.

6.2.9 *Theorem.* $T(G(T)) = \langle T \rangle$ for any diagonalizable toral k-subcoring T of $H(K/k)$. The set $\mathbf{T}$ of diagonalizable toral k-subbirings of $H(K/k)$ is mapped bijectively to the set $\mathbf{G}$ of splittings of K/k by the mapping $T \mapsto G(T)$, the inverse being the mapping $G \mapsto T(G)$.

Proof. Let T be a diagonalizable toral k-subcoring of $H(K/k)$ and let $t \in T$. Then the element t' of $(kG(T))^*$ such that $t'(g) = \alpha(t)$ for $g = K_\alpha(T) \in G(T)$ satisfies the equations $\bar{t}'(x) = t'(g)x = \alpha(t)x = t(x)$ for $x \in g$ and $g = K_\alpha(T) \in G(T)$. Thus, $t = \bar{t}' \in T(G(T))$. It follows that $T \subset T(G(T))$, whence $\langle T \rangle \subset T(G(T))$. Moreover, $G(\langle T \rangle) = G(T) = G(T(G(T)))$, so that $K^{\langle T \rangle} = K^{T(G(T))}$. It follows from the Jacobson-Bourbaki Correspondence Theorem that $K\langle T \rangle = KT(G(T))$. But then

$$\langle T \rangle : k = K\langle T \rangle : K = KT(G(T)) : K = T(G(T)) : k,$$

by 6.2.1. It follows that $\langle T \rangle = T(G(T))$. In particular, $T = T(G(T))$ for $T \in \mathbf{T}$. Since we have seen that $G = G(T(G))$ for $G \in \mathbf{G}$, $T \mapsto G(T)$ maps $\mathbf{T}$ bijectively to G, the inverse mapping being $G \mapsto T(G)$. □

We describe now the connection between splittings of K/k (and therefore diagonalizable toral k-subcoalgebras of $H(K/k)$) and tensor product decompositions of K/k.

6.2.10 *Proposition.* Let G be a splitting of K/k, let $g_1, \ldots, g_n$ be a basis for G as finite Abelian group and let x_i be a nonzero element of g_i for $1 \le i \le n$. Then $K = K^G(x_1) \cdots K^G(x_n)$ (internal tensor product of K^G-algebras) and $x_i^{e_i} \in K^G$ where e_i is the order of g_i for $1 \le i \le n$. Conversely, if $K \supset$

$k' \supset k$ and $K = k'(x_1) \cdots k'(x_n)$ (internal tensor product of k'-algebras) where $x_i \in K$, $e_i \geq 0$ and $x_i^{e_i} \in k$ for $1 \leq i \leq n$, then

$$G = \{kx_1^{f_1} \cdots x_n^{f_n} \mid 0 \leq f_i \leq e_i - 1 \text{ for } 1 \leq i \leq n\}$$

is a splitting for K/k with basis $g_1 = k'x_1, \ldots, g_n = k'x_n$ and identity $K^G = k'$.

Proof. The $x_1^{f_1} \cdots x_n^{f_n}$ form a basis for K over K^G since $Kx_1^{f_1} \cdots x_n^{f_n} = g_1^{f_1} \cdots g_n^{f_n}$ and $K = \sum_{g \in G} g$ (direct). □

6.2.11 Theorem. Let $p > 1$ and let T be a diagonalizable toral k-subcoring/k-subbiring of $H(K/k)$. Then T_π is a π-subcoring/π-subbiring of $H(K)$ and $H(K/K^T) = K\langle T\rangle = K\langle T_\pi\rangle / H(K/K^T) = KT = KT_\pi$.

Proof. Suppose first that T is a diagonalizable toral k-subbiring of $H(K/k)$, so that $T = T(G) = \{\bar{t} \mid t \in (kG)^*\}$ for some splitting G of K/k. Then $T_\pi = \{\bar{t} \mid t \in (\pi G)^*\}$ where $(\pi G)^* = \{t \in (kG)^* \mid t(\pi G) \subset \pi\}$. It follows that a π-basis for T_π is a k-basis for T and that $\varepsilon(T_\pi) = \pi$. Next, suppose only that T is a diagonalizable toral k-subcoring of $H(K/k)$. Then a π-basis for $\langle T\rangle_\pi$ is a k-basis for $\langle T\rangle$ and $\varepsilon(\langle T_\pi\rangle) = \pi$. Since $T = kT_\pi$ (see 8.1.13), it follows that a π-basis for T_π is a k-basis for T and $\varepsilon(T_\pi) = \pi$. Let $t_1, \ldots, t_n$ be a π-basis for T_π and let $t \in T_\pi$. Then $\Delta t = \sum_1^n {}_it \otimes_k t_i$ where ${}_1t, \ldots, {}_nt$ are elements of T. It follows from this that

$$\sum_1^n {}_it \otimes_k t_i = \Delta t = \Delta t^p = \sum_1^n {}_it^p \otimes_k t_i^p = \sum_1^n {}_it^p \otimes_k t_i$$

(see E.8.2). Thus, ${}_it = {}_it^p$ and ${}_it \in T_\pi$ for $1 \leq i \leq n$. Thus, T_π is a π-subcoring of $H(K)$. That $H(K/K^T) = K\langle T\rangle = K\langle T_\pi\rangle$ follows from the Jacobson-Bourbaki Correspondence Theorem, since $K^T = K^{T_\pi} = K^{\langle T\rangle} = K^{\langle T_\pi\rangle} = K^{K\langle T\rangle} = K^{K\langle T_\pi\rangle}$. If T is closed under multiplication, then T_π is closed under multiplication and $T = \langle T\rangle, T_\pi = \langle T_\pi\rangle$. □

6.3 Coradical toral k-subcorings of $H(K/k)$

We begin with a short discussion about subcorings of $H(K)$.

6.3.1 Definition. Let C be a subcoring of $H(K)$. Then a *filtration* for C is a sequence C_i of K_C-subspaces such that $C_0 = K_C I$, $C_i \subset C_{i+1}$ and each element x of C_{i+1}^0 is primitive modulo C_i for all i. If C has a filtration, then C is *coradical* (see 5.3.17, 5.3.18).

6.3.2 Theorem. Let C be a subcoring of $H(K)$ such that $C{:}K_C < \infty$. Then C is coradical if and only if $K^{p^n} \subset K^C$ for some n.

Proof. Suppose first that C is coradical and let C_i be a filtration for C. Then $C = \bigcup_0^n C_i$ for some n. As in the proof of 5.3.20, we have $K = K^{C_0}$ and $(K^{C_i})^p \subset K^{C_{i+1}}$ for all i, so that $K^{p^n} \subset K^{C_n} = K^C$.

Suppose, conversely, that $K^{p^n} \subset K^C$. Let $\hat{K} = \{\hat{a} \mid a \in K\}$ where $\hat{a}\colon KC \to K$ is defined by $\hat{a}(x) = x(a)$ for $x \in KC$. Since $\hat{K}$ separates the points of KC, the K-dual space A of KC is $A = K\hat{K}$ (K-span of $\hat{K}$). Since KC is a K-coalgebra,

A has a unique K-algebra product fg and identity 1_A such that $(fg)(x) = \sum_i f({}_ix)g(x_i)$ and $1_A(f) = f(1)$ for $f, g \in A$ and $x \in KC$. Letting $a, b \in K$ and $x \in KC$, we have $(\hat{a}\hat{b}(x) = \sum_i \hat{a}({}_ix)\hat{b}(x_i) = \sum_i {}_ix(a)x_i(b) = x(ab) = \widehat{ab}(x)$ and $1_A(x) = x(1) = \hat{1}(x)$. It follows that $\hat{a}\hat{b} = \widehat{ab}$ and $1_A = \hat{1}$, so that the mapping $a \mapsto \hat{a}$ is a K^C-algebra homomorphism from K into A with image $\hat{K}$. The ideal $M = \{\sum_1^m a_i\hat{b}_i \mid m \geq 1, a_i, b_i \in K \text{ for } 1 \leq i \leq m \text{ and } \sum_1^m a_i b_i = 0\}$ of A consists of nilpotent elements, since $a_i^{p^n} \in K^C$ for $1 \leq i \leq m$ and

$$\left(\sum_1^m a_i\hat{b}_i\right)^{p^n} = \sum_1^m a_i^{p^n}\hat{b}_i^{p^n} = \text{etc.} = \hat{0}$$

as in the proof of 5.3.30. Since $A = K1_A + M$, A is a split local K-algebra (see A.3) and KC is coradical by C.4. To see that C is coradical, we consider $A' = \{f \in A \mid f(C) \subset K_C\}$. Letting $x_1, \ldots, x_d$ be a K_C-basis for C, $x_1, \ldots, x_d$ is a K-basis for KC and we let $f_1, \ldots, f_d$ be a dual basis for A. Then clearly the $f_1, \ldots, f_d$ are in A' and $f_1, \ldots, f_d$ is a K_C-basis for A'. Thus, A' can be identified with the K_C-dual space of C. Clearly A' is a subring of A and $(fg)(x) = \sum_i f({}_ix)g(x_i)$, $1_A(x) = x(1)$ for $x \in C$. To show that C is coradical, it therefore suffices to show that A' is a split local K_C-algebra (see C.4). For this, note that $M = \{f \in A \mid f(I) = 0\}$ and let $M' = \{f \in A' \mid f(I) = 0\} = A' \cap M$. Then M' is an ideal of A' consisting of nilpotent elements and $A' = K_C 1_A + M'$. Thus, A' is split local and C is coradical. □

Suppose that T is a coradical toral k-subcoring of $H(K/k)$ and let T_i be a filtration of T. Letting $\langle S\rangle_p$ be the k-span of $\{s^{p^e} \mid e \geq 0, s \in S\}$ for $S \subset H(K/k)$, we obtain a filtration $\langle T_i\rangle_p$ of $\langle T\rangle_p = T$ consisting of toral k-subcorings of $H(K/k)$ (see E.6.6). Thus, any coradical toral k-subcoring T of $H(K/k)$ has a p-filtration in the sense of the following definition.

6.3.3 Definition. A *p-filtration* for a toral k-subcoring T of $H(K/k)$ is filtration T_i for T such that $T_i^p \subset T_i$ for all i.

Suppose next that K/k is a radical extension. We know that $\overline{T} = L \otimes_k T$ is a diagonalizable $\bar{k}$-torus on $\overline{K} = L \otimes_k K$ where L is the separable closure of k in the algebraic closure $K_{\text{Alg}} = k_{\text{Alg}}$ of K and $\bar{k} = L \otimes_k k$ (see 2.2.12 and 6.1.8). The extension L/k is Galois with Galois group $G = \operatorname{Aut}_k L$ and $\overline{K}$ is a radical field extension of $\bar{k}$. For any k-vector space T, we let $\overline{T} = L \otimes_k T$ and regard $\overline{T}$ as $\bar{k}$-vector space. For any K-vector space V, we let $\overline{V} = L \otimes_k V$ and regard $\overline{V}$ as $\overline{K}$-vector space. Each element g of G induces k-linear mappings (also denoted g for convenience) $g: \overline{K} \to \overline{K}$, $g: \bar{k} \to \bar{k}$, $g: \overline{T} \to \overline{T}$, $g: \overline{V} \to \overline{V}$ such that $g(a \otimes b) = g(a) \otimes b$ $(a \in L, b \in K)$, $g(a \otimes b) = g(a) \otimes b$ $(a \in L, b \in k)$, $g(a \otimes t) = g(a) \otimes t$ $(a \in L, t \in T)$, $g(a \otimes v) = g(a) \otimes v$ $(a \in L, v \in V)$. In this way, G acts as a group of transformations of $\overline{K}$, $\bar{k}$, $\overline{T}$ and $\overline{V}$. Moreover, $\overline{K}^G = K$, $\bar{k}^G = k$, $\overline{T}^G = T$ and $\overline{V}^G = V$ where we identify K and $k \otimes_{\bar{k}} K$, k and $k \otimes_k k$, T and $k \otimes_k T$, V and $k \otimes_k V$. If K/k is finite dimensional, then we can identify $\overline{H(K/k)}$ and $H(\overline{K}/\bar{k})$, since $\overline{H(K/k)} = \overline{\operatorname{End}_k K}$ and $H(\overline{K}/\bar{k}) = \operatorname{End}_{\bar{k}} \overline{K}$ (see E.6.3). Thus, G acts as a group of transformations on $H(\overline{K}/\bar{k}) =$

$\overline{H(K/k)}$ and $H(\bar{K}/\bar{k})^G = \overline{H(K/k)}^G = H(K/k)$. Moreover, $g(\bar{x}(\bar{a})) = g(\bar{x})(g(\bar{a}))$ for $\bar{x} \in H(\bar{K}/\bar{k})$, $\bar{a} \in \bar{K}$ and $g \in G$.

The importance of the next theorem is that it enables us to pass from a toral k-subcoring T of $H(K/k)$ to the diagonalizable (see 8.1.8) toral $\bar{k}$-subcoring $\bar{T}$ of $H(\bar{K}/\bar{k})$, study $\bar{T}$ and then transfer the conclusions about $\bar{T}$ to $T = \bar{T}^G$. The passage from T to $\bar{T}$ is referred to as *ascent*, that from $\bar{T}$ to T *descent*.

6.3.4 Theorem. Let K/k be a finite dimensional radical extension. Then the set of toral k-subcorings/k-subbirings of $H(K/k)$ is mapped bijectively to the set of G-stable (necessarily diagonalizable) toral $\bar{k}$-subcorings/$\bar{k}$-subbirings of $H(\bar{K}/\bar{k})$ by the mapping $T \mapsto \bar{T}$, the inverse mapping being the mapping $\bar{T} \mapsto \bar{T}^G$. If T is a k-torus on K containing I such that $\varepsilon(T) \subset k$ and for each $t \in T$ there exist ${}_it, t_i \in T$ such that $\Delta t = \sum_i {}_it \otimes_K t_i$, then T is a coradical toral k-subcoring of T.

Proof. Let T be a k-torus on K containing I such that $\varepsilon(T) \subset k$ and for each $t \in T$, there exist ${}_it, t_i \in T$ such that $\Delta t = \sum_i {}_it \otimes_K t_i$. Then $\bar{T}$ is a diagonalizable $\bar{k}$-torus on $\bar{K}$ containing I such that $\varepsilon(\bar{T}) \subset \bar{k}$ and for each $t \in \bar{T}$, there exist ${}_it, t_i \in \bar{T}$ such that $\Delta t = \sum_i {}_it \otimes_K t_i$. Thus, $\bar{T}$ is a toral $\bar{k}$-subcoring of $H(\bar{K}/\bar{k})$ (see 6.2.1 and 6.2.3). If $t_1, \ldots, t_n$ is a k-basis for T, then $t_1, \ldots, t_n$ is a $\bar{k}$-basis for $\bar{T}$, hence a $\bar{K}$-basis for $\bar{K}\bar{T}$, hence a K-basis for KT. Thus, T is a toral k-subcoring of $H(K/k)$, and T is coradical by 6.3.2. Furthermore, $\bar{T}$ is closed under products if T is closed under products. Suppose next that $\hat{T}$ is a G-stable toral $\bar{k}$-subcoring of $H(\bar{K}/\bar{k})$ and let $T = \hat{T}^G$. Then T is a k-form of $\hat{T}$ since $k = \bar{k}^G$ (see 3.2.5). Clearly T contains I and we have $t^p \in T = \bar{T}^G$ for $t \in T = \bar{T}^G$. Letting $t \in T$, we have $\varepsilon(t) \in \bar{k}$ and $g(\varepsilon(t)) = g(t(1)) = g(t)(g(1)) = t(1) = \varepsilon(t)$ for $g \in G$, so that $\varepsilon(t) \in \bar{k}^G = k$. Thus, $\varepsilon(T) \subset k$. Let $t_1, \ldots, t_n$ be a k-basis for T, hence a $\bar{k}$-basis for $\bar{T}$, hence a $\bar{K}$-basis for $\bar{K}^{\bar{T}}$. For $t \in T$, there exist uniquely determined elements ${}_1t, \ldots, {}_nt$ of $\bar{T}$ such that $\Delta t = \sum_i {}_it \otimes t_i$. Since $\sum_i {}_it \otimes t_i = \Delta t = \Delta g(t) = \sum_i g({}_it) \otimes g(t_i) = \sum_i g({}_it) \otimes t_i$ so that $g({}_it) = {}_it$ for $g \in G$, so that we have ${}_it \in T = \bar{T}^G$ for $1 \leq i \leq n$. Thus, T is a toral k-subcoring of $H(K/k)$. If $\bar{T}$ is closed under products, then so is T. □

6.3.5 Definition. For $S \subset H(K/k)$ and $H \subset H(K/k)$, $H^S = \{x \in H \mid sx = xs \text{ for all } s \in S\}$ and $\langle S \rangle$ is the subring of $H(K/k)$ generated by S.

6.3.6 Proposition. Let K/k be a finite dimensional radical extension and let S be a toral subcoring of $H(K/k)$. Then $H(K/K^S)^S = K^S\langle S \rangle$ and every semisimple element of $K^S\langle S \rangle$ over k is contained in $\langle S \rangle$.

Proof. Let $T = K^S\langle S \rangle$. Then $K^S = K^T$, $H(K/K^S)^S = H(K/K^T)^T$ and T is a toral K^T-subbiring of $H(K/K^T)$ (see 8.3.4). We have $H(K/K^T) = KT$ and $K:K^T = H(K/K^T):K = T:K^T$ (see 8.2.11). Now $\bar{T}$ is a diagonalizable $\bar{K}^{\bar{T}}$-space of linear transformations of $\bar{K}/\bar{K}^{\bar{T}}$ and $\bar{T}:\bar{K}^{\bar{T}} = \bar{K}:\bar{K}^{\bar{T}}$. It follows by linear algebra that $H(\bar{K}/\bar{K}^{\bar{T}})^{\bar{T}} = \bar{T}$. Thus, $H(K/K^T)^T = T$, so that $H(K/K^S)^S = K^S\langle S \rangle$. Since K^S/k is radical, $(K^S)p^e \subset k$ and $(K^S\langle S \rangle)^{p^e} \subset \langle S \rangle$ for some e. It follows that $\langle S \rangle$ contains every semisimple element of $K^S\langle S \rangle$ over k (see E.6.5). □

6.3.7 Theorem. Let K/k be a finite dimensional radical extension and let S and T be toral k-subcorings of $H(K/k)$ such that $S \subset T$ and every element of T^0 is primitive modulo S (see 5.3.16). Then $T \cap \langle S \rangle = \{t \in T \mid t(a) = \varepsilon(t)a \text{ for all } a \in K^S\}$.

Proof. One direction is easy (see E.6.7). For the other, let $t \in T^0$ and suppose that $t(a) = \varepsilon(t)a$ for all $a \in K^S$. Let

$$\Delta t = \sum_1^n {}_it \otimes t_i = t \otimes I + I \otimes t + \sum_3^n {}_it \otimes t_i$$

where the ${}_it$, t_i are elements of $S^0 = S \cap \text{Kernel } \varepsilon$ for $3 \leq i \leq n$. For $a \in K^S$ and $b \in K$, we have

$$t(ab) = t(a)b + at(b) + \sum_3^n {}_it(a)t_i(b) = \varepsilon(t)ab + at(b) + \sum_3^n \varepsilon({}_it)at_i(b)$$

$$= a\left(\sum_1^n \varepsilon({}_it)t_i\right)(b) = at(b),$$

so that $t \in T \cap H(K/K^S) = T \cap H(K/K^S)^S = T \cap K^S\langle S \rangle$ (see 6.3.6). Since t is semisimple, $t \in \langle S \rangle$ by 6.3.6. □

6.3.8 Theorem. Let K/k be a finite dimensional radical extension and let S and T be diagonalizable toral k-subcorings of $H(K/k)$ such that $S \subset T$ and every element of T^0 is primitive modulo S (see 5.3.16). Let $\mathbf{R} = \{\alpha \in T^* \mid K_\alpha(T) \neq \{0\}\}$ and let $\mathbf{R}^S = \{\alpha \in \mathbf{R} \mid K_\alpha{}^S(T) \neq \{0\}\}$ where $K_\alpha{}^S(T) = K^S \cap K_\alpha(T)$ for $\alpha \in \mathbf{R}$. Then $\mathbf{R}^S = \{\alpha \in T_\pi^* \mid (\alpha - \varepsilon_T)(t) = 0 \text{ for } t \in T_\pi \cap \langle S \rangle\}$ where $T_\pi^* = \{\alpha \in T^* \mid \alpha(T_\pi) \subset T_\pi\}$. And $K^S = \sum_{\alpha \in \mathbf{R}^S} K_\alpha{}^S(T)$ (direct), $K_\alpha{}^S(T)K_\beta{}^S(T) = K_{\alpha*\beta}^S(T)$ for $\alpha, \beta \in \mathbf{R}^S$ and $K_\alpha{}^S(T) = K^T a_\alpha$ for $a_\alpha \in K_\alpha{}^S - \{0\}$ and $\alpha \in \mathbf{R}^S$.

Proof. We have $K = \sum_{\alpha \in \mathbf{R}} K_\alpha(T)$ (direct), $K_\alpha(T)K_\beta(T) = K_{\alpha*\beta}(T)$ for $\alpha, \beta \in \mathbf{R}$ and $K_\alpha(T) = K^T a_\alpha$ for $a_\alpha \in K_\alpha(T) - \{0\}$ and $\alpha \in \mathbf{R}$ (see 6.2.6 and 6.2.9). Since K^S is a T-stable K^T-subspace of K, it follows that

$$K^S = \sum_{\alpha \in \mathbf{R}^S} K_\alpha{}^S(T) \quad \text{and} \quad K_\alpha{}^S(T) = K_\alpha(T) = K^T a_\alpha \qquad \text{for } \alpha \in \mathbf{R}^S.$$

Moreover, $\mathbf{R}^S = \{\alpha \in \mathbf{R} \mid (\alpha - \varepsilon_T)(t) = 0 \text{ for } t \in T \cap \langle S \rangle\}$, since

$$\alpha \in \mathbf{R}^S \Leftrightarrow K_\alpha{}^S(T) \neq \{0\} \Leftrightarrow a_\alpha \in K^S \Leftrightarrow \alpha(t)a_\alpha = t(a_\alpha) = \varepsilon_T(t)a_\alpha$$
$$\text{for } t \in T \cap \langle S \rangle \Leftrightarrow (\alpha - \varepsilon_T)(t) = 0$$
$$\text{for } t \in T \cap \langle S \rangle.$$

Next, let $\alpha, \beta \in \mathbf{R}^S$. Letting $t \in T^0 = T \cap \text{Kernel } \varepsilon$ and $\Delta t = t \otimes I + I \otimes t + \sum_n^3 {}_it \otimes t_i$ where the ${}_it$, t_i are elements of S^0, we have $(\alpha * \beta)(t) = \sum_1^n \alpha({}_it)\beta(t_i) = \alpha(t) + \beta(t) + \sum_3^n \varepsilon({}_it)\varepsilon(t_i) = (\alpha + \beta)(t) + 0$. It follows that $\alpha * \beta = \alpha + \beta - \varepsilon_T$ for $\alpha, \beta \in \mathbf{R}^S$. Since $\mathbf{R}^S$ is a subset of T_π^* and $\alpha * \beta \in \mathbf{R}^S$

for $\alpha, \beta \in \mathbf{R}^S$, it follows that $\mathbf{R}^S - \varepsilon_T = \{\alpha - \varepsilon_T \mid \alpha \in \mathbf{R}^S\}$ is a π-subspace of T_π^*. Thus, $\mathbf{R}^S - \varepsilon_T = (\mathbf{R}^S - \varepsilon_T)^{\perp\perp}$ where

$$\begin{aligned}(\mathbf{R}^S - \varepsilon_T)^\perp &= \{t \in T_\pi \mid (\alpha - \varepsilon_T)(t) = 0 \text{ for } \alpha \in \mathbf{R}^S\}\\ &= \{t \in T_\pi \mid t \in H(K/K^S)\}\\ &= T_\pi \cap \langle S\rangle \quad \text{(see 6.3.6)}\end{aligned}$$

and $(\mathbf{R}^S - \varepsilon_T)^{\perp\perp} = ((\mathbf{R}^S - \varepsilon_T)^\perp)^\perp = (T_\pi \cap \langle S\rangle)^\perp = \{\beta \in T_\pi^* \} \beta(t) = 0$ for $t \in T_\pi \cap \langle S\rangle\}$. Thus, $\mathbf{R}^S - \varepsilon_T = \{\beta \in T_\pi^* \mid \beta(t) = 0$ for $t \in T_\pi \cap \langle S\rangle\}$ and $\mathbf{R}^S = \{\alpha \in T_\pi^* \mid (\alpha - \varepsilon_T)(t) = 0$ for $t \in T_\pi \cap \langle S\rangle\}$. □

6.3.9 Definition. For any k-subspaces H and T of $H(K/k)$, we let $H_\alpha(T) = \{x \in H \mid tx - xt = \alpha(t)x$ for all $t \in T\}$ for $\alpha \in T^*$ (k-dual space of T) and $H^T = H_0(T) = \{x \in H \mid tx = xt$ for all $t \in T\}$.

6.3.10 Theorem. Let K/k be a finite dimensional radical extension and let S and T be diagonalizable toral k-subcorings of $H(K/k)$ such that $S \subset T$ and every element of T^0 is primitive modulo S. Let $H = H(K/k)$ and $H_\alpha{}^S(T) = H_\alpha(T) \cap H^S$ for $\alpha \in T^*$. Let $\mathbf{R}^S = \{\alpha \in T^* \mid K_\alpha{}^S(T) \neq \{0\}\}$ and note that $\alpha * 0 = \alpha - \varepsilon_T$ for $\alpha \in T^*$ where 0 is the zero element of T^* (see 6.3.8). Then $H^S = \sum_{\alpha\in\mathbf{R}^S} H_{\alpha*0}^S(T)$ (direct), $K_\alpha{}^S(T)H_\beta{}^S(T) = H_{\alpha*\beta}^S(T)$ for $\alpha \in \mathbf{R}^S$, $\beta \in \mathbf{R}^S * 0$ and $H_{\alpha*0}^S = a_\alpha H^T$ for $a_\alpha \in K_\alpha{}^S - \{0\}$ and $\alpha \in \mathbf{R}^S$.

Proof. For $t \in T$, let $ad\,t: H^S \to H^S$ be the linear transformation of H^S defined by $ad\,t(x) = tx - xt$ for $x \in H^S$ (see E.4.9). Then $ad\,s\,ad\,t = ad\,t\,ad\,s$ for $s, t \in T$ and $ad\,t^p = (ad\,t)^p$ for $t \in T$, so that $adT = \{ad\,t \mid t \in T\}$ is a k-torus on H^S (see E.4.9). Since T is the k-span of T_π and since $ad(T_\pi) \subset (adT)_\pi$, adT is the k-span of $(adT)_\pi$ and adT is diagonalizable on H^S (see 6.1.13). Letting $\mathbf{Q} = \{\beta \in T^* \mid H_\beta{}^S(T) \neq \{0\}\}$, we therefore have $H^S = \sum_{\beta\in\mathbf{Q}} H_\beta{}^S(T)$ (direct). For $a \in K_\alpha{}^S(T)$, $x \in H_\beta{}^S(T)$, $t \in T^0$ and

$$\Delta t = \sum_i {}_it \otimes t_i = t \otimes I + I \otimes t + \sum_3^n {}_it \otimes t_i$$

where the ${}_it$, t_i are elements of S^0 for $i \geq 3$, we have

$$\begin{aligned} ad\,t(ax) &= t(ax) - axt = t(a)x + atx + \sum_3^n {}_it(a)t_ix - axt\\ &= t(a)x + a(ad\,t(x)) + \sum_3^n 0\\ &= \alpha(t)ax + \beta(t)ax + \left(\sum_3^n \alpha({}_it)\beta(t_i)\right)ax\\ &= (\alpha * \beta)(t)ax.\end{aligned}$$

It follows easily that $K_\alpha{}^S(T)H_\beta{}^S(T) \subset H_{\alpha*\beta}^S(T)$ for $\alpha, \beta \in T^*$. In particular, $\mathbf{R}^S * \beta \subset \mathbf{Q}$ for any $\beta \in \mathbf{Q}$, where $\mathbf{R}^* * \beta = \{\alpha * \beta \mid a \in \mathbf{R}^*\}$ for $\beta \in T^*$. Next, note that for any $t \in T_\pi \cap \langle S\rangle$, we have $0 = ad\,t(x) = \beta(t)x$ for all $x \in H_\beta{}^S(T)$, so that $\beta(t) = 0$ for $\beta \in \mathbf{Q}$. Thus, $\mathbf{Q} \subset \mathbf{R}^S - \varepsilon_T = \mathbf{R}^S * 0$ (see 6.3.8). The number of elements in the sets $\mathbf{R}^S$, $\mathbf{R}^S * 0$ and $\mathbf{R}^S * \beta$ ($\beta \in \mathbf{Q}$) are the same, since $\alpha * \beta = \alpha + \beta - \varepsilon_T$ for $\alpha \in \mathbf{R}^S$. Thus, $\mathbf{R}^S * \beta = \mathbf{Q} = \mathbf{R}^S * 0$ for $\beta \in \mathbf{Q}$, and $H^S = \sum_{\alpha \in \mathbf{R}^S} H^S_{\alpha * 0}$ (direct). Letting $a_\alpha \in K_\alpha{}^S(T) - \{0\}$ and $\alpha \in \mathbf{R}^S$, we have

$$H^S = a_\alpha H^S = \sum_{\beta \in \mathbf{R}^S * 0} a_\alpha H_\beta{}^S(T) \subset \sum_{\beta \in \mathbf{R}^S * 0} H^S_{\alpha * \beta}(T) = H^S.$$

It follows that $a_\alpha H_\beta{}^S(T) = H^S_{\alpha*\beta}(T)$, hence that $K_\alpha{}^S(T) H_\beta{}^S(T) = H^S_{\alpha*\beta}(T)$ for all $\beta \in \mathbf{R}^S * 0$. In particular, $a_\alpha H^T = a_\alpha H_0{}^S(T) = H^S_{\alpha*0}(T)$. $\square$

6.3.11 Theorem. Let K/k be a finite dimensional radical extension and let T be any toral k-subcoring in $H(K/k)$. Then $H(K/k)^T$ is a K^T-subbiring of $H(K)$ and $H(K/k) = KH(K/k)^T$ (K-span of $H(K/k)^T$).

Proof. Let $H = H(K/k)$. We first consider the case in which T is diagonalizable. Since K/k is radical, T is coradical (see 6.3.2) and has a p-filtration T_i (see 6.3.3). We prove by induction on n that H^{T_n} is a K^{T_n}-subbiring of $H(K)$ and $H = KH^{T_n}$, which suffices since $T = T_n$ for some n. If $n = 0$, this is clear. Next, let $n > 0$ and $S = T_{n-1}$, and suppose that H^S is a K^S-subbiring of $H(K)$ and $H = KH^S$. Following the notation of 6.3.8 and 6.3.10 with T_n in the place of T, we have $K^S = \sum_{\alpha \in \mathbf{R}^S} K^{T_n} a_\alpha$ (direct) and $H^S = \sum_{\alpha \in \mathbf{R}^S} a_\alpha H^{T_n}$ (direct), so that H^{T_n} is a K^{T_n}-form of H^S as K^S-vector space (see 3.2.3 and 6.2.2). Since H^S is a K^S-form of H as K-vector space, it follows that H^{T_n} is a K^{T_n}-form of H as K-vector space. Obviously, H^{T_n} is a subring of $H(K)$. We now let $x \in H$ and describe ${}_ix$, x_i in H^{T_n} such that $\Delta x = \sum_i {}_ix \otimes x_i$. For this, let $a_1, \ldots, a_m$ be a k-basis for K contained in $\bigcup_{\alpha \in T_n^*} K_\alpha(T_n)$ and let $x_1, \ldots, x_m$ be the K-basis for $H = \mathrm{End}_k\, K$ such that $x_i(a_j) = \delta_{ij} a_j$ ($1 \leq i, j \leq m$). One sees easily that the $K_\alpha(T_n)$ ($\alpha \in T_n^*$) are stable under the $x_1, \ldots, x_m$, so that the $x_1, \ldots, x_m$ are elements of H^{T_n}. Since H^{T_n} is a K^{T_n}-form of H, $x_1, \ldots, x_m$ is a K^{T_n}-basis for H^{T_n}. Let $x \in H^{T_n}$ and let ${}_1x, \ldots, {}_mx$ be elements of H such that $\Delta x = \sum_i {}_ix \otimes x_i$. Since $H^{T_n} \subset H^S$, we have $x \in H^S$. It follows that the ${}_1x, \ldots, {}_mx$ are elements of H^S (see E.5.4). Let $t \in T_n{}^0$ and $\Delta t = t \otimes I + I \otimes t + \sum_3^\infty {}_it \otimes t_i$ where the ${}_it$, t_i are elements of S^0 for $i \geq 3$. Then

$$\begin{aligned} t(x(ba_i)) &= \sum_j t({}_jx(b) x_j(a_i)) = t({}_ix(b) a_i) \\ &= t({}_ix(b)) a_i + {}_ix(b) t(a_i) + \sum_r {}_rt({}_ix(b)) t_r(a_i) \end{aligned}$$

and

$$\begin{aligned} t(x(ba_i)) &= x(t(ba_i)) = x(t(b)a_i) + x(bt(a_i)) + \left(\sum_r {}_rt(b) t_r(a_i)\right) \\ &= \sum_j {}_jxt(b) x_j(a_i) + \sum_j {}_jx(b) x_j t(a_i) + \sum_r \sum_j {}_jx_r t(b) x_j t_r(a_i) \\ &= {}_ixt(b) a_i + {}_ix(b) t(a_i) + \sum_r {}_rt_i x(b) t_r(a_i) \end{aligned}$$

for all $b \in K$ and $1 \leq i \leq m$, since $x_j t = t x_j$, $x_{jr} t = {}_rt x_j$ and ${}_jx_r t = {}_rt_j x$ for all j, r. It follows that $t_i x = {}_ixt$ for all $t \in T_n$ and $1 \leq i \leq m$. Thus, the ${}_1x, \ldots, {}_mx$

are elements of H^{T_n} as desired. It remains only to show that $\varepsilon(H^{T_s}) \subset K^{T_n}$. For this, let $x \in H^{T_n}$, $t \in T_n$ and $b \in K$. Then

$$\begin{aligned} t(x(1)b) &= \sum_i {}_it(x(1))t_i(b) = \sum_i x({}_it(1))t_i(b) \\ &= \sum_i x(1)_it(1)t_i(b) = x(1)t(b) \end{aligned}$$

since ${}_it(1) \in k$ for all i. Thus, $x(1) \in K^{T_n}$ for $x \in H^{T_n}$, so that $\varepsilon(H^{T_n}) \subset K^{T_n}$. We have now shown that H^{T_n} is a K^{T_n}-subbiring of $H(K)$ for all n, hence that H^T is a K^T-subbiring of $H(K)$ for any diagonalizable toral k-subcoring T of $H(K/k)$.

We finally drop the assumption that T be diagonalizable. Following the terminology of 6.3.4, $\overline{T}$ is a diagonalizable $\overline{k}$-subcoring of $\overline{H} = H(\overline{K}/\overline{k})$, $\overline{H}^{\overline{T}}$ is therefore a $\overline{K}^{\overline{T}}$-subbiring of $H(\overline{K})$ and $\overline{H} = \overline{K}\overline{H}^{\overline{T}}$. Since $H^T = (\overline{H}^{\overline{T}})^G$ and $K^T = (\overline{K}^{\overline{T}})^G$ where $G = \operatorname{Aut}_k L = \operatorname{Aut}_k \overline{k} = \operatorname{Aut}_K \overline{K}$, H^T is a K^T-form of $\overline{H}^{\overline{T}}$ as $\overline{K}^{\overline{T}}$-space. Consequently, H^T is a K^T-form of $\overline{H}$ as $\overline{K}$-space, hence a K^T-form of H as K-space. Let $x_1, \ldots, x_m$ be a K^T-basis for H^T, hence a K-basis for H, hence a $\overline{K}$-basis for $\overline{H}$. Let $x \in H^T$ and choose ${}_1x, \ldots, {}_mx \in \overline{H}^{\overline{T}}$ such that $\Delta x = \sum_i {}_ix \otimes x_i$. For $g \in G$, we then have

$$\sum_i {}_ix \otimes x_i = \Delta x = \Delta g(x) = \sum_i g({}_ix) \otimes g(x_i) = \sum_i g({}_ix) \otimes x_i$$

(see E.6.8). It follows that ${}_ix = g({}_ix)$ for $g \in G$, so that ${}_ix \in (\overline{H}^{\overline{T}})^G = H^T$ for $1 \le i \le m$. Next note that $\varepsilon(H^T) \subset \varepsilon(\overline{H}^{\overline{T}}) \cap K \subset \overline{K}^{\overline{T}} \cap K = K^T$. Since H^T is a subring of $H(K)$, we have therefore proved that H^T is a K^T-subbiring of $H(K)$ and $H = KH^T$. $\square$

6.4 Radical extensions

The theory of toral subbirings developed thus far can be used in studying the structure of radical field extensions. It is clear from 6.2.10 that a finite dimensional radical field extension K/k splits as a tensor product over k of simple extensions of k if and only if $K^T = k$ for some diagonalizable toral k-subbiring T of $H(K/k)$. For an arbitrary finite dimensional radical extension K/k and an arbitrary toral k-subbiring T of $H(K/k)$, we have seen that $H(K/k)^T$ is a K^T-subbiring of $H(K/k)$ and that $H(K/k) = KH(K/k)^T$ (see 6.3.11). Also, $H(K/k^T) = KT$ (see 6.2.11). Finally, the K^T-biring $H(K^T/k)$ is a homomorphic image of $H(K/k)^T$, as we show in 6.4.1. Thus, $H(K/k)$ is completely determined by the K^T-subbiring $H(K/k)^T$ and its action on K, $H(K/K^T)$ is completely determined by T and its action on K and $H(K^T/k)$ is completely determined by $H(K/k)^T$ and its action on K^T. This shows that K/k can be effectively studied by studying K/K^T in terms of T (see 6.2 and 6.3.4) and K^T/k in terms of $H(K/k)^T$. Since a nontrivial radical extension K/k has some nonzero toral k-subbiring T (see E.6.9), these observations enable one to study K/k inductively in terms of the proper subextensions K/K^T and

K^T/k. This approach to studying radical extensions is illustrated in the proof of 6.4.2.

In the extreme case where a finite dimensional radical extension K/k has a toral k-subbiring T such that $K^T = k$, we know from 6.2.10 and 6.3.4 that $\bar{K}$ is a tensor product of certain simple extensions of $\bar{k}$ where $\bar{K} = L \otimes_k K$, $\bar{k} = \mathrm{L} \otimes_k k$ and L is the separable closure of k. Not every finite dimensional radical extension K/k has this property (see E.6.30). Thus not every finite dimensional radical extension K/k has a toral k-subbiring T such that $K^T = k$. It is not known whether every finite dimensional radical extension K/k has some k-subbiring H_k such that $K^{H_k} = k$, hence such that $H(K/k) = KH_k$. It has been shown by Moss Sweedler, however, that there exists a K-measuring k-bialgebra $H_k(K)$ such that $K^{H_k(K)} = k$. (See E.6.18, E.6.22.)

We now compare the K^T- birings $H(K^T/k)$ and $H(K/k)^T$.

6.4.1 Theorem. Let K/k be a finite dimensional radical extension and let T be a toral k-subbiring of $H(K/k)$. Then K^T is stable under $H(K/k)^T$ and the restriction mapping $x \mapsto x|_{K^T}$ from $H(K/k)^T$ to $H(K^T/k)$ is a surjective homomorphism of k-algebras and K^T-coalgebras.

Proof. It is clear that K^T is stable under $H(K/k)^T$ and that $x \mapsto x|_{K^T}$ is a K^T-linear k-algebra homomorphism. Let Δ_{K^T} be the coproduct for $H(K/k)^T$ as K^T-coalgebra. Then if $x \in H(K/k)^T$ and $\Delta_{K^T}(x) = \sum_i {}_ix \otimes_{K^T} x_i$, we have $x(ab) = \sum_i {}_ix(a)x_i(b)$ for all $a, b \in K$. Thus $x|_{K^T}(ab) = \sum_i {}_ix|_{K^T}(a)x_i|_{K^T}(b)$ for all $a, b \in K^T$, so that $\Delta(x|_{K^T}) = \sum_i {}_ix|_{K^T} \otimes_{K^T} x_i|_{K^T}$. Thus, $x \mapsto x|_{K^T}$ is a K^T-coalgebra homomorphism. To show that $x \mapsto x|_{K^T}$ is surjective from $H(K/k)^T$ to $H(K^T/k)$, let H be the subbiring $H = \{x|_{K^T} \mid x \in H(K/k)^T\}$ of $H(K^T/k)$. Since $H(K/k) = KH(K/k)^T$, we have $K^{H(K/k)^T} = k$. It follows that $(K^T)^H = k$, so that $H = H(K^T/k)$, by 5.3.12. □

We conclude with a generalization of the theorem of Jacobson which states that if K is a field and $\mathscr{D}$ a K-sub Lie ring of Der K which is finite dimensional over K, then $K/K^{\mathscr{D}}$ is finite dimensional (see 5.2). The proof illustrates how the foregoing ideas can be applied to the study of radical extensions.

6.4.2 Theorem. Let K be a field and let C be a coradical K-subcoring of $H(K)$ such that $xy - yx \in C$ and $x^p \in C$ for all $x, y \in C$. Then if C is finite dimensional over K, K is finite dimensional over K^C.

Proof. Suppose that $C:K < \infty$ and let $K^C = k$. Since C is coradical, K/k is radical (see 6.3.2). Let L be the separable closure of k, $\bar{K} = L \otimes_k K$, $\bar{k} = L \otimes_k k$, $\bar{C} = L \otimes_k C$. Then $\bar{K}$ is a field, $\bar{C}$ is a coradical $\bar{K}$-subcoring of $H(\bar{K})$ such that $[x, y] = xy - yx \in \bar{C}$ and $x^p \in \bar{C}$ for all $x, y \in \bar{C}$ (see E.6.31), $\bar{C}:\bar{K} < \infty$ and $\bar{K}^{\bar{C}} = \bar{k}$. To show that K/k is finite dimensional, it suffices to show that $\bar{K}/\bar{k}$ is finite dimensional. Thus, we may replace K/k by $\bar{K}/\bar{k}$, that is, we may suppose that k is separably closed.

We proceed now by induction on the dimension of C over K. If $C = \{0\}$, then the theorem is certainly true for C. Suppose next that $C \neq \{0\}$ and that the theorem is true for dimensions less than the dimension of C over K. Since C is nonzero and coradical, C has a nonzero primitive element x. Since $\Delta x = x \otimes I + I \otimes x$, we have $x \in \mathrm{Der}\, K$. Choose $a \in K$ such that $x(a) \neq 0$ and let $y = ax(a)^{-1}x$, so that $y(a) = a$. Then $y \in \mathrm{Der}\, K$. Since $y \in C$ and $C:K < \infty$, the K-sub Lie ring $\mathscr{D}$ of Der K generated by y is finite dimensional over K and the extension $K/K^{\mathscr{D}}$ corresponding to $\mathscr{D}$ under the Jacobson Differential Correspondence is finite dimensional (see 5.2). Let $k' = K^{\mathscr{D}}$ and choose e such that y^{p^e} is a semisimple linear transformation of K/k' (see 6.1.6). Note that $t = y^{p^e}$ is nonzero, since $t(a) = a \neq 0$. Let T be the k'-span of the powers $t, t^p, t^{p^2}, \ldots$ of t, so that T is a k'-torus on the finite dimensional vector space K over k'. Since k is separably closed, k' is separably closed. Thus, T is diagonalizable over k' (see 6.1.8), so that T is the k'-span of $T_\pi = \{t \in T \mid t^p = t\}$ (see 6.1.13). We now imitate some of the material in 6.3 in order to show that the centralizer C^{T_π} of T_π in C is a K^{T_π}-form of C. We have $K = \sum_{\alpha \in \mathbf{R}} K_\alpha(T_\pi)$ where $\mathbf{R} = T_\pi^*$ (π-dual space of the π-space T_π) (see 6.1.12). The set $\mathbf{Q} = \{\alpha \in \mathbf{R} \mid K_\alpha(T_\pi) \neq \{0\}\}$ is nonempty. One shows easily that $K_\alpha(T_\pi)K_\beta(T_\pi) \subset K_{\alpha+\beta}(T_\pi)$ for $\alpha, \beta \in \mathbf{R}$, since $T_\pi \subset \mathrm{Der}\, K$. It follows that $\mathbf{Q}$ is closed under $+$ and therefore that $\mathbf{Q}$ is a π-subspace of $\mathbf{R} = T_\pi^*$. Since $\mathbf{Q}$ separates the points of T_π, it follows that $\mathbf{Q} = \mathbf{R}$ and $K_\alpha(T_\pi) \neq 0$ for all $\alpha \in \mathbf{R}$. Moreover, we have $K_\alpha(T_\pi) = K^{T_\pi}x_\alpha$ for $\alpha \in \mathbf{R}$ and $x_\alpha \in K_\alpha(T_\pi) - \{0\}$ (compare with 6.3.8). For $u \in C$, let $adu\colon C \to C$ be defined by $adu(v) = [u, v] = uv - vu$. Then adu is a π-linear transformation of C and $(adu)^p = adu^p$ for all $u \in C$ (see E.4.9). Since $t^p = t$ for $t \in T_\pi$, it follows that $(adt)^p = adt$ for $t \in T_\pi$. Since the elements of adT_π commute pairwise, the elements of $adT_\pi = \{adt \mid t \in T_\pi\}$ commute pairwise. It follows easily from 6.1.12 that $C = \sum_{\beta \in \mathbf{R}} C_\beta(T_\pi)$ (direct) where $C_\beta(T_\pi) = \{v \in C \mid adt(v) = \beta(t)v$ for all $t \in T_\pi\}$ for $\beta \in \mathbf{R}$. Let $\alpha, \beta \in \mathbf{R}$, $x_\alpha \in K_\alpha(T_\pi) - \{0\}$ and $v_\beta \in C_\beta(T_\pi)$. Then $x_\alpha v_\beta \in C_{\alpha+\beta}(T_\pi)$, for we have

$$\begin{aligned}[t, x_\alpha v_\beta] &= t(x_\alpha v_\beta) - (x_\alpha v_\beta)t = t(x_\alpha)v_\beta + x_\alpha t v_\beta - x_\alpha v_\beta t \\ &= t(x_\alpha)v_\beta + x_\alpha[t, v_\beta] = \alpha(t)x_\alpha v_\beta + x_\alpha\beta(t)v_\beta \\ &= (\alpha + \beta)(t)x_\alpha v_\beta.\end{aligned}$$

It follows easily that $x_\alpha C_\beta(T_\pi) = C_{\alpha+\beta}(T_\pi)$ for all $\alpha, \beta \in \mathbf{R}$, thus that $C_\beta(T_\pi) \neq \{0\}$ for all $\beta \in \mathbf{R}$, thus that $C_\beta(T_\pi) = x_\beta C^{T_\pi}$ for all $\beta \in \mathbf{R}$, thus that C^{T_π} is a K^{T_π}-form of C as vector space over K (compare with 6.3.10). One now shows, as in the proof of 6.3.11, that C^{T_π} is a K^{T_π}-subcoring of C such that $C = KC^{T_\pi}$. Furthermore, we have $[u, v] \in C^{T_\pi}$ and $u^p \in C^{T_\pi}$ for all $u, v \in C^{T_\pi}$. Letting $K' = K^{T_\pi}$, the mapping $f(u) = u|_{K'}$ maps the K'-subcoring C^{T_π} of $H(K)$ homomorphically to a K'-subcoring

$$C' = \{u|_{K'} \mid u \in C^{T_\pi}\}$$

of $H(K')$ such that $[u, v] \in C'$ and $u^p \in C'$ for all $u, v \in C'$ (compare with 6.4.1). Since $T_\pi \subset C^{T_\pi} \cap \mathrm{Der}_{K'} K$, we have $f(T_\pi) = 0$. Thus,

$$C':K' < C^{T_\pi}:K' = C^{T_\pi}:K^{T_\pi} = C:K.$$

We may therefore apply the induction hypothesis to C' and K' and conclude that K' is finite dimensional over $K'^{C'}$. Since $K/K' = K/K^{T_\pi}$ is finite dimensional, it follows that K is finite dimensional over $K'^{C'}$. But $K'^{C'} = K^C$, since $C = KC^{T_\pi} = KC'$. Thus, K is finite dimensional over K^C. □

E.6 Exercises to Chapter 6

E.6.1. Let t be a linear transformation of a vector space V over an algebraically closed field k of characteristic $p > 0$. Show that t^{p^e} is diagonalizable for some e. (Hint: Consider the Jordan Canonical Form.)

E.6.2. Show that if S is a set of pairwise commuting semisimple linear transformations of a finite dimensional vector space V over a field k of characteristic $p > 0$, then $\langle S \rangle_p$ (defined following 6.1.7) is a k-torus on V. Show that $\langle S \rangle_p$ is diagonalizable if every element of S is diagonalizable. (Hint: The eigenspaces of any element s of S are invariant under S.)

E.6.3. Let K/k be a finite dimensional radical extension. Show that for any separable algebraic extension L of k, there is a bijective k-linear mapping from $L \otimes_k \operatorname{End}_k K$ to $\operatorname{End}_{L\otimes_k k} L \otimes_k K$ mapping $x \otimes T$ to $(x id_L) \otimes T$ for $x \in L$, $T \in \operatorname{End}_k K$.

E.6.4. Under what conditions on a finite dimensional extension K/k is KI (K-span of $I \in \operatorname{End}_k K$) a k-torus? Show that the diagonalizability of T in 6.2.1 is essential.

E.6.5. Let V be a finite dimensional vector space over a field k of characteristic $p > 0$. Let T be a k-subspace of $\operatorname{Hom}_k V$ consisting of pairwise commuting linear transformations of V and suppose that $t^p \in T$ for all $t \in T$. Then any subspace S of T such that for each $t \in T$, $t^{p^e} \in S$ for some e contains every semisimple element of T.

E.6.6. Verify that the $\langle T_i \rangle_p$ following 6.3.2 are toral k-subcorings.

E.6.7. Let C be a subcoring of $H(K/k)$ and let $t \in C$, $a \in K$. Show that if $t(ab) = at(b)$ for all $b \in K$, then $t(a) = \varepsilon(t)a$.

E.6.8. Let K/k be a finite dimensional radical extension and let L/k be the separable closure of k. Following the identifications and notation introduced following 6.3.3 and letting $g \in \operatorname{Aut}_k L$, $x \in \overline{H(K/k)} = H(\bar{K}/\bar{k})$, show that $\Delta x = \sum_i {}_ix \otimes x_i$ if and only if $\Delta g(x) = \sum_i g({}_ix) \otimes g(x_i)$.

E.6.9. Show that if K/k is a radical extension of dimension greater than one, then there is a nonzero diagonalizable toral k-subbiring in $H(K/k)$.

E.6.10. Describe an algebraic field extension K/k which is radical, but which does not satisfy the condition $\bigcap_i K^{p^i} \subset k$.

E.6.11. A *higher derivation* of K/k is a sequence $D = (D_0, D_1, \ldots, D_m)$ of elements D_i of $\operatorname{End}_k K$ such that $D_0 = I$ and $D_s(ab) = \sum_r D_{s-r}(a) D_r(b)$ for all $a, b \in K$ and $0 \leq s \leq m$. Prove the following for any higher derivation $(D_0, \ldots, D_m)$ of K/k.

(a) The mapping $a \mapsto \sum_0^m D_r(a)\bar{X}^r$ is a k-homomorphism from K into the k-algebra $K[\bar{X}] = K[X]/(X^m K[X])$, $\bar{X}$ being the coset $\bar{X} = X + X^m K[X]$.

(b) The pth power mapping in K and in $K[\bar{X}]$ is a ring homomorphism.
(c) $D_r(a^p) = (D_{r/p}(a))^p$ if $p|r$ and $D_r(a^r) = 0$ if $p \nmid r$. (Hint: Use parts (*a*) and (*b*).)

E.6.12 (Sequence of Divided Powers). In a coalgebra C over k, a *sequence of divided powers* is a sequence $D_0, \ldots, D_m$ of elements of C such that $\Delta D_s = \sum_0^s {}_{s-r}D \otimes D_r$ for $0 \leq s \leq m$. Show that if $D_0, \ldots, D_m$ is a sequence of divided powers, then the k-span of $D_0, \ldots, D_m$ is a colocal cocommutative subcoalgebra D of C and for any measuring representation $f: D \to \mathrm{End}_k K$ of D on a field extension K/k mapping D_0 to I, the sequence $f(D_0), f(D_1), \ldots, f(D_m)$ is a higher derivation on K.

E.6.13. Let K/k be a finite dimensional field extension and suppose that k is the subfield of constants of the set of higher derivations of K over k, that is, for each $a \in K - k$, there exists a higher derivation $D = (D_0, D_1, \ldots, D_m)$ such that $D_i(a) \neq 0$ for some i with $1 \leq i \leq m$. Show that K/k is a radical extension. (Hint: Show for a fixed higher derivation $D = (D_0, D_1, \ldots, D_m)$ that K is a radical extension of the field $K^D = \{c \in K \mid D_i(c) = 0 \text{ for } 1 \leq i \leq m\}$ by showing that the K-span C of $D_0, D_1, \ldots, D_m$ is a coradical colocal subcoring of $H(K/k)$. Or simply analyze the proof of 5.3.20 and prove directly that K/K^D is radical).

E.6.14 (Sweedler). Let K/k be a finite dimensional field extension and suppose that k is the subfield of constants of the set of higher derivations of K/k. Show that K is the tensor product over k of simple extensions of k by proving the following.

(a) Show that if $D = (D_0, \ldots, D_m)$ is a higher derivation of K/k, then the D_j map the subfield $K^{p^i} = \{x^{p^i} \mid x \in K\}$ into itself for $i \geq 0$. (Hint: Use E.6.11.)
(b) The fields k and K^{p^i} are linearly disjoint in K over $k \cap K^{p^i}$ for all $i \geq 0$. (Hint: Suppose the contrary, and take linearly independent elements $c_1, \ldots, c_t$ of k over $k \cap K^{p^i}$ and elements $a_1, \ldots, a_t$ of K^{p^i} (not all zero) such that $c_1a_1 + \ldots + c_ta_t = 0$. Do this in such a way that t is as small as possible. Note that all the a_i are nonzero and that a_t can be taken to be 1. Take j such that $a_j \notin k$ and choose a higher derivation $D = (D_0, \ldots, D_m)$ such that $D_i(a_j) \neq 0$ for some $i \geq 1$. Show that $c_1D_i(a_1) + \cdots + c_{t-1}D_i(a_{t-1}) = 0$ contradicts the initial supposition.)
(c) If elements $x_1, \ldots, x_r$ of K are linearly independent over $\sqrt[p^i]{k} = \{x \in K \mid x^{p^i} \in k\}$, then $x_1^{p^i}, \ldots, x_r^{p^i}$ are linearly independent over k. (Hint: If $x_1^{p^i}, \ldots, x_r^{p^i}$ are linearly dependent over k, then they are linearly dependent over $k \cap K^{p^i}$, by part (*b*)).
(d) If $y_1, \ldots, y_s$ are p-independent elements of $\sqrt[p^{i+1}]{k}$ over $\sqrt[p^i]{k}$, then $y_1^{p^i}, \ldots, y_s^{p^i}$ are p-independent elements of $\sqrt[p]{k}$ over k. (Hint: Using 4.3.5, 5.2.1 or E.4.7, relate the p-independence of $y_1, \ldots, y_s$ over k' to the k'-independence of the monomials $x_r = y_1^{e_1} \cdots y_s^{e_r}$ with $0 \leq e_i \leq p - 1$ for $1 \leq i \leq s$. Then use part (c).)

(e) If $y_1, \ldots, y_s$ are p-independent elements of $\sqrt[p^{i+1}]{k}$ over $\sqrt[p^i]{k}$, then $y_1{}^p, \ldots, y_s{}^p$ are p-independent elements of $\sqrt[p^i]{k}$ over $\sqrt[p^{i-1}]{k}$. (Hint: Use part (d).)

(f) Choosing m such that $K = \sqrt[p^m]{k} \gneqq \sqrt[p^{m-1}]{k} \gneqq \cdots \gneqq \sqrt[p]{k} \gneqq k$, there exist elements $y_1, \ldots, y_{s_m}, y_{s_m+1}, \ldots, y_{s_{m-1}}, \ldots, y_{s_2+1}, \ldots, y_{s_1}$ of K such that

$y_1, \ldots, y_{s_m}$ is a p-basis for K over $\sqrt[p^{m-1}]{k}$;

$y_1{}^p, \ldots, y_{s_m}{}^p, y_{s_m+1}, \ldots, y_{s_{m-1}}$ is a p-basis for $\sqrt[p^{m-1}]{k}$ over $\sqrt[p^{m-2}]{k}$;

$\vdots$

$y_1^{p^{m-1}}, \ldots, y_{s_m}^{p^{m-1}}, y_{p_m+1}^{p^{m-2}}, \ldots, y_{p_{m-1}}^{p^{m-2}}, \ldots, y_{s_2+1}, \ldots, y_{s_1}$,

is a p-basis for $\sqrt[p]{k}$ over k.

(Hint: Use part (e).)

(g) $K = k(y_1) \otimes_k \cdots \otimes_k k(y_{s_1})$ (internal tensor product over k) where the $y_1, \ldots, y_{s_1}$ are as described in part (f).

E.6.15 (Dual of Measuring Representation). Let $\rho: C \to \operatorname{End}_k K$ be a measuring representation of a finite dimensional colocal k-coalgebra C with grouplike element e on an extension K/k and suppose that $\rho(e) = I$ (identity of $\operatorname{End}_k K$). Let A be the dual k-algebra of C, let 1_A be the identity of A and let Nil $A = e^\perp$ be the nil radical of A. Show that

(a) ρ induces a unique k-algebra homomorphism $\alpha: K \to A \otimes_k K$ such that for $y \in K$, $\alpha(y) = \sum_j a_j \otimes y_j$ if and only if $\rho(c)(y) = \sum_j a_j(c) y_j$ for all $c \in C$;

(b) for each $y \in K$, $\alpha(y) = 1_A \otimes y + \sum_j u_j \otimes y_j$ for suitable $u_j \in$ Nil A and $y_j \in K$. (Hint: A is spanned by 1_A and Nil A, so that an expression for $\alpha(y)$ of the above kind exists, the only problem being that the first term is $1_A \otimes z$. Show that $y = z$ by applying $\rho(e)$.)

E.6.16 (Dual Measuring Representation). Let $\alpha: K \to A \otimes_k K$ be a k-algebra homomorphism from a field extension K of k to the tensor product k-algebra of a finite dimensional k algebra A and K. Let C be the dual k-coalgebra of A. Show that α induces a unique measuring representation $\rho: C \to \operatorname{End}_k K$ of C on K such that for $c \in C$ and $y \in K$, $\rho(c)(y) = \sum_j c\,(a_j) y_j$ if and only if $\alpha(y) = \sum_j a_j \otimes y_j$.

E.6.17 (Sweedler). Let K/k be a finite dimensional field extension and let $\sqrt{K}$ be a finite dimensional extension of K such that $K = k((\sqrt{K})^p)$. Suppose that C is a finite dimensional cocommutative colocal k-coalgebra with grouplike element e and let $\rho: C \to \operatorname{End}_k K$ be a measuring representation of C on K such that $\rho(e) = I$. Show that there is a finite dimensional cocommutative colocal k-coalgebra $\sqrt{C}$ with grouplike element $\sqrt{e}$ and measuring representation $\sqrt{\rho}: \sqrt{C} \to \operatorname{End}_k \sqrt{K}$ of $\sqrt{C}$ on $\sqrt{K}$ such that $\sqrt{\rho}\,(\sqrt{e}) = I$ which lifts the action of C on K to the action of $\sqrt{C}$ on $\sqrt{K}$ in the sense that there is a surjective k-coalgebra homomorphism $\tau: \sqrt{C} \to C$ such that for $\sqrt{c} \in \sqrt{C}$ and $y \in K$, $\sqrt{\rho}(\sqrt{c})(y) = \rho(\tau(\sqrt{c}))(y)$. Do this by proving the following.

(a) Let $x_1, \ldots, x_m$ be a p-basis for $\sqrt{K}/K$. Let $X = \{X_1, \ldots, X_m\}$ (set of m commuting indeterminants) and let I be the ideal of $K[X]$ generated by $\{X_i{}^p - x_i{}^p \mid 1 \leq i \leq m\}$. Let $\bar{X}_i = X_i + I$ for $1 \leq i \leq m$ and let $K[\bar{X}] = K[X]/I$. Show that there is a K-linear isomorphism from $K[\bar{X}]$ to $\sqrt{K}$ sending $\bar{X}_i$ to x_i for $1 \leq i \leq m$.

(b) Show that there exists an integer n and elements $u_{ij} \in \operatorname{Nil} A$, $x_{ij} \in \sqrt{K}$ $(1 \leq i \leq m, 1 \leq j \leq n)$ such that $\alpha(x_i{}^p) = 1_A \otimes x_i{}^p + \sum_j u_{ij} \otimes x_{ij}{}^p$ for $1 \leq i \leq m$, A being the dual k-algebra of C and $\alpha: K \to A \otimes_k K$ being the k-algebra homomorphism dual to $\rho: C \to \operatorname{End}_k K$ described in E.6.15.

(c) Let $Y = \{Y_{ij} \mid 1 \leq i \leq m, 1 \leq j \leq n\}$ (set of mn commuting indeterminants) and let J be the ideal of $A[Y]$ generated by $\{Y_{ij}{}^p - u_{ij} \mid 1 \leq i \leq m, 1 \leq j \leq n\}$. Let $\sqrt{u_{ij}} = Y_{ij} + J$ for $1 \leq i \leq m$, $1 \leq j \leq n$ and let $\sqrt{A} = A[Y]/J$. Show that $\beta: A \to \sqrt{A}$, defined by $\beta(a) = a + J$, is an injective k-algebra homomorphism. Identifying A and $\beta(A)$, show that $(\sqrt{u_{ij}})^p = u_{ij}$ for all i, j and show that $\sqrt{A}$ is a split local finite dimensional commutative algebra.

(d) Show that there is an extension of $\alpha: K \to A \otimes_k K$ to a k-algebra homomorphism from $K[X]$ to $\sqrt{A} \otimes_k \sqrt{K}$ mapping X_i to $1_A \otimes x_i + \sum_j \sqrt{u_{ij}} \otimes x_{ij}$ for $1 \leq i \leq m$ and therefore vanishing on I. Thus, show that $\alpha: K \to A \otimes_k K$ has an extension to a k-algebra homomorphism $\sqrt{\alpha}: \sqrt{K} \to \sqrt{A} \otimes_k \sqrt{K}$, by applying part (a).

(e) Let $\sqrt{C}$ be the dual k-coalgebra of $\sqrt{A}$ and let $\tau: \sqrt{C} \to C$ be the surjective k-coalgebra homomorphism induced by $\beta: A \to \sqrt{A}$ and the equations $\tau(\sqrt{c}) = d$ if and only if $a(d) = \sqrt{c}(\beta(a))$ for all $a \in A (\sqrt{c} \in C, d \in C)$. Let $\sqrt{\rho}: \sqrt{C} \to \operatorname{End}_k \sqrt{K}$ be the measuring representation of $\sqrt{C}$ on $\sqrt{K}$ dual to $\sqrt{\alpha}$ (see E.6.16). Show that $\sqrt{\rho}(\sqrt{c})(y) = \rho(\tau(\sqrt{c}))(y)$ for $\sqrt{c} \in \sqrt{C}$ and $y \in K$. Show in particular that $\sqrt{\rho}(\sqrt{e}) = \rho(e) = I$ where $\sqrt{e}$ is the grouplike element of $\sqrt{C}$.

E.6.18. Let K/k be a finite dimensional field extension and let $\sqrt{K}$ be a finite dimensional extension of K such that $K = k((\sqrt{K})^p)$. Let $x_1, \ldots, x_m$ be a p-basis for $\sqrt{K}/K$. Show that

(a) there are derivations $t_1, \ldots, t_m$ of $\sqrt{K}$ over k such that $t_j(x_i) = \delta_{ji} x_i$ for $1 \leq i, j \leq m$;

(b) the k-span T of $I, t_1, \ldots, t_m$ can be regarded as a cocommutative colocal k-coalgebra with grouplike element I such that the inclusion mapping $\delta: T \to \operatorname{End}_k \sqrt{K}$ is a measuring representation;

(c) the commutant $\sqrt{K}^T$ is K.

E.6.19 (Sweedler). Let K/k be finite dimensional radical extension. Show that there exists a finite dimensional cocommutative colocal k-coalgebra C

with grouplike element e and measuring representation $\rho\colon C \to \operatorname{End}_k K$ such that $\rho(e) = I$ and the commutant $K^C = \{x \in K \mid \rho(c)(x) = \varepsilon(c)c \text{ for all } c \in C\}$ is k. Show this as follows.

(a) Let $K_i = k(K^{p^i})$ for $i = 0, 1, 2, \ldots$ and let $\sqrt{K_i} = K_{i-1}$ for $i \geq 1$. Show that $K = K_0 \supset K_1 \supset \cdots \supset K_m = k$ for some m and $K_i = k((\sqrt{K_i})^p)$ for all i.

(b) For each i, there exists a finite dimensional cocommutative k-coalgebra C_i with grouplike element e_i and measuring representation $\rho_i\colon C_i \to \operatorname{End}_k \sqrt{K_i}$ such that $\rho(e_i) = I$ and $\sqrt{K_i}^{C_i} = K_i$. (Use E.6.18.)

(c) For each i, there exists a finite dimensional cocommutative colocal k-coalgebra $\sqrt{\sqrt{C_i}}$ with grouplike element $\sqrt{\sqrt{e_i}}$ and measuring representation $\sqrt{\sqrt{\rho_i}}\colon \sqrt{\sqrt{C_i}} \to \operatorname{End}_k K$ such that $\sqrt{\sqrt{\rho_i}}(\sqrt{\sqrt{e_i}}) = I$ having the property that $K^{\sqrt{\sqrt{C_i}}} \cap \sqrt{K_i} = K_i$. (Use (*b*) and apply E.6.17 over and over again.)

(d) Show that if D_1, D_2 are cocommutative colocal k-coalgebras with grouplike elements f_1, f_2, then the k-span $P = k(f_1 - f_2)$ is a coideal of $D_1 + D_2$ (direct sum k-coalgebra) and $D = (D_1 + D_2)/P$ is a cocommutative colocal k-coalgebra with grouplike element $f = f_1 + P = f_2 + P$. Show that if $\rho_i\colon D_i \to \operatorname{End}_k K$ $(i = 1, 2)$ are measuring representations of D_1 and D_2 such that $\rho_i(f_i) = I$ $(i = 1, 2)$, then $\rho\colon D \to \operatorname{End}_k K$, defined by $\rho(d_1 + d_2) = \rho_1(d_1) + \rho_2(d_2)$ for $d_1 \in D_1$, $d_2 \in D_2$, is a measuring representation of D such that $\rho(f) = I$.

(e) Using parts (c) and (d), show that there is a finite dimensional cocommutative colocal k-coalgebra C and measuring representation $\rho\colon C \to \operatorname{End}_k K$ such that $K^C = k$.

E.6.20 (Sweedler). Let K/k be a finite dimensional normal extension. Prove that there exists a finite dimensional cosplit cocommutative k-coalgebra C and measuring representation $\rho\colon C \to \operatorname{End}_k K$ such that $K^C = k$. (Hint: Let $K = K_1 K_2$ (internal tensor product over k) where K_1/k is Galois and K_2/k is radical. Let C_1 be the k-span of $\operatorname{Aut}_k K_1$ and regard C_1 as group k-coalgebra measuring K_1. Let C_2 be the k-coalgebra measuring K_2 constructed in E.6.19 and take $C = C_1 \otimes_k C_2$.)

E.6.21 (Sweedler). Let A be a k-algebra. A *measuring k-coalgebra* on A is a pair (C, ρ) where C is a k-coalgebra and $\rho\colon C \to \operatorname{End}_k A$ a measuring representation. A *morphism* from a measuring k-coalgebra (C, ρ) to a measuring k-coalgebra (C', ρ') is a k-coalgebra homomorphism $d\colon C \to C'$ such that the diagram

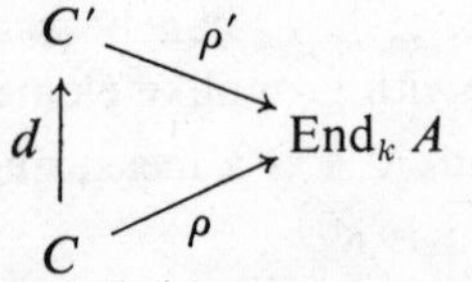

is commutative. A *universal* measuring k-coalgebra on A is a measuring k-coalgebra (C', ρ') on K such that for any measuring k-coalgebra (C, ρ) on

A, there exists a unique morphism φ from (C, ρ) to (C', ρ'). A *universal* cosplit cocommutative k-coalgebra on A is a cosplit cocommutative k-coalgebra (C', ρ') on A such that for any cosplit cocommutative k-coalgebra (C, ρ) on A, there is a unique morphism φ from (C, ρ) to $(', \rho')$.

(a) Given any vector space V over k, show that there is a pair (C', ρ'), called the *free k-coalgebra* on V, where C' is a k-coalgebra and $\rho'\colon C' \to V$ a k-linear mapping such that for any pair (C, ρ) where C is a k-coalgebra and $\rho\colon C \to V$ a k-linear mapping, there is a unique k-coalgebra homomorphism $\varphi\colon C \to C'$ such that the diagram

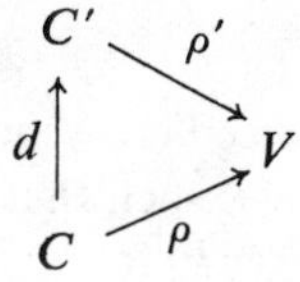

is commutative. (Hint: Let $T(*)$ be the tensor algebra on the dual space V^* of V. Then let C' be the sum of the subcoalgebras D of the dual k-coalgebra $T(V^*)^0$ of $T(V^*)$ such that $i^*(D) \subset V$ where i^* is the transpose $i^*\colon T(V^*)^* \to V^{**}$ of the embedding $i\colon V^* \to T(V^*)$ and where V is embedded canonically in V^{**}.)

(b) There exists a universal measuring k-coalgebra $(M_k(A), \rho_M)$ on A. (Hint: Let $M_k(A)$ be the sum of the subcoalgebras D of the free k-coalgebra (C', ρ') on $\mathrm{End}_k K$ such that $\rho_M|_D\colon D \to \mathrm{End}_k K$ is a measuring representation of D on K).

(c) There exists a universal cosplit cocommutative measuring k-coalgebra $(H_k(A), \rho_H)$ on A. (Hint: Let $H_k(A)$ be the sum of the cosplit cocommutative k-subcoalgebras of $M_k(A)$.)

(d) There is a k-linear mapping $\pi\colon H_k(A) \otimes_k H_k(A) \to H_k(A)$ such that $H_k(A)$ together with π is k-algebra and $H_k(A)$ as k-algebra and k-coalgebra is k-bialgebra. (Hint: Describe a measuring representation of the cosplit cocommutative k-coalgebra $H_k(A) \otimes_k H_k(A)$ and use the universality of $(H_k(A), \rho_H)$ to obtain π.)

E.6.22 (Sweedler). Let K/k be a finite dimensional normal extension. Show that $K^{H_k(K)} = k$. (Hint: Use E.6.20 and the universality of $H_k(K)$).

E.6.23. Let K/k be a finite dimensional normal field extension. Show that

(a) $\rho_H\colon H_k(K) \to \mathrm{End}_k K$ is a measuring representation of $H_k(K)$ as k-bialgebra;

(b) letting $KH_k(K)$ denote the semidirect product K/k-bialgebra of K and $H_k(K)$ with respect to ρ_H, and letting I be the K/k-biideal of elements of $KH_k(K)$ which map into 0 under $I \otimes \rho_H$, the K-measuring K/k-bialgebras $KH_k(K)/I$ and $H(K/k)$ are isomorphic.

E.6.24 (Pickert). Let K/k be any finite dimensional radical extension and let S be a minimal subset such that $K = k(S)$. Show that

(a) S is a p-basis for K/k;

(b) S has subsets S_i such that $S = S_0 \supset S_1 \supset \cdots$ and $S_i^{p^i} = \{s^{p^i} \mid s \in S_i\}$ is a p-basis for $k(K^{p^i})/k$ for all i;
(c) the subsets $T_i = S_{i-1} - S_i$ satisfy the conditions
   (1) $S = \bigcup_i T_i$
   (2) $T_i^{p^i} \subset k(\bigcup_{j>i} T_j^{p^i})$.

The sets $T_1, T_2 \cdots$ are a *canonical generating system* of K/k.

E.6.25 (Mordeson and Vinograd). Let K/k be a radical extension of infinite dimension. Assuming that $K^{p^n} \subset k$ for some n, prove (a), (b), (c) of the preceding exercise in the present context. What can be said for arbitrary radical extensions K/k?

E.6.26 (Haddix, Mordeson, Sweedler, Vinograd). A *subbase* of a radical extension K/k is a subset S of K-k such that $K = k(S)$ and $k(S_0) = k(s_1) \cdots k(s_n)$ (internal tensor product over k) for any finite subset $S_0 = \{s_1, \ldots, s_n\}$ of S. Show that if $K^{p^n} \subset k$ for some n, then the following conditions are equivalent, using the ideas described in E.6.14.

(a) K/k has a subbase;
(b) K^{p^i} and k are linearly disjoint over $K^{p^i} \cap k$ for all i;
(c) for any canonical generating system $T_1, T_2, \cdots$ of K/k,

$$T_i^{p^i} \subset (K^{p^i} \cap k)\left(\bigcup_{j>i} T_j^{p^i}\right) \quad \text{for all } i.$$

An arbitrary extension K/k which satisfies conditions (b) is a *modular* extension.

E.6.27. A finite dimensional radical extension K/k is *toral/diagonalizable* if there exists a toral/diagonalizable k-subbiring T of $H(K/k)$ such that $K^T = k$. Show that the following conditions are equivalent:

(a) K/k is diagonalizable;
(b) K/k has a subbase;
(c) K/k is modular;
(d) $k = K^{\mathscr{D}}$ where $\mathscr{D}$ is the set of higher derivations of K/k
   (Hint: Use E.6.14).

Using the equivalence of (a) and (d), show that K has a unique minimal subfield k' such that K/k' is diagonalizable.

E.6.28. Show that if K/k is a finite dimensional radical extension, L is the separable closure of k, $\bar{K} = L \otimes_k K$, $\bar{k} = L \otimes_k k$, then $\bar{K}/\bar{k}$ is diagonalizable if K/k is toral.

E.6.29 (Weisfeld). Let $k = \mathbb{Z}_p(X, Y, Z)$ (purely transcendental extension of $\mathbb{Z}_p$ with transcendency basis X, Y, Z), choose $a \in k_{\text{Alg}}$ such that $a^{p^2} = X$, choose $b \in k_{\text{Alg}}$ such that $b^p = a^p Y + Z$ and let $K = k(a, b)$. Show that $K = k(a)k(b)$, $k(a) \cap k(b) = k(a^p) \supsetneqq k$ and that the dimensions indicated by the following diagram are correct:

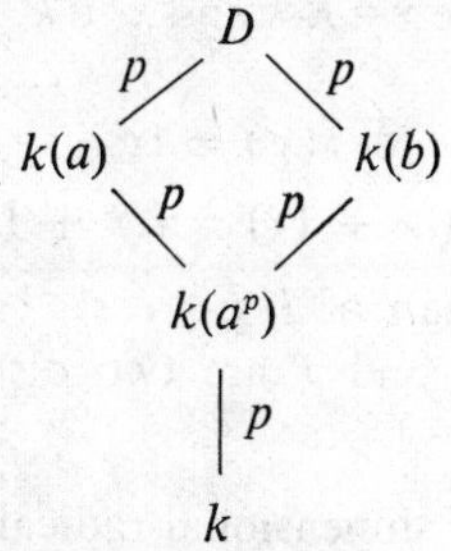

Finally, show that K/k is not diagonalizable.

E.6.30. Show that there is a finite dimensional radical extension which is not toral. (Hint: Let L be the separable closure of k and let $\bar{K} = L \otimes_k K$, $\bar{k} = L \otimes_k k$ where K/k is the extension constructed in E.6.28. Show that the dimensions indicated by the diagram

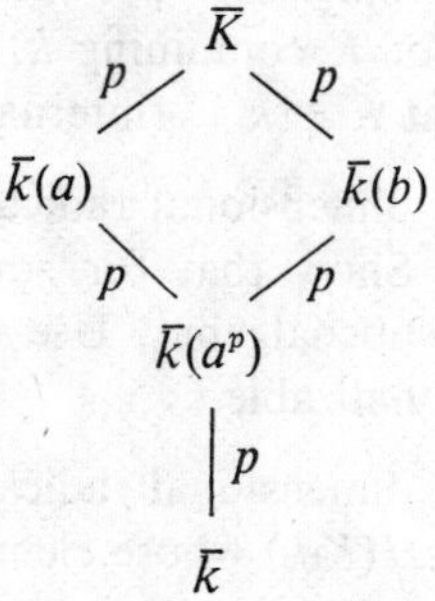

are correct. Finally, using E.6.28, show that K/k is not toral).

E.6.31. Let K/k be a radical extension. Let C be a finite dimensional K-subspace of $\text{End}_k K$ such that $[x, y] \in C$ and $x^p \in C$ for all $x, y \in C$. Letting L be a separable extension of k, $\bar{K} = L \otimes_k K$, $\bar{k} = L \otimes_k k$, $\bar{C} = L \otimes_k C$, show that $\bar{C}$ is a finite dimensional $\bar{K}$-subspace of $\text{End}_{\bar{k}} \bar{K}$ such that $[x, y] \in \bar{C}$ and $x^p \in \bar{C}$ for all $x, y \in \bar{C}$. (Hint: After showing that the $\bar{k}$-span $[\bar{C}, \bar{C}]$ of $\{[x, y] \mid x, y \in \bar{C}\}$ is a subset of $\bar{C}$, show that if $x, y \in \bar{C}$ and $x^p, y^p \in \bar{C}$, then $(x + y)^p \in \bar{C}$ because $(x + y)^p \equiv x^p + y^p \pmod{[\bar{C}, \bar{C}]}$).

E.6.32. Show that if K/k is an Abelian extension of finite degree not divisible by p, then the k-span T of $\text{Aut}_k K$ is toral k-subbiring of $H(K/k)$.

E.6.33. Let K/k be a Galois extension of degree n not divisible by p, and suppose that k contains all of the nth roots of unity in k_{Alg}. Show that $H(K/k)$ has at most one diagonalizable k-subbiring T such that $K^T = k$, namely, the k-span of $\text{Aut}_k K$.

E.6.34. Let K/k be a Galois extension of degree n and suppose that $H(K/k)$ has a diagonalizable toral cosplit k-subbiring T such that $K^T = k$. Show that K/k is an Abelian extension and that k contains all of the nth roots of unity in k_{Alg}. (Hint: Show that T is the k-span of $\text{Aut}_k K$).

E.6.35. Let $K = k(x)$ where $x \in K\text{-}k$ and $x^p \in k$. Let s and t be the derivations of K/k such that

$$s(x^i) = ix^i$$
$$t((x + 1)^i) = i(x + 1)^i$$

for all i. Let S be the k-span of $I, s, \ldots, s^{p-1}$ and let T be the k-span of $I, t, \ldots, t^{p-1}$. Show that S and T are two distinct cosplit k-forms of the K/k-bialgebra $H(K/k)$.

E.6.36. Let K/k be a finite dimensional radical extension of exponent one. Prove the following.

(a) If T is a maximal finite dimensional diagonalizable toral k-subring of $H(K/k)$, then $K^T = k$.
(b) If S is a diagonalizable k-torus in $\mathrm{Der}_k K$, then there exists a diagonalizable k-torus S' in $\mathrm{Der}_k K$ such that $S \cap S' = \{0\}$, $T = S + S'$ (direct) is a maximal k-torus in $\mathrm{Der}_k K$ and, consequently, $K^T = k$ and $K = K^S K^{S'}$ (internal tensor product over k).
(c) For any subfield k' of K containing k, there is a subfield k'' of K containing k such that $K = k'k''$ (internal tensor product over k).

E.6.37. Let K/k be a finite dimensional radical extension of characteristic $p > 0$ and let $D \in \mathrm{Der}_k K$. Show that for some e, the k-span $\langle D^{p^e} \rangle$ of $D^{p^e}, D^{p^{e+1}}, \ldots$ is a toral k-subcoalgebra. Use this to show that not every toral k-subcoalgebra is diagonalizable.

E.6.38. Let K/k be a finite dimensional radical extension of characteristic $p > 0$. Let A be a subset of $H(K/k)$ whose elements commute pairwise, and suppose that $\varepsilon(A) \subset k$ and for each $x \in A$, there exist ${}_ix, x_i \in A$ such that $\Delta x = \sum_i {}_ix \otimes x_i$. Show that there is a toral k-subcoalgebra T of $H(K/k)$ such that $K^A = K^T$. (Hint: Consider the k-span of all p^eth powers of elements of A, e being chosen sufficiently large).

E.6.39 (Toral Descent). Let K/k be a (possibly infinite dimensional) field extension of characteristic $p > 0$. A *toral* k-subcoring of $H(K/k)$ is a finite dimensional k-subspace T of $H(K/k)$ containing I and consisting of pairwise commuting elements such that T is the k-span of $T^p = \{t^p \mid t \in T\}$ and T is a k-subcoring of $H(K/k)$. Let T be a toral k-subcoring of $H(K/k)$ and let V be a vector space V over K. Then a *T-product* on V is a mapping $T \times V \to V$, denoted $(t, v) \mapsto t_L(v) = t(v)$, such that

1. $t \mapsto t_L$ is a k-linear mapping from T to $\mathrm{Hom}_k V \ni I_L = id_V$;
2. $s(t(v)) = t(s(v))$ for $s, t \in T$, $v \in V$;
3. $(t^p)_L(v) = (t_L)^p(v)$ for all $t \in T$, $v \in V$;
4. $t(av) = \sum_i {}_it(a)t_i(v)$ for all $t \in T$, $a \in K$, $v \in V$.

Suppose that K/k is radical, let L be the separable closure, let $\bar{K} = L \otimes_k K$, $\bar{k} = L \otimes_k k$, $\bar{T} = L \otimes_k T$, $\bar{V} = L \otimes_k V$ and make the usual identifications. Describe how $\bar{T}$ can be regarded as a toral $\bar{k}$-subcoring of $H(\bar{K}/\bar{k})$ and show that for any T-product $T \times V \to V$, there is an extension of $T \times V \to V$ to a $\bar{T}$-product $\bar{T} \times \bar{V} \to \bar{V}$ on $\bar{V}$. Assuming that $T \times V \to V$ is a T-product on V

having the property that if $t \in T^0$ and if t is contained in the subring $\langle S \rangle$ of $\text{End}_k K$ generated by a k-subspace S of T, then $t(v) = 0$ for all $v \in V^S = \{v \in V \mid s(v) = 0 \text{ for } s \in S^0\}$, prove the following. (Here, $T^0 = T \cap \text{Kern}\, \varepsilon$ as usual).

(a) $\overline{T}_\pi = \{t \in \overline{T} \mid t^p = t\}$ is a π-form of the $\bar{k}$-space $\overline{T}$, π being the prime field of $\bar{k}$.
(b) $\overline{V} = \sum_{\alpha \in \overline{T}_\pi^*} \overline{V}_\alpha$ (direct) where $\overline{T}_\pi^*$ is the π-dual space of $\overline{T}_\pi$ and $t(v) = \alpha(t)v$ for $t \in \overline{T}_\pi$, $v \in V$, $\alpha \in \overline{T}_\pi^*$. (Hint: Use part (a).)
(c) For each $v \in \overline{V}$, v is contained in a finite dimensional $\overline{T}$-stable $\overline{K}$-subspace of $\overline{V}$. (Hint: Use part (b).)
(d) For each $v \in V$, v is contained in a finite dimensional T-stable K-subspace of V. (Hint: Use part (c).)
(e) $\overline{V}^{\overline{T}} = \{v \in \overline{V} \mid t(av) = t(a)v \text{ for all } t \in \overline{T}, a \in \overline{K}\}$ is a $\overline{K}^{\overline{T}}$-form of the $\overline{K}$-space $\overline{V}$. (Hint: Use part (c) to reduce the case where $\overline{V}:\overline{K}$ is finite. Then follow the ideas used in Chapter 6 in proving that $H(K/k)^T$ and C^T are K^T-forms of $H(K/k)$ and C.)
(f) $V^T = \{v \in V \mid t(av) = t(a)v \text{ for all } t \in T, a \in K\}$ is a K^T-form of V. (Hint: Use part (e).)

(The relationship of "torus" as used here to "torus" as used in Chapter 6 is described in Chapter 4 of [22].)

S Set theory

Roughly speaking, sets are collections of objects or elements and set theory is a precise language in which sets and their elements are, ultimately, the sole object of discussion. This language is extremely well suited as a language within which to develop most mathematics. In this appendix, we briefly describe the set theory needed for this book. We describe the basic concepts and the most meaningful consequences of the usual axioms or basic assumptions of set theory, but do not attempt to describe the axioms themselves. The reader who is interested in further pursuing set theory is referred to [4], [7].

S.1 Sets and elements

Sets are usually denoted by upper case letters $A, B, C, \ldots$. Familiar sets are the set $\mathbb{Z}$ of integers, $0, \pm 1, \pm 2, \ldots$, the set $\mathbb{Q}$ of rational numbers m/n (m, n integers, $n \neq 0$), the set $\mathbb{R}$ of real numbers, the set $\mathbb{C}$ of complex numbers and the set $\mathbb{Z}_n$ of residues $\bar{0}, \bar{1}, \ldots, \overline{n-1}$ modulo a positive integer n.

Given x and a set A, either x is an element of A, written $x \in A$, or x is not an element of A, written $x \notin A$. A set A is a *subset* of a set B, written $A \subset B$, if $x \in A$ implies that $x \in B$ for all x. If A is not a subset of B, we write $A \not\subset B$. For any two sets A and B, $A = B$ if and only if $A \subset B$ and $B \subset A$. We sometimes write $A \subseteqq B$ for $A \subset B$. If $A \subset B$ and $A \neq B$, we write $A \subsetneqq B$ and say that A is a *proper* subset of B. We also write $A \ni x$, $A \not\ni x$, $B \supset A$, $B \not\supset A$, $B \supseteqq A$, $B \supsetneqq A$ for $x \in A$, $x \notin A$, $A \subset B$, $A \not\subset B$, $A \subseteqq B$, $A \subsetneqq B$ respectively. Note that $x \in A$ and $A \subset B$ implies $x \in B$, and $A \subset B$ and $B \subset C$ implies $A \subset C$.

Sets can be constructed in a number of axiomatically prescribed ways. If A is a set and $S(x)$ a statement about x, then $\{x \in A \mid S(x)\}$ is a set (called the *set of* $x \in A$ *such that* $S(x)$) such that $y \in \{x \in A \mid S(x)\}$ if and only if $y \in A$ and the statement $S(y)$ is true. There is a (unique) set $\varnothing$ (called the *empty set* or *null set*) such that $y \notin \varnothing$ for all y. For any x, there is a set $\{x\}$ (called the *one point set containing* x or the *singleton* of x) such that $y \in \{x\}$ if and only if $y = x$. If A and B are sets, there are sets $A - B$ (called the *difference* of A and B), $A \cup B$ (called the *union* of A and B) and $A \cap B$ (called the *intersection* of A and B) such that $y \in A - B$ if and only if $y \in A$ and $y \notin B$, $y \in A \cup B$ if and only if $y \in A$ or $y \in B$, $y \in A \cap B$ if and only if $y \in A$ and $y \in B$. If $A \cap B = \varnothing$, we say that A and B are *disjoint*. Note that $A \cup B = B \cup A$, $A \cap B = B \cap A$, $(A \cup B) \cup C = A \cup (B \cup C)$, $(A \cap B) \cap C = A \cap (B \cap C)$, $A \cap (B \cup C) = (A \cap B) \cup (A \cap C)$. We let $A_1 \cup A_2 \cup A_3 \cup \cdots \cup A_n = (\cdots((A_1 \cup A_2) \cup A_3) \cup \cdots) \cup A_n$ and $A_1 \cap A_2 \cap \cdots \cap A_n =$

$(\cdots((A_1 \cap A_2) \cap A_3) \cap \cdots) \cap A_n$. For any $x_1, \ldots, x_n$, we let $\{x_1, \ldots, x_n\} = \{x_1\} \cup \cdots \cup \{x_n\}$ and call it the *set consisting of* $x_1, \ldots, x_n$.

For any set A, there is a set $\mathbf{P}(A)$ (called the *set of subsets of A* or the *power set* of A) such that $S \in \mathbf{P}(A)$ if and only if $S \subset A$. The set $\mathbf{P}(A)$ is an example of a *collection of sets*, that is, a set all of whose elements are sets. If $\mathbf{A}$ is a collection of sets, then there is a set $\bigcup_{S \in \mathbf{A}} S$ (called the *union* of the sets in $\mathbf{A}$) such that $y \in \bigcup_{S \in \mathbf{A}}$ if and only if $y \in S$ for some $S \in \mathbf{A}$. And there is a set $\bigcap_{S \in \mathbf{A}} S$ (called the *intersection* of the elements of $\mathbf{A}$) such that $y \in \bigcap_{S \in \mathbf{A}} S$ if and only if $y \in S$ for all $S \in \mathbf{A}$. For any x and y, (x, y) denotes the set $\{\{x\}, \{x, y\}\}$ and is called the *ordered pair* of x and y. The importance of this asymmetric definition of ordered pair is that it leads to the familiar property $(x, y) = (x', y')$ if and only if $x = x'$ and $y = y'$. For any $x_1, \ldots, x_n$, we let $(x_1, \ldots, x_n)$ denote $((\cdots((x_1, x_2), x_3), \cdots), x_n)$ and call it the *ordered n-tple* of $x_1, \ldots, x_n$. Given sets $A_1, \ldots, A_n$, there is a set $A_1 \times \cdots \times A_n$ (called the *Cartesian product* of $A_1, \ldots, A_n$) such that $y \in A_1 \times \cdots \times A_n$ if and only if $y = (x_1, \ldots, x_n)$ where $x_1 \in A_1, \ldots, x_n \in A_n$. For any set A, A^n denotes $A \times \cdots \times A$ (n times).

S.2 Functions

A *function* from A to B is a subset f of $A \times B$ such that for each $a \in A$, there exists precisely one $b \in B$ such that $(a, b) \in f$. Suppose that f is a function from A to B. For $a \in A$, we denote the element $b \in B$ such that $(a, b) \in f$ by $f(a)$ and call $f(a)$ the *image* of a under b. Thus, $f = \{y \in A \times B \mid y = (a, f(a))$ for some $a \in A\}$ or, in a more abbreviated notation, $f = \{(a, f(a)) \mid a \in A\}$. To indicate that f is a function from A to B, we often write $f: A \to B$.

The *domain* of $f: A \to B$ is Domain $f = A$. The *image* of f is Image $f = \{b \in B \mid b = f(a)$ for some $a \in A\}$ or, in a more abbreviated notation, Image $f = \{f(a) \mid a \in A\}$. We say that the function $f: A \to B$ is *surjective* from A to B if Image $f = B$. If $f(a) = f(a')$ implies that $a = a'$ for all $a, a' \in A$, we say that f is *injective* from A to B. If $f: A \to B$ is surjective and injective from A to B, we say that f is *bijective* from A to B.

If $f: A \to B$ is bijective from A to B, then $f^{-1} = \{(b, a) \mid (a, b) \in f\}$ is a bijective function from B to A such that $f^{-1}(b) = a$ if and only if $f(a) = b$ for all $b \in B$. If f is a function from A to B and g is a function from B to C, then $g \circ f = \{(a, g(f(a))) \mid a \in A\}$ is a function from A to C such that $(g \circ f)(a) = g(f(a))$ for all $a \in A$. The function $g \circ f$ is called the *composite* of f and g. If, furthermore, h is a function from C to D, then $(h \circ g) \circ f = h \circ (g \circ f)$. Letting $\mathrm{id}_A = \{(a, a) \mid a \in A\}$, $\mathrm{id}_A(a) = a$ for all $a \in A$. If f is a function from A to B, then $f \circ \mathrm{id}_A = f$ and $\mathrm{id}_B \circ f = f$. The function $\mathrm{id}_A: A \to A$ is called the *identity function* on A.

Suppose that f is a function from A to B and g is a function from B to C. If $g \circ f$ is surjective, then g is surjective. If $g \circ f$ is injective, then f is injective. Suppose, furthermore, that $C = A$, so that f is a function from A to B and g a function from B to A. Then f and g are bijective if and only if $g \circ f$ is bijective. In particular, f and g are bijective and $g = f^{-1}$ if and only if $g \circ f = \mathrm{id}_A$ and $f \circ g = \mathrm{id}_B$.

If f is a function from A to B and $S \subset A$, then $f|_S = \{(a, f(a)) \mid a \in S\}$ is a function from S to A such that $f|_S(a) = f(a)$ for all $a \in S$. The function $f|_S$ is called the *restriction* of f to S. If f is a function from A to B and if $A' \supset A$, $B' \supset B$, an *extension* of f to A' is a function $f': A' \to B'$ such that $f'(a) = f(a)$ for all $a \in A$.

If f is a function from A to B, g a function from A to C and h a function from A to C, then we say that the diagram

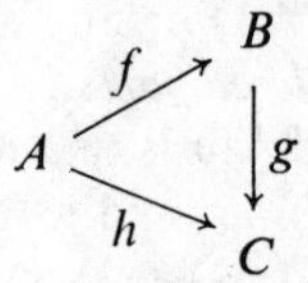

is *commutative* if $h = g \circ f$. In a more complicated diagram of sets and arrows labelled by functions, any directed path from one point in the diagram to another point in the diagram corresponds to a composite function $f_m \circ \cdots \circ f_1$. The diagram is *commutative* if for any two points in the diagram and any two directed paths from the first point to the second point, the corresponding composite functions $f_m \circ \cdots \circ f_1$ and $g_n \circ \cdots \circ g_1$ are equal.

S.3 Cardinality

Two sets A and B have the same *cardinality* if there exists a bijective function f from A to B. If A and B have the same cardinality, we write $|A| = |B|$. If A and B do not have the same cardinality, we write $|A| \neq |B|$. If A and C are sets and if C has a subset B such that $|A| = |B|$, we write $|A| \leq |C|$. If $|A| \leq |B|$ and $|A| \neq |B|$, we write $|A| \lneqq |B|$. For any set A, we have $|A| \lneqq |\mathbf{P}(A)|$. For any two sets A and B, precisely one of the following is the case:

1. $|A| \gneqq |B|$;
2. $|A| = |B|$;
3. $|B| \lneqq |A|$.

This theorem is dependent upon the socalled *Axiom of Choice*, which states that for any set A, there exists a function f from $\mathbf{P}(A)$ to A such that $f(S) \in S$ for every nonempty subset S of A. Such a function f is called a *choice function* and may be regarded as "choosing" an element $f(S)$ from each nonempty subset S of A.

If A is similar to the set $\{1, \ldots, n\}$, then n is uniquely determined by A, we write $|A| = n$ and we say that A is *finite* and that *A has n elements*. If A is not finite, we say that A is *infinite*. If A is similar to the set $\{n \in \mathbb{Z} \mid n > 0\}$ of positive integers $1, 2, 3, \ldots$, then A is infinite and we say that A is *countably infinite*. Every infinite set contains a countably infinite subset. (This theorem is dependent on the Axiom of Choice). If A is either finite or countably infinite, we say that A is *countable*. If A is not countable, we say that A is *uncountable*. For any infinite set A, the power set $\mathbf{P}(A)$ is uncountable.

S.4 Zorn's Lemma

A *partial order* on a set A is a subset $\preceq$ of $A \times A$ such that

1. $a \preceq a$ for all $a \in A$;
2. $a \preceq b$ and $b \preceq a$ implies that $a = b$ for all $a, b \in A$;
3. $a \preceq b$ and $b \preceq c$ implies that $a \preceq c$ for all $a, b, c \in A$;

where "$a \preceq b$" is the statement "(a, b) is an element of the subset $\preceq$" for $a, b \in A$. We write $a \precneqq$ if $a \preceq b$ and $a \neq b$. Set inclusion $X \subset Y$ is a partial order on the set $\mathbf{A} = \mathbf{P}(B)$ of subsets X of a set B. Let A be a *partially ordered set*, that is, a set together with a partial order $\preceq$. A *chain* in A is a subset C of A such that for any $a, b \in C$, either $a \preceq b$ or $b \preceq a$. An *upper bound* for a chain C in A is an element $b \in A$ such that $c \preceq b$ for all $c \in C$. A *least upper bound* for a chain C in A is an upper bound b for C such that $b \preceq b'$ for every upper bound b' of C. A *maximal element* in A is an element $m \in A$ such that $m \precneqq a$ is false for all $a \in A$. *Zorn's Lemma* (also dependent on the Axiom of Choice) states that if every chain C in a partially ordered set A has a least upper bound, then A has a maximal element.

S.5 Transfinite induction

A *total order* on a set A is a partial order $\preceq$ on A such that for any $a, b \in A$, either $a \preceq b$ or $b \preceq a$. In $\mathbb{Z}$, $a \leq b$ is a total order. A *first element* of a subset S of a set A with total order $\preceq$ is an element $s \in S$ such that $s \preceq t$ for all $t \in S$. A *well ordering* of a set A is a total order $\preceq$ on A such that every nonempty subset S of A has a first element. The total order $\preceq$ on $\mathbb{Z}$ is not a well ordering of $\mathbb{Z}$, since $\mathbb{Z}$ has no first element. However, $\mathbb{Z}$ has a well ordering $\preceq$, defined by $m \preceq n$ if m and n are both nonnegative and $m \leq n$, $m \preceq n$ if m and n are both negative and $-m \leq -n$, and $m \preceq n$ if m is negative and n is nonnegative. In fact, any set A has a well ordering on A. (This theorem is dependent upon the Axiom of Choice.)

Suppose now that A is a *well ordered set*, that is, a set together with a well ordering $\preceq$. Let $S(a)$ be a statement about the elements $a \in A$. To prove by *Transfinite Induction* that $S(a)$ is true for all $a \in A$, it suffices to prove that $S(a)$ is true if a is the first element of A and that for any element a of A, if $S(b)$ is true for all $b \precneqq a$, then $S(a)$ is true. To see why, suppose that $S(a)$ is not true for all $a \in A$ and let $S = \{a \in A \mid S(a) \text{ is not true}\}$. Let a be the first element of S. Then $S(a)$ is not true, but $S(b)$ is true for all $b \precneqq a$, so that the conditions just described are not met.

T Tensor products

In this appendix, we construct the tensor product $V \otimes W$ of two vector spaces and discuss its properties. In T.1, we construct $V \otimes W$ and relate the bases and generating subsets of V and W and the bases and generating subsets of $V \otimes W$. In T.2, we discuss the tensor product $A \otimes B \in \text{Hom}(V \otimes W, V' \otimes W')$ of $A \in \text{Hom}(V, V')$ and $B \in \text{Hom}(W, W')$. In T.3, we prove that $k \otimes V$, $V \otimes W$, $(V \otimes W) \otimes X$ and $(V_1 \oplus V_2) \otimes W$ are naturally isomorphic to V, $W \otimes V$, $V \otimes (W \otimes X)$ and $(V_1 \otimes W) \oplus (V_2 \otimes W)$ respectively. In T.4, we use tensor products to introduce the processes of ascent and descent for passing from a vector space V_k over a subfield k of a field K to a certain vector space $(V_k)_K$ over K, and from a vector space V_K over K to a vector space V_k over k such that V_K and $(V_k)_K$ are naturally isomorphic.

T.1 Preliminaries

Let V and W be vector spaces over a field k. A *pairing* of V and W is an ordered pair (f, X) where X is a vector space over k and f is a function from the Cartesian product $V \times W$ to X which is bilinear in the sense that

1. $f(v_1 + v_2, w) = f(v_1, w) + f(v_2, w)$,
2. $f(v, w_1 + w_2) = f(v, w_1) + f(v, w_2)$,
3. $f(cv, w) = f(v, cw) = cf(v, w)$ for $v, v_1, v_2 \in V$, $w, w_1, w_2 \in W$ and $c \in k$.

If (f, X) and (g, Y) are pairings of V and W, then a *morphism/isomorphism* from (f, X) to (g, Y) is a linear transformation/bijective linear transformation $\alpha: X \to Y$ such that the diagram

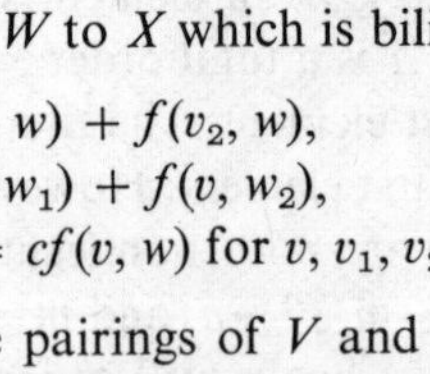

is commutative. A *universal* bilinear pairing of V and W is a pairing (f, X) of V and W such that for each bilinear pairing (g, Y) of V and W, there exists a unique morphism α from (f, X) to (g, Y).

If (f, X) and (g, Y) are universal pairings of V and W, then (f, X) and (g, Y) are isomorphic. For let $\alpha: X \to Y$ be a morphism from (f, X) to (g, Y) and let $\beta: Y \to X$ be a morphism from (g, Y) to (f, X). Then $\beta \circ \alpha: X \to X$ and $\text{id}_X: X \to X$ are morphisms from (g, X) to (g, X). Thus, $\beta \circ \alpha = \text{id}_X$. Similarly, $\alpha \circ \beta = \text{id}_Y$. Thus, α and β are inverses and α is an isomorphism from (f, X) to (g, Y).

We now let $\{v_i \mid i \in S\}$ and $\{w_j \mid j \in T\}$ be bases for V and W respectively, and describe a particular universal pairing of V and W which we refer to as the *standard* universal pairing of V and W with respect to the bases $\{v_i \mid i \in S\}$ and $\{w_j \mid j \in T\}$. Let $F(S \times T, k)$ be the vector space of functions from $S \times T$ to k. For each $(i, j) \in S \times T$, let x_{ij} be the function such that $x_{ij}(i', j') = 0$ if $(i, j) \neq (i', j')$ and $x_{ij}(i, j) = 1$. Let X be the subspace of $F(S \times T, k)$ with basis $\{x_{ij} \mid i \in S, j \in T\}$. Let $f\colon V \times W \to X$ be defined by $f(\sum_i a_i v_i, \sum_j b_j w_j) = \sum_{i,j} a_i b_j x_{ij}$. Then (f, X) is a pairing of V and W. Moreover, (f, X) is universal. For if (g, Y) is a pairing of V and W, then the linear mapping $\alpha\colon X \to Y$ such that $\alpha(x_{ij}) = g(v_i, w_j)$ for $e \in S, j \in T$ is a (unique) morphism from (f, X) to (g, Y).

We now know that V and W have a universal pairing, and that any two universal pairings of V and W are isomorphic. For this reason, we speak of *the* universal pairing of V and W. The universal pairing of V and W is called the *tensor product* of V and W over k.

We usually denote the tensor product of V and W over k by $(\otimes, V \otimes W)$, so that $V \otimes W$ is a vector space over k and $\otimes\colon V \times W \to V \otimes W$ is a bilinear mapping such that for any pairing (g, Y) of V and W, there exists a linear mapping $\alpha\colon V \otimes W \to Y$ such that the diagram

$$\begin{array}{ccc} V \times W & \xrightarrow{\otimes} & V \otimes W \\ & \searrow^{g} & \downarrow \alpha \\ & & Y \end{array}$$

is commutative. We denote $\otimes(v, w)$ by $v \otimes w$ for $v \in V, w \in W$, and call $v \otimes w$ the *tensor product* of v and w. The commutativity of the above diagram now says that $g(v, w) = \alpha(v \otimes w)$ for all $v \in V, w \in W$. We let $v \otimes W = \{v \otimes w \mid w \in W\}$ and $V \otimes w = \{v \otimes w \mid v \in V\}$ for $v \in V$, $w \in W$. At times, it is convenient to display the field k when using tensor products, in which case we may write $V \otimes_k W, v \otimes_k w, v \otimes_k W, V \otimes_k w$ for $V \otimes W, v \otimes w, v \otimes W$, $V \otimes w$ respectively.

If $\{v_i \mid i \in S\}$ and $\{w_j \mid j \in T\}$ are basis for V and W respectively, then $\{v_i \otimes w_j \mid i \in S, j \in T\}$ is a basis for $V \otimes W$. To see this, let (f, X) be the standard universal pairing with respect to the given bases, so that $\{x_{ij} \mid i \in S, j \in T\}$ is a basis for X where $x_{ij} = f(v_i, w_j)$ for $i \in S, j \in T$. Let $\alpha\colon V \otimes W \to X$ be the isomorphism such that the diagram

$$\begin{array}{ccc} V \times W & \xrightarrow{\otimes} & V \otimes W \\ & \searrow^{f} & \downarrow \alpha \\ & & X \end{array}$$

is commutative. Then we have $\alpha(v_i \otimes w_j) = f(v_i, w_j) = x_{ij}$ for $i \in S, j \in T$. Thus, $\{v_i \otimes w_j \mid i \in S, j \in T\}$ is a basis for $V \otimes W$.

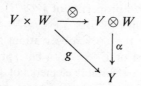

The vector space $V \otimes W$ is generated by $\{v \otimes w \mid v \in V, w \in W\}$, as we see from the preceding paragraph. More generally, if $\{v_i \mid i \in S\}$ and $\{w_j \mid j \in T\}$ generate V and W respectively, then $\{v_i \otimes w_j \mid i \in S, j \in T\}$ generates $V \otimes W$. To see this, it suffices to observe that each $v \otimes w$ ($v \in V$, $w \in W$) is a linear combination of the $v_i \otimes w_j$. However, for $v = \sum_i c_i v_i$ and $w = \sum_j d_j w_j$, we have $v \otimes w = \sum_{i,j} c_i d_j (v_i \otimes w_j)$.

If $\{v_i \mid i \in S'\}$ and $\{w_j \mid j \in T'\}$ are linearly independent subsets of V and W respectively, then $\{v_i \otimes v_j \mid i \in S', j \in T'\}$ is a linearly independent subset of $V \otimes W$. For we may extend these linearly independent subsets to bases $\{v_i \mid i \in S\}$ and $\{w_j \mid j \in T\}$ for V and W respectively, where $S' \subset S$ and $T' \subset T$, and then $\{v_i \otimes w_j \mid i \in S, j \in T\}$ is a basis for $V \otimes W$.

If $\{w_j \mid j \in T\}$ is a linearly independent subset of W and if $\sum_j {}_jv \otimes w_j = 0$ where the ${}_jv$ are in V, then ${}_jv = 0$ for all j. To see this, let $\{v_i \mid i \in S\}$ be a basis for V and ${}_jv = \sum_i c_{ji} v_i$ for $j \in T$. Then

$$0 = \sum_i \left(\sum_j c_{ji} v_i \right) \otimes w_j = \sum_{i,j} c_{ji} (v_i \otimes w_j),$$

so that $c_{ji} = 0$ for all $i \in S, j \in T$ and ${}_jv = 0$ for all $j \in T$, by the linear independence of the $v_i \otimes w_j$. Similarly if $\{v_i \mid i \in S\}$ is a linearly independent subset of V and if $\sum_i v_i \otimes {}_iw = 0$ where the ${}_iw$ are elements of W, then ${}_iw = 0$ for all i. These two properties of $V \otimes W$ are referred to as the *linear disjointness* of V and W in $V \otimes W$.

It follows easily from the above observations that if $\{w_j \mid j \in T\}$ is a basis for W, then each element of $V \otimes W$ can be expressed uniquely in the form $\sum_j {}_jv \otimes w_j$, where the ${}_jv$ are suitable elements of V. Similarly, if $\{v_i \mid i \in S\}$ is a basis for V, each element of $V \otimes W$ can be expressed uniquely in the form $\sum_i v_i \otimes {}_iw$, where the ${}_iw$ are suitable elements of W.

For a pairing (f, X) of V and W to be a tensor product of V and W, any one of the following conditions is sufficient:

1. V and W have bases $\{v_i \mid i \in S\}$ and $\{w_j \mid j \in T\}$ respectively such that $\{f(v_i, w_j) \mid i \in S, j \in T\}$ is a basis for X;
2. W has a basis $\{w_j \mid j \in T\}$ such that for each element $x \in X$, there exist unique elements ${}_jv \in V (j \in T)$ such that $x = \sum_j f({}_jv, w_j)$;
3. V has a basis $\{v_i \mid i \in S\}$ such that for each element $x \in X$, there exist unique elements ${}_iw (i \in S)$ such that $x = \sum_i f(v_i, {}_iw)$.

To see this, let $\alpha: V \otimes W \to X$ be the linear mapping such that the diagram

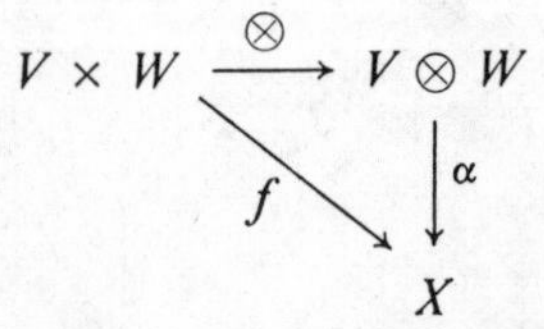

is commutative. If V and W have bases $\{v_i \mid i \in S\}$ and $\{w_j \mid j \in T\}$ such that $\{f(v_i, w_j) \mid i \in S, j \in T\}$ is a basis for X, then $\{v_i \otimes w_j \mid i \in S, j \in T\}$ is a basis for $V \otimes W$ and $\alpha(v_i \otimes w_j) = f(v_i, w_j)$, so that α is an isomorphism and (f, X)

is a tensor product. This proves that (1) is sufficient. Suppose next that $\{w_j \mid j \in T\}$ is a basis for W such that for each element $x \in X$, there exist unique elements $_jv \in V$ $(j \in T)$ such that $x = \sum_j f(_jv, w_j)$. Each element t of $V \otimes W$ can be written uniquely as $t = \sum _jv \otimes w_j$ where the $_jv(j \in T)$ are elements of V, and $\alpha(\sum_j {}_jv \otimes w_j) = \sum_j f(_jv, w_j)$. Thus α is an isomorphism and (f, x) is a tensor product. This proves that (2) is sufficient. The proof that (3) is sufficient is similar.

For a pairing (f, X) of V and W such that X is generated by $\{f(v, w) \mid v \in V, w \in W\}$ to be a tensor product of V and W, any one of the following conditions is sufficient:

1. if $\{v_i \mid i \in S\}$ and $\{w_j \mid j \in T\}$ are linearly independent subsets of V and W respectively, then $\{f(_i, w_j) \mid i \in S, j \in T\}$ is linearly independent;
2. if $\{w_j \mid j \in T\}$ is a linearly independent subset of W and if $\sum_j f(_jv, w_j) = 0$ where the $_jv(j \in T)$ are elements of V, then $_jv = 0$ for all $j \in T$;
3. if $\{v_i \mid i \in S\}$ is a linearly independent subset of V and if $\sum_i f(v_i, {}_iw) = 0$ where the $_iw(i \in S)$ are elements of W, then $_iw = 0$ for all $i \in S$.

This follows easily from the preceding paragraph.

T.2   Tensor products of linear transformations

Let V, V', V'', W, W', W'' be vector spaces over k.

For $A \in \operatorname{Hom}(V, V')$ and $B \in \operatorname{Hom}(W, W')$, we obtain a bilinear mapping $A \times B : V \times W \to V' \otimes W'$ defined by $A \times B(v, w) = A(v) \otimes B(w)$ for $v \in V, w \in W$. The pair $(A \times B, V' \otimes W')$ is a pairing of V and W, so there exists a unique linear transformation $A \otimes B : V \otimes W \to V' \otimes W'$ such that the diagram

$$
\begin{array}{ccc}
V \times W & \xrightarrow{\;\otimes\;} & V \otimes W \\
& \searrow^{\!A \times B} & \downarrow{\scriptstyle A \otimes B} \\
& & V' \otimes W'
\end{array}
$$

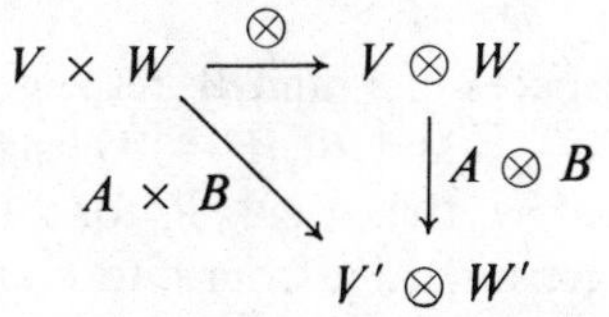

is commutative, that is, such that $(A \otimes B)(v \otimes w) = A(v) \otimes B(w)$ for $v \in V$, $w \in W$. Note that $A \otimes B \in \operatorname{Hon}(V \otimes W, V' \otimes W')$.

Let X be the subspace of $\operatorname{Hom}(V \otimes W, V' \otimes W')$ spanned by $\{A \otimes B \mid A \in \operatorname{Hom}(V, V'), B \in \operatorname{Hom}(W, W')\}$ and let $f : \operatorname{Hom}(V, V') \times \operatorname{Hom}(W, W') \to X$ be defined by $f(A, B) = A \otimes B$. Then (f, X) is a tensor product of $\operatorname{Hom}(V, V')$ and $\operatorname{Hom}(W, W')$. For obviously (f, X) is a pairing of $\operatorname{Hom}(V, V')$ and $\operatorname{Hom}(W, W')$. It therefore suffices to show that for any linearly independent subset $\{B_j \mid j \in T\}$ of $\operatorname{Hom}(W, W')$ and for $\sum_j {}_jA \otimes B_j = 0$ where the $_jA(j \in T)$ are elements of $\operatorname{Hom}(V, V')$, $_jA = 0$ for all $j \in T$. For this, let $v \in V$. Then $\sum_j {}_jA(v)B_j(w) = 0$ for all $w \in W$, so that

$$
\sum_j {}_jA(v)B_j = 0.
$$

Thus, $_jA(v) = 0$ for all $j \in T$, by the linear independence of the B_j. Since this is true for all $v \in V$, $_jA = 0$ for all $j \in T$.

What we have shown above is that the tensor product $\mathrm{Hom}(V, V') \otimes \mathrm{Hom}(W, W')$ can be taken to be the span of $\{A \otimes B \mid A \in \mathrm{Hom}(V, V'), B \in \mathrm{Hom}(W, W')\}$ together with the mapping $(A, B) \mapsto A \otimes B$, where the $A \otimes B$ are as defined earlier in the section.

Letting V^* and W^* be the dual spaces $V^* = \mathrm{Hom}(V, k)$ and $W^* = \mathrm{Hom}(W, k)$ of V and W respectively, it follows that $V^* \otimes W^*$ is the span of $\{f \otimes g \mid f \in V^*, g \in W^*\}$. The linear mapping $f \otimes g\colon V \otimes W \to k \otimes k$ can be identified with a linear mapping from $V \otimes W$ to k, since $k \otimes k$ and k are one dimensional and are therefore isomorphic under the mapping $\pi\colon k \otimes k \to k$ defined by $\pi(a \otimes b) = ab$ for $a, b \in k$. Then $f \otimes g$ is an element of $(V \otimes W)^* = \mathrm{Hom}(V \otimes W, k)$ and $(f \otimes g)(v \otimes w) = f(v)g(w)$ for $v \in V, w \in W$. By this identification, we have imbedded $V^* \otimes W^*$ in $(V \otimes W)^*$. If V and W are finite dimensional, we have $V^* \otimes W^* = (V \otimes W)^*$, and otherwise, $V^* \otimes W^* \subsetneqq (V \otimes W)^*$.

If $A \in \mathrm{Hom}(V, V')$, $C \in \mathrm{Hom}(V', V'')$, $B \in \mathrm{Hom}(W, W')$ and $D \in \mathrm{Hom}(W', W'')$, then we have $(C \otimes D) \circ (A \otimes B) = (C \circ A) \otimes (D \circ B)$ in $\mathrm{Hom}(V \otimes W, V'' \otimes W'')$, as is easily verified.

If $A \in \mathrm{Hom}(V, V')$ and $B \in \mathrm{Hom}(W, W')$ are injective/surjective/bijective, then $A \otimes B \in \mathrm{Hom}(V \otimes W, V' \otimes W')$ is injective/surjective/bijective, as one easily verifies.

Let V_0 and W_0 be subspaces of V and W respectively, and let $A\colon V_0 \to V$ and $B\colon W_0 \to W$ be the inclusion mappings, that is, the mappings defined by $A(v) = v$ for $v \in V_0$ and $B(w) = w$ for $w \in W_0$. Since A and B are injective, $A \otimes B$ is injective. Thus, we may use $A \otimes B$ to identify $V_0 \otimes W_0$ with the span of $\{v \otimes w \mid v \in V_0, w \in W_0\}$ of $V \otimes W$. Henceforth, we will tacitly assume this identification to have been made, so that we may write $V_0 \otimes W_0 = \{v \otimes w \in V \otimes W \mid v \in V_0, w \in W_0\}$.

Let V_0 and W_0 be subspaces of V and W respectively, let $\overline{V} = V/V_0$ and $\overline{W} = W/W_0$ and let $A\colon V \to \overline{V}$ and $B\colon W \to \overline{W}$ be the quotient mappings, that is, the mappings defined by $A(v) = v + V_0$ for $v \in V$ and $B(w) = w + W_0$ for $w \in W$. Since A and B are surjective from V to $\overline{V}$ and W to $\overline{W}$ respectively, $A \otimes B$ is surjective from $V \otimes W$ to $\overline{V} \otimes \overline{W}$. One easily shows that the kernel of $A \otimes B$ is Kernel $A \otimes B = V_0 \otimes W + V \otimes W_0$. It follows that $V \otimes W/(V_0 \otimes W + V \otimes W_0)$ is isomorphic to $V/V_0 \otimes W/W_0$.

T.3 The identity, commutative, associative and distributive laws for tensor products

The vector spaces V and $k \otimes V$ are isomorphic under the linear mapping $f\colon V \to k \otimes V$ defined by $f(v) = 1 \otimes v$. For f maps a basis $\{v_i \mid i \in S\}$ for V to the basis $\{1 \otimes v_i \mid i \in S\}$ for $k \in V$. Similarly, V and $V \otimes k$ are isomorphic. The property $V \simeq k \otimes V \simeq V \otimes k$ is called the *identity law* for tensor products, where we write "$\simeq$" for "is isomorphic to." Note that this law can also be expressed by the equalities $k \otimes V = 1 \otimes V$ and $V \otimes k = V \otimes 1$.

The vector spaces $V \otimes W$ and $W \otimes V$ are isomorphic under a linear mapping $\alpha\colon V \otimes W \to W \otimes V$ such that $\alpha(v \otimes w) = w \otimes v$ for $v \in V$ and $w \in W$, called the *twist* from $V \otimes W$ to $W \otimes V$. To see this, let $f\colon V \times W \to$

$W \otimes V$ be defined by $f(v, w) = w \otimes v$. Then $(f, W \otimes V)$ is a pairing of V and W, and there exists a unique linear mapping $\alpha: V \otimes W \to W \otimes V$ such that the diagram

$$\begin{array}{ccc} V \times W & \xrightarrow{\otimes} & V \otimes W \\ & {\scriptstyle f} \searrow & \downarrow {\scriptstyle \alpha} \\ & & W \otimes V \end{array}$$

is commutative. Clearly, $\alpha(v \otimes w) = w \otimes v$ for all $v \in V$ and $w \in W$. Similarly, there exists a unique linear mapping $\beta: W \otimes V \to V \otimes W$ such that $\beta(w \otimes v) = v \otimes w$ for all $w \in W$ and $v \in V$. Clearly, $\beta \circ \alpha = \mathrm{id}_{V \otimes W}$ and $\alpha \circ \beta = \mathrm{id}_{W \otimes V}$, so that α and β are inverses and α is an isomorphism. The property $V \otimes W \simeq W \otimes V$ is called the *commutative law* for tensor products.

The vector spaces $(V \otimes W) \otimes X$ and $V \otimes (W \otimes X)$ are isomorphic under a linear mapping $\alpha: (V \otimes W) \otimes X \to V \otimes (W \otimes X)$ such that $\alpha((v \otimes w) \otimes x) = v \otimes (w \otimes x)$ for $v \in V$, $w \in W$ and $x \in X$. To see this, we fix an element $x \in X$ and define $f_x: V \times W \to V \otimes (W \otimes X)$ by $f_x(v, w) = v \otimes (w \otimes x)$ for $v \in V$ and $w \in W$. Then $(f_x, V \otimes (W \otimes X))$ is a pairing of V and W, and there exists a linear mapping $\alpha_x: V \otimes W \to V \otimes (W \otimes X)$ such that the diagram

$$\begin{array}{ccc} V \times W & \xrightarrow{\otimes} & V \otimes W \\ & {\scriptstyle f_x} \searrow & \downarrow {\scriptstyle \alpha_x} \\ & & V \otimes (W \otimes X) \end{array}$$

is commutative. Next, define $f: (V \otimes W) \times X \to V \otimes (W \otimes X)$ by $f(t, x) = \alpha_x(t)$. Then $(f, V \otimes (W \otimes X))$ is a pairing of $V \otimes W$ and X, and there exists a linear mapping $\alpha: (V \otimes W) \otimes X \to V \otimes (W \otimes X)$ such that the diagram

$$\begin{array}{ccc} (V \otimes W) \times X & \xrightarrow{\otimes} & (V \otimes W) \otimes X \\ & {\scriptstyle f} \searrow & \downarrow {\scriptstyle \alpha} \\ & & V \otimes (W \otimes X) \end{array}$$

is commutative. Clearly, we have $\alpha((v \otimes w) \otimes x) = v \otimes (w \otimes x)$ for all $v \in V$, $w \in W$ and $x \in X$. Similarly, there exists a linear mapping $\beta: V \otimes (W \otimes X) \to (V \otimes W) \otimes X$ such that $\beta(v \otimes (w \otimes x)) = (v \otimes w) \otimes x$ for all $v \in V$, $w \in W$ and $x \in X$. Since $(\beta \circ \alpha)((v \otimes w) \otimes x) = (v \otimes w) \otimes x$ for all $v \in V$, $w \in W$ and $x \in X$, the linear mappings $\beta \circ \alpha$ and $\mathrm{id}_{(V \otimes W) \otimes X}$ are equal on a set of generators of $(V \otimes W) \otimes X$ and $\beta \circ \alpha = \mathrm{id}_{(V \otimes W) \otimes X}$. Similarly we have $\alpha \circ \beta = \mathrm{id}_{V \otimes (W \otimes X)}$. Thus, α and β are inverses and α is an isomorphism. The property $(V \otimes W) \otimes X = V \otimes (W \otimes X)$ is called the *associative law* for tensor products. We may now disregard or omit parentheses accordingly, when convenient.

Finally, suppose that $V = V_1 + V_2$ (inner direct sum of subspaces). Then $V \otimes W \simeq V_1 \otimes W + V_2 \otimes W$ (inner direct sum of subspaces). For let $\{v_i^1 \mid i \in S^1\}$, $\{v_i^2 \mid i \in S^2\}$, $\{w_j \mid j \in T\}$ be bases for V_1, V_2 and W respectively. Then $\{v_i^1 \otimes w_j \mid i \in S^1, j \in T\}$ and $\{v_i^2 \otimes w_j \mid i \in S^2, j \in T\}$ are bases for $V_1 \otimes W$ and $V_2 \otimes W$ respectively, and $\{v_i^1 \otimes w_j \mid i \in S^1, j \in T\} \dot\cup \{v_i^2 \otimes w_j \mid i \in S^2, j \in T\}$ is a basis for $V \otimes W$. Similarly, $W \otimes V = W \otimes V_1 + W \otimes V_2$ (inner direct sum of subspaces).

It follows from the preceding paragraph that for $V = V_1 \oplus V_2$ (outer direct sum), there is an isomorphism α from $V \otimes W$ to $(V_1 \otimes W) \oplus (V_2 \otimes W)$ (outer direct sum) such that $\alpha((v_1 \oplus v_2) \otimes w) = (v_1 \otimes w) \oplus (v_2 \otimes w)$ for $v_1, v_2 \in V$ and $w \in W$. Similarly, there exists an isomorphism β from $W \otimes V$ to $(W \otimes V_1) \oplus (W \otimes V_2)$ (outer direct sum) such that

$$\beta(w \otimes (v_1 \oplus v_2)) = (w \otimes v_1) \oplus (w \otimes v_2) \qquad \text{for } w \in W \text{ and } v_1, v_2 \in V.$$

The property $(V_1 \oplus V_2) \otimes W \simeq V_1 \otimes W \oplus V_2 \otimes W$ and $W \otimes (V_1 \oplus V_2) \simeq W \otimes V_1 \oplus W \otimes V_2$ is called the *distributive law* for tensor products.

T.4 Ascent and descent

Let K be a field and let k be a subfield of K. Let V be a vector space over k. Fix an element $c \in K$ and consider the function $f_c: K \times V \to K \otimes_k V$ defined by $f_c(d, v) = (cd) \otimes v$ for $d \in K, v \in V$. Then $(f_c, K \otimes_k V)$ is a pairing of K and V, so that there exists a linear mapping $\alpha_c: K \otimes_k V \to K \otimes_k V$ such that the diagram

$$\begin{array}{ccc} K \times V & \longrightarrow & K \otimes_k V \\ & \searrow^{f_c} & \downarrow \alpha_c \\ & & K \otimes_k V \end{array}$$

is commutative. Let $V_K = K \otimes_k V$ and define cv for $c \in K$ and $v \in V_K$ by $cv = \alpha_c(v)$. Then we have $c(\sum_{i,j} d_i \otimes v_j) = \sum_{i,j} cd_i \otimes v_j$ for any $d_1, \ldots, d_m \in K$ and $v_1, \ldots, v_n \in V$. One verifies easily from this that V_K as additive group together with the scalar product $cv (c \in C, v \in V_K)$ is a vector space over K. It is clear from T.1 that for any basis $\{v_i \mid i \in S\}$ for V over k, $\{1 \otimes v_i \mid i \in S\}$ is a basis for V_K over K. Thus, a basis for $1 \otimes V$ over k is a basis for V_K over K. We refer to V_K as the vector space over K obtained from the vector space V over k by *ascent* from k to K.

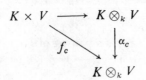

Suppose, conversely, that V_K is an arbitrary vector space over K. Since k is a subfield of K, we can also regard V_K as a vector space over k. Let V_k be a k-subspace of V_K, that is, a subspace of V_K as vector space over k. Let $(V_k)_K = K \otimes_k V_k$ be the vector space over K obtained from V_k by ascent from k to K. Let $f: K \times V_k \to V_K$ be defined by $f(c, v) = cv$ (scalar product in V_K) for $c \in K$, $v \in V_k$. Then (f, V_K) is a pairing of K and V_k, so that there exists a k-linear mapping (linear mapping of vector spaces over k) $\alpha: K \otimes_k V_k \to V_K$ such that the diagram

$$\begin{array}{ccc} K \times V_k & \xrightarrow{\otimes} & K \otimes_k V_k \\ & \searrow^{f} & \downarrow \alpha \\ & & V_K \end{array}$$

is commutative. Now $\alpha(d \otimes v) = dv$ for $d \in K$, $v \in V_k$. It follows that the mapping $\alpha: (V_k)_K \to V_K$ is in fact K-linear, for we have

$$\alpha c\left(\sum_{i,j} d_i \otimes v_j\right) = \alpha\left(\sum_{i,j} cd_i \otimes v_j\right) = \sum_{i,j} (cd_i)v_j = c \sum_{i,j} d_i v_j = c\alpha\left(\sum_{i,j} d_i \otimes v_j\right)$$

for $d_1, \ldots, d_m \in K$ and $v_1, \ldots, v_n \in V_k$.

We say that V_k is a *k-form* of V_K if the mapping $\alpha: (V_k)_K \to V_K$ is an isomorphism of vector spaces over K. The passage from a vector space V_K over K to a k-form V_k of V_K is called *descent* from K to k.

It is easily seen that a k-subspace V_k of a vector space V_K over K is a k-form of V_K if and only if a k-basis for V_k is a K-basis for V_K. In particular, for any vector space V over k, $1 \otimes_k V$ is a k-form of the vector space $K \otimes_k V$ over K obtained from V by ascent from k to K.

W Witt vectors

Our objective here is to define the ring W_eK of Witt vectors and establish the properties of W_eK needed in 3.9. The original account by Ernst Witt [24] is quite nice, and we develop the material along similar lines.

We begin with a positive integer e, a prime number p and an integral domain A containing the field $\mathbb{Q}$ of rational numbers as subring, and consider the e-fold Cartesian product A^e of A.

For $x \in A^e$, we let $x_0, \ldots, x_{e-1}$ be the elements of A such that $x = (x_0, \ldots, x_{e-1})$. We call x_i the *ith major component* of x. We let $\pi(x) = x^\pi = ({x_0}^p, \ldots, x_{e-1}^p)$ for $x \in A^e$ and call the mapping $\pi: A^e \to A^e$ the *pth power operator* on A^e.

Elements $x_0, \ldots, x_{e-1}$ of A define elements $x^{(0)}, \ldots, x^{(e-1)}$ of A by the recursion formulas

$$x^{(0)} = x_0 \quad \text{and} \quad x^{(i+1)} = (x^\pi)^{(i)} + p^{i+1}x_{i+1} \qquad (0 \leq i \leq e-2)$$

where $x = (x_0, \ldots, x_{e-1})$. We call $x^{(i)}$ the *ith minor component* of x for $x = (x_0, \ldots, x_{e-1})$ in A^e.

Elements $x^{(0)}, \ldots, x^{(e-1)}$ of A define elements $x_0, \ldots, x_{e-1}$ of A by the inverse recursion formulas

$$x_0 = x^{(0)} \quad \text{and} \quad x_{i+1} = \frac{1}{p^{i+1}}(x^{(i+1)} - (x^\pi)^{(i)}) \qquad (0 \leq i \leq e-2)$$

where $(x^\pi)^{(i)}$ denotes the ith minor component of $({x_0}^p, \ldots, {x_i}^p) \in A^{i+1}$. We let $[x^{(0)}, \ldots, x^{(e-1)}] = (x_0, \ldots, x_{e-1})$ where the $x_0, \ldots, x_{e-1}$ are defined in terms of the $x^{(0)}, \ldots, x^{(e-1)}$ in this manner.

Upon writing out the above recursion formulas, we have the following proposition.

W.1 Proposition. The following conditions on elements $x_0, \ldots, x_{e-1}$ and $x^{(0)}, \ldots, x^{(e-1)}$ of A are equivalent:

1. $(x_0, \ldots, x_{e-1}) = [x^{(0)}, \ldots, x^{(e-1)}]$;
2. $x^{(0)} = x_0$ and
   $x^{(i+1)} = (x^\pi)^{(i)} + p^{i+1}x_{i+1} \qquad (0 \leq i \leq e-2)$ where
   $x = (x_0, \ldots, x_{e-1})$;
3. $x^{(0)} = x_0$ and
   $x^{(i+1)} = x_0^{p^{i+1}} + p{x_1}^{p^i} + \cdots + p^{i+1}x_{i+1} \qquad (0 \leq i \leq e-2)$;
4. $x_0 = x^{(0)}$ and
   $$x_{i+1} = \frac{1}{p^{i+1}}(x^{(i+1)} - (x_0^{p^{i+1}} + p{x_1}^{p^i} + \cdots + p^i{x_i}^p)). \qquad \square$$

W.2 ***Definition.*** Let $x = [x^{(0)}, \ldots, x^{(e-1)}]$ and $y = [y^{(0)}, \ldots, y^{(e-1)}]$. Then we let $x \circ y = [x^{(0)} \circ y^{(0)}, \ldots, x^{(e-1)} \circ y^{(e-1)}]$ where $\circ$ is addition, subtraction or multiplication in the ring A.

The above definition makes $A^e = \{(x_0, \ldots, x_{e-1}) \mid x_i \in A \text{ for } 0 \le i \le e-1\}$ into a commutative ring. We wish to study the major components of $x + y$ and xy for x, y in this ring A^e. The major zeroth and 1st components are given by

$$\begin{aligned}
(x + y)_0 &= x_0 + y_0,\\
(x + y)_1 &= \frac{1}{p}((x + y)^{(1)} - (x + y)_0{}^p) = \frac{1}{p}(x^{(1)} + y^{(1)} - (x_0 + y_0)^p)\\
&= \frac{1}{p}(x_0{}^p + px_1 + y_0{}^p + py_1 - (x_0 + y_0)^p),\\
&= x_1 + y_1 - \frac{1}{p}(x_0{}^p + y_0{}^p - (x_0 + y_0)^p),\\
(xy)_0 &= y_0 y_0,\\
(xy)_1 &= \frac{1}{p}((xy)^{(1)} - (xy)_0{}^p) = \frac{1}{p}(x^{(1)}y^{(1)} - (x_0 y_0)^p)\\
&= \frac{1}{p}((x_0{}^p + px_1)(y_0{}^p + py_1) - x_0{}^p y_0{}^p)\\
&= x_0{}^p y_1 + x_1 y_0{}^p + px_1 y_1.
\end{aligned}$$

Note that the above zeroth and 1st major components of $x + y$ and xy are polynomials in x_0, x_1, y_0, y_1. We show shortly that the other major components of $x + y$ and xy are also polynomials in the x_i, y_i with integer coefficients. This we then use to define $x + y$ and xy for $x, y \in A_p{}^e$ (e-fold Cartesian product of A_p), where A_p is any commutative ring containing the field $\mathbb{Z}_p$ of p elements as subring. The resulting ring structure in $A_p{}^e$ cannot be introduced directly in the same manner as in A^e, because an element $(x_0, \ldots, x_{e-1})$ of $A_p{}^e$ cannot be expressed as $[x^{(0)}, \ldots, x^{(e-1)}]$ as it can be in A^e. In fact, the resulting ring $A_p{}^e$ is not isomorphic to the ring $A_p \times \cdots \times A_p$ (outer direct product of rings) whereas the ring A^e is isomorphic to the ring $A \times \cdots \times A$ (outer direct product of rings) under the mapping

$$[x^{(0)}, \ldots, x^{(e-1)}] \mapsto (x^{(0)}, \ldots, x^{(e-1)}).$$

For studying the major components of $x + y$ and xy for $x, y \in A^e$, it is convenient to introduce the *shift operator* $V: A^e \mapsto A^e$, defined as follows.

W.3 ***Definition.*** Let $x = (x_0, \ldots, x_{e-1})$. Then $V(x) = (0, x_0, \ldots, x_{e-2})$.

W.4 ***Proposition.*** Let $x = [x^{(0)}, \ldots, x^{(e-1)}]$. Then

$$V^i(x) = [0, \ldots, 0, p^i x^{(0)}, \ldots, p^i x^{(e-1-i)}]$$

for $0 \le i \le e - 1$ and $V^i(x) = 0$ for $i \ge e$.

Proof. One sees directly from W.3 that $V(x) = [0, px^{(0)}, \ldots, px^{(e-2)}]$. Now the above formula for $V^i(x)$ is obtained by iteration. □

W.5 Corollary. $V(x + y) = V(x) + V(y)$ for $x, y \in A^e$.

W.6 Definition. For $a \in A$, we let

$$\{a\} = (a, 0, \ldots, 0) = [a, a^p, \ldots, a^{p^{e-1}}].$$

W.7 Proposition. $V^i\{a\} = [0, \ldots, 0, p^i a, \ldots p^i a^{p^{e-1-i}}]$ for $0 \leq i \leq e - 1$.

Proof. Apply W.4. □

W.8 Corollary. For $x = (x_0, \ldots, x_{e-1}) \in A^e$ and $a \in A$, have

$$x = \sum_0^{e-1} V^i\{x_i\} \quad \text{and} \quad \{a\}x = (ax_0, a^p x_1, \ldots, a^{p^{e-1}} x_{e-1}) = \sum_0^{e-1} V^i\{a^{p^i} x_i\}.$$

Proof. By W.7, we have

$$\sum_0^{e-1} V^i\{x_i\} = \sum_0^{e-1} [0, \ldots, 0, p^i x_i, \ldots, p^i x_i^{p^{e-1-i}}].$$

But the latter sum is $[x^{(0)}, \ldots, x^{(e-1)}] = x$, by W.2 and W.1. Next, we compute $\{a\}x$. By W.2, this is

$$\{a\}\, x = [a, a^p, \ldots, a^{p^{e-1}}][x^{(1)}, \ldots, x^{(e-1)}] = [ax^{(0)}, \ldots, a^{p^{e-1}} x^{(e-1)}].$$

By W.1, the latter is $(ax_0, \ldots, a^{p^{e-1}} x_{e-1})$. □

W.9 Corollary. The zero element and identity element of A^e are $\mathbf{0} = (0, 0, \ldots, 0)$ and $\mathbf{1} = (1, 0, \ldots, 0)$ respectively.

W.10 Corollary. Let $x, y \in A^e$ be *disjoint*, that is, $x_i = 0$ or $y_i = 0$ for $1 \leq i \leq e - 1$. Then $x + y = (x_0 + y_0, \ldots, x_{e-1} + y_{e-1})$.

Proof. We have

$$\begin{aligned} x + y = \sum_0^{e-1} V^i\{x_i\} + \sum_{i=0}^{e-1} V^i\{y_i\} &= \sum_0^{e-1} V^i(\{x_i\} + \{y_i\}) \\ &= \sum_0^{e-1} V^i(\{x_i + y_i\}) \end{aligned}$$

since $x_i = 0$ or $y_i = 0$ for $0 \leq i \leq e - 1$. □

W.11 Proposition. $(x_1, \ldots, x_{e-1}) + (y_1, \ldots, y_{e-1}) = (z_1, \ldots, z_{e-1})$ in A^{e-1} if and only if $(0, x_1, \ldots, x_{e-1}) + (0, y_1, \ldots, y_{e-1}) = (0, z, \ldots, z_{e-1})$ in A^e.

Proof. Let $x' = (x_1, \ldots, x_{e-1}) = [{}^1x, \ldots, {}^{e-1}x]$ in A^{e-1} and $x = (0, x_1, \ldots, x_{e-1}) = [0, x^{(1)}, \ldots, x^{(e-1)}]$ in A^e. Then ${}^{i+1}x = x_1^{p^i} + \cdots + p^i x_{i+1}$ and $x^{(i+1)} = 0^{p^{i+1}} + px_1^{p^i} + \cdots + p^{i+1} x_{i+1}$ for $0 \leq i \leq e - 2$, by W.1. Thus, ${}^{i+1}x = px^{(i+1)}$ for $0 \leq i \leq e - 2$. Generalizing this notation, we now have

$$\begin{aligned} x' + y' = (x_1, \ldots, x_{e-1}) + (y_1, \ldots, y_{e-1}) &= [{}^1x + {}^1y, \ldots, {}^{e-1}x + {}^{e-1}y] \\ &= [px^{(1)} + py^{(1)}, \ldots, px^{(e-1)} + py^{(e-1)}] \end{aligned}$$

and $x + y = [0, x^{(1)} + y^{(1)}, \ldots, x^{(e-1)} + y^{(e-1)}]$. Letting $z = (x + y) = (0, z_1, \ldots, z_{e-1})$ and $z' = (z_1, \ldots, z_{e-1}) = [{}^1z, \ldots, {}^{e-1}z]$, we then have ${}^{i+1}z = pz^{(i+1)} = p(x + y)^{(i+1)} = p(x^{(i+1)} + y^{(i+1)}) = px^{(i+1)} + py^{(i+1)}$ for $0 \le i \le e - 2$ and $z' = x' + y'$, by the above description of $x' + y'$. ▯

Our entire discussion applies to the ring of polynomials over $\mathbb{Q}$ in commuting indeterminants $X_0, \ldots, X_{e-1}, Y_0, \ldots, Y_{e-1}$. Letting $X = (X_0, \ldots, X_{e-1})$ and $Y = (Y_0, \ldots, Y_{e-1})$, we denote this polynomial ring by $\mathbb{Q}[X, Y]$. We let $\mathbb{Z}[X, Y]$ be the subring of $\mathbb{Q}[X, Y]$ generated by the X_r, Y_s $(0 \le r, s \le e - 1)$, that is, the ring of polynomials in the X_r, Y_s with coefficients in $\mathbb{Z}$. The major components $(X + Y)_i = a_i(X_r, Y_s)$ and $(XY)_i = m_i(X_r, Y_s)$ of $X + Y$ and XY have the property that $(x + y)_i = a_i(x_r, y_s)$ and $(xy)_i = m_i(x_r, y_s)$ for $x, y \in A^e$.

We claim that the coefficients of $a_i(X_r, Y_s)$ and $m_i(X_r, Y_s)$ are integers, that is, that $(X + Y)_i$ and $(XY)_i$ are elements of $\mathbb{Z}[X, Y]$. The proof is by induction on e and is obvious if $e = 1$. We now take $e > 1$, assume the assertion for $f = e - 1$ and proceed to prove it for e.

W.12 Proposition. Let $x, y \in A^f$. Then $((x + y)^\pi)_j \equiv (x^\pi + y^\pi)_j \pmod{pA}$ and $((xy)^\pi)_j \equiv (x^\pi y^\pi)_j \pmod{pA}$ for $0 \le j \le f - 1$.

Proof. By the induction hypothesis, the coefficients of the $a_j(X_r, Y_s)$, $m_j(X_r, Y_s)$ are in $\mathbb{Z}$. Thus, using the formula $(a + b)^p \equiv a^p + b^p \pmod{pA}$ for $a, b \in A$ (see E.1.4), we have $((x + y)^\pi)_j - (x^\pi + y^\pi)_j = (x + y)_j^p - (x^\pi + y^\pi)_j = (a_j(x_r, y_s))^p - a_j(x_r^p, y_s^p) \equiv 0 \pmod{pA}$. Similarly, $((xy)^\pi)_j - (x^\pi y^\pi)_j = (m_j(x_r, y_s))^p - m_j(x_r^p, y_s^p) \equiv 0 \pmod{pA}$. ▯

W.13 Lemma. Let $x, y \in A^e$ and let $j > 0$. Then $x_i \equiv y_i \pmod{p^jA}$ for $0 \le i \le e - 1$ if and only if $x^{(i)} \equiv y^{(i)} \pmod{p^{i+j}A}$ for $0 \le i \le e - 1$.

Proof. The proof is by induction on e and is trivial if $e = 1$. Next, let $e > 1$ and assume the assertion for $f = e - 1$. Letting $x_i \equiv y_i \pmod{p^jA}$ and $x^{(i)} \equiv y^{(i)} \pmod{p^{i+j}A}$ for $0 \le i \le f - 1$, we have $x_i^p \equiv y_i^p \pmod{p^{1+j}A}$ for $0 \le i \le f - 1$ since $j > 1$ (see E.3.6). Applying induction to x^π, y^π with $i = e - 2$, we now have $(x^\pi)^{(e-2)} \equiv (y^\pi)^{(e-2)} \pmod{p^{e-2+1+j}A}$. By condition 2 of W.1, it follows that $x^{(e-1)} - p^{e-1}x_{e-1} \equiv y^{(e-1)} - p^{e-1}x_{e-1} \pmod{p^{e-1+j}A}$, so that $x^{(e-1)} \equiv y^{(e-1)} \pmod{p^{e-1+j}A}$ if and only if $x_{e-1} \equiv y_{e-1} \pmod{p^jA}$. ▯

W.14 Theorem. The polynomials $(X + Y)_i = a_i(X_r, Y_s)$ and $(XY)_i = m_i(X_r, Y_s)$ have integer coefficients for $0 \le i \le e - 1$.

Proof. What we must show is that $(X + Y)_i$ and $(XY)_i$ are in $\mathbb{Z}[X, Y]$. It suffices for this to verify the following sequence of equalities and congruences for $1 \le i \le e - 1$, in view of part 2 of W.1:

$$\begin{aligned} p^i(X + Y)_i &= (X + Y)^{(i)} - ((X + Y)^\pi)^{(-1)} \\ &\equiv (X + Y)^{(i)} - (X^\pi + Y^\pi)^{(i-1)} \pmod{p^i\mathbb{Z}[X, Y]} \\ p^i(XY)_i &= (XY)^{(i)} - ((XY)^\pi)^{(i-1)} \\ &\equiv (XY)^{(i)} - (X^\pi Y^\pi)^{(i-1)} \pmod{p^i\mathbb{Z}[X, Y]} \end{aligned}$$

We verify only the first sequence, the verification of the second being precisely the same. The equality $p^i(X + Y)_i = (X + Y)^{(i)} - ((X + Y)^{\pi})^{(i-1)}$ is just condition 2 of W.1. For the following congruence, simply note that $((X + Y)^{\pi})_j \equiv (X^{\pi} + Y^{\pi})_j \pmod{p\mathbb{Z}[X, Y]}$ for $0 \leq j \leq i - 1$, by W.12, so that $((X + Y)^{\pi})^{(i-1)} \equiv (X^{\pi} + Y^{\pi})^{(i-1)} \pmod{p^i\mathbb{Z}[X, Y]}$ by W.13. For the last congruence, note that $(X + Y)^{(i)} = X^{(i)} + Y^{(i)} = (X^{\pi})^{(i-1)} = p^iX_i + (Y^{\pi})^{(i-1)} + p^iY_i \equiv (X^{\pi})^{(i-1)} + (Y^{\pi})^{(i-1)} = (X^{\pi} + Y^{\pi})^{(i-1)} \pmod{p^i\mathbb{Z}[X, Y]}$, by condition 2 of W.1. □

For any positive integer n, we let $\pm\mathbf{n} = \pm(\mathbf{1} + \cdots + \mathbf{1})$ (n times), so that $\pm\mathbf{n}x = \pm(x + \cdots + x)$ (n times) for $x \in A$. Note that $\mathbf{n} = [n, \ldots, n]$ and that $\mathbf{n} + x = [n + x^{(0)}, \ldots, n + x^{(e-1)}]$ and $\mathbf{n}x = [nx^{(0)}, \ldots, nx^{(e-1)}]$ for $x \in A$ and $n \in \mathbb{Z}$. Now let $X = (X_0, \ldots, X_{e-1})$ as before, let $\mathbb{Q}[X]$ and $\mathbb{Z}[X]$ be the rings of polynomials in the X_r with coefficients in $\mathbb{Q}$ and $\mathbb{Z}$ respectively, and note that the polynomials $(\mathbf{p}X)_i$ and $V(X^{\pi})_i$ $(0 \leq i \leq e - 1)$ have coefficients in $\mathbb{Z}$.

W.15 Proposition. $(\mathbf{p}X)_i \equiv V(X^{\pi})_i \pmod{p\mathbb{Z}[X]}$ for $0 \leq i \leq e - 1$.

Proof. We have $(\mathbf{p}X)^{(i)} = \mathbf{p}X^{(i)} = \mathbf{p}(x^{\pi(i-1)} + \mathbf{p}^ix_i) \equiv \mathbf{p}x^{\pi(i-1)} = V(x^{\pi})^{(i)} \pmod{p^{i+1}\mathbb{Z}[X]}$ for $0 \leq i \leq e - 1$, by W.1 and W.3. Thus, $(\mathbf{p}X)_i \equiv V(x^{\pi})_i \pmod{p\mathbb{Z}[X]}$ for $0 \leq i \leq e - 1$, by W.13. □

We now drop the assumption that the commutative ring A contain $\mathbb{Q}$ as subring. Since the coefficients of the polynomials $a_i(X_r, Y_s)$, $m_i(X_r, Y_s)$ are integers, these polynomials determine polynomials $\bar{a}_i(X_r, Y_s)$, $\overline{m}_i(X_r, Y_s)$ obtained from $a_i(X_r, Y_s)$, $m_i(X_r, Y_s)$ by replacing each coefficient $c \in \mathbb{Z}$ by the corresponding $\bar{c}$ in the prime subring of A. We now define $x + y$, xy for $x, y \in A^e$ by the conditions $(x + y)_i = \bar{a}_i(x_r, y_s)$, $(xy)_i = \overline{m}_i(x_r, y_s)$ for $0 \leq i \leq e - 1$. If A contains $\mathbb{Q}$ as subring, these are the same operations as before.

W.16 Definition. The set A^e together with the addition $x + y$ and multiplication xy defined above is denoted W_eA and is called the *ring of Witt vectors over A*.

Letting $\mathbb{Z}[X, Y, Z] = \mathbb{Z}[X_r, Y_s, Z_t]$ be the ring of polynomials in commuting indeterminants $X_0, \ldots, X_{e-1}, Y_0, \ldots, Y_{e-1}, Z_0, \ldots, Z_{e-1}$, and letting $X = (X_0, \ldots, X_{e-1})$, $Y = (Y_0, \ldots, Y_{e-1})$, $Z = (Z_0, \ldots, Z_{e-1})$, the properties of $W_e\mathbb{Z}[X, Y, Z]$ go over to the properties of W_eA by specializing X, Y, Z to elements x, y, z of W_eA. In particular, the axioms for a commutative ring go over from $\mathbb{Z}[X, Y, Z]$ to W_eA. Now suppose that the ring A has characteristic p, that is, that $\bar{p} = 0$ in A. Then we have $(x + y)^{\pi} = x^{\pi} + y^{\pi}$ and $(xy)^{\pi} = x^{\pi}y^{\pi}$ for $x, y \in W_eA$, by W.12. And we have $\mathbf{p}x = V(x^{\pi})$ for $x \in W_eA$, by W.15. Iterating the latter, we have $\mathbf{p}^ix = V^i(x^{\pi^i})$. Consequently, $V^i(x^{\pi^i})V^j(y^{\pi^j}) = (\mathbf{p}^ix)(\mathbf{p}^jy) = \mathbf{p}^i\mathbf{p}^jxy = \mathbf{p}^{i+j}xy = V^{i+j}(x^{\pi^{i+j}}y^{\pi^{i+j}})$. If A is a field of characteristic p, we can extend the operations to the algebraic closure A_{Alg} of A. Then, letting $x = u^{\pi^i}$ and $y = v^{\pi^i}$ with $u, v \in A_{\mathrm{Alg}}$, the formula just derived yields $V^i(x)V^j(y) = V^{i+j}(x^{\pi^j}y^{\pi^i})$ when applied to u, v.

At this point, we have established most of the following theorem.

W.17 Theorem. Let A be a commutative ring. Then W_eA is a commutative ring and the major components of $x + y$, xy are polynomials in the x_r, y_s with coefficients in the prime subring of A. The ring W_eA has the properties described in W.5, W.8, W.9, W.10, W.11, W.12. If A has characteristic p and prime subring π_p, then π is a homomorphism from W_eA to W_eA, $\mathbf{p}x = V(x^\pi)$ for $x \in W_eA$ and the prime subring of W_eA is $W_e\pi_p$ and is naturally isomorphic to $\mathbb{Z}_{p^e}$. If A is a field of characteristic p, then

$$V^i(x)V^j(y) = V^{i+j}(x^{\pi^j}y^{\pi^i}) \qquad \text{for } x, y \in W_eA.$$

Proof. It remains only to show that if A has characteristic p, then the mapping $\varphi: \mathbb{Z} \to W_eA$ defined by $\varphi(n) = \mathbf{n}$ for $n \in \mathbb{Z}$ is a homomorphism with Kernel $\varphi = p^e\mathbb{Z}$ and Image $\varphi = W_e\pi_p$, so that the prime subring of W_eA is $W_e\pi_p$ and is naturally isomorphic to $\mathbb{Z}_{p^e}$. We noted in Chapter 0 that φ is a homomorphism. Obviously, Image φ is the prime subring of A. We next show that $\varphi(1) = \mathbf{1}$ has additive order p^e, so that Image φ has p^e elements and Kernel $\varphi = p^e\mathbb{Z}$. Since $\mathbf{p}^e\mathbf{1} = V^e(\mathbf{1}) = \mathbf{0}$ and $\mathbf{p}^f\mathbf{1} = V^f(\mathbf{1}) \neq 0$ for $f < e$, $\mathbf{1}$ has additive order p^e. Now Image φ has p^e elements and is contained in $W_e\pi_p$. The latter also has p^e elements, so that $W_e\pi_p =$ Image φ and is therefore the prime subring of W_eA. □

A. Algebras

The material in Chapters 5 and 6 is best understood when the reader is familiar with certain material on algebras, coalgebras and bialgebras. In this appendix, we introduce the language of algebras, discuss tensor products of algebras and describe the structure of finite dimensional commutative algebras. This leads us into the remaining two appendices, C and B, where we discuss coalgebras and bialgebras respectively.

A.1 Preliminaries

Let k be a field. Then a *k-algebra* is a k-vector space A together with k-linear mappings $\pi: A \otimes_k A \to A$ and $\iota: k \to A$, called the *product* and *identity* mappings of A respectively, such that the following associativity and identity diagrams are commutative:

$$\begin{array}{ccc}
A \otimes_k A \otimes_k A & \xrightarrow{\pi \otimes I_A} & A \otimes_k A \\
{\scriptstyle I_A \otimes \pi}\downarrow & & \downarrow{\scriptstyle \pi} \\
A \otimes_k A & \xrightarrow[\pi]{} & A
\end{array}$$

$$\begin{array}{ccccc}
A \otimes_k A & \xrightarrow{\pi} & A & \xleftarrow{\pi} & A \otimes_k A \\
{\scriptstyle \iota \otimes I_A}\uparrow & & \uparrow & & \uparrow{\scriptstyle I_A \otimes \iota} \\
k \otimes_k A & \longrightarrow & A & \longleftarrow & A \otimes_k k
\end{array}$$

Here, I_A denotes the identity mapping from A to A. We denote $\pi\,(x \otimes y)$ by xy and $\iota(1_k)$ by 1_A where $x, y \in A$ and where 1_k is the identity element of k. Then A can be regarded as k-vector space and as ring with product xy and identity element 1_A. Moreover, $(cx)y = x(cy) = c(xy)$ for $c \in k$ and $x, y \in A$. Conversely, any k-vector space and ring A whose product xy satisfies $(cx)y = x(cy) = c(xy)$ for $c \in k$, $x, y \in A$ can be regarded as a k-algebra.

Any extension field K of k can be regarded as a k-algebra. And any k-algebra A which is a field can be regarded as a field extension of k by identifying k and the subfield $k1_A = \{c1_A \mid c \in k\}$ of A.

If G is a monoid and if $k[G]$ is a vector space over k having G as a basis, then the product $(\sum_{g \in G} c_g g)(\sum_{h \in G} d_h h) = \sum_{g \in G} \sum_{h \in G} (c_g d_h)(gh)$ (all but finitely many of the c_g, d_h being 0) makes $k[G]$ into a k-algebra with identity 1_G (the identity of G). This k-algebra is called the *monoid algebra* of G over k or, if G is a group, the *group algebra* of G over k.

If $A_1, \ldots, A_n$ are k-algebras, then the direct sum $A = A_1 + \cdots + A_n$ of the A_i as vector spaces can be regarded as a k-algebra with product $(a_1 + \cdots + a_n)(b_1 + \cdots + b_n) = a_1b_1 + \cdots + a_nb_n$ and identity $1_A = 1_{A_1} + \cdots + 1_{A_n}$. This k-algebra A is called the *direct product k-algebra* of $A_1, \ldots, A_n$.

A *homomorphism/isomorphism* from a k-algebra A to a k-algebra B is a k-linear mapping/bijective k-linear mapping $f: A \to B$ such that $f(xy) = f(x)f(y)$ for $x, y \in A$ and $f(1_A) = 1_B$. A *subalgebra/ideal* of a k-algebra A is a subalgebra/ideal of A as ring. Any subalgebra of a k-algebra A can be regarded as a k-algebra. If I is an ideal of a k-algebra A, then A/I can be regarded as a k-algebra.

A.2 Tensor products of algebras

If A and B are k-algebras, there are k-linear mappings $\pi_{A\otimes B}: (A \otimes_k B) \otimes_k (A \otimes_k B) \to A \otimes_k B$ and $\iota_{A\otimes B}: k \to A \otimes_k B$ such that $\pi_{A\otimes B}((x \otimes u) \otimes (y \otimes v)) = xy \otimes uv$ for $x, y \in A$, $u, v \in B$ and $\iota_{A\otimes B}(1) = 1_A \otimes 1_A$ for $c \in k$. Together with $\pi_{A\otimes B}$ and $\iota_{A\otimes B}$, $A \otimes_k B$ is a k-algebra. Note that $A \otimes 1 = \{x \otimes 1_B \mid x \in A\}$ and $1 \otimes B = \{1_A \otimes u \mid u \in B\}$ are subalgebras of $A \otimes_k B$.

Suppose that C is a k-algebra and that A and B are subalgebras of C. Let AB denote the k-span of $\{xu \mid x \in A, u \in B\}$. Then there is a k-linear mapping $f: A \otimes_k B \to C$ such that $f(A \otimes_k B) = AB$. If f is injective, we say that A and B are *linearly disjoint* over k. It is clear from T.1 that A and B are linearly disjoint over k if and only if for any linearly independent set $u_1, \ldots, u_n$ of elements of B over k, $x_1u_1 + \cdots + x_nu_n = 0$ with $x_1, \ldots, x_n$ in A only if $x_1 = \cdots = x_n = 0$. Similarly, A and B are linearly disjoint over k if and only if for any linearly independent set $x_1, \ldots, x_n$ of elements of A, $x_1u_1 + \cdots + x_nu_n = 0$ with $u_1, \ldots, u_n$ in B if and only if $u_1 = \cdots = u_n = 0$. It is also clear from T.1 that A and B are linearly disjoint over k if and only if B has a basis $\{u_\beta \mid \beta \in S\}$ such that each element of AB can be written as $\sum_{\beta\in\mathbf{S}} x_\beta u_\beta$ where the x_β are uniquely determined elements of A. Similarly, A and B are linearly disjoint over k if and only if $\{x_\alpha u_\beta \mid \alpha \in \mathbf{R}, \beta \in \mathbf{S}\}$ is a basis for AB over k for any basis $\{x_\alpha \mid a \in \mathbf{R}\}$ for A over k and any basis $\{u_\beta \mid \beta \in \mathbf{S}\}$ for B over k. If the elements of A commute with the elements of B, it is clear that A and B are linearly disjoint over k if and only if B and A are linearly disjoint over k.

If $f: A \otimes_k B \to C$ is an isomorphism of k-algebras, we say that C is the *internal tensor product* of A and B over k and we write $C = AB$ (internal tensor product of k-algebras). For C to be the internal tensor product of the subalgebras A and B, it is necessary and sufficient that the elements of A commute with the elements of B, $C = AB$ and A and B are linearly disjoint over k. If the elements of A commute with the elements of B, then C is the internal tensor product of A and B if and only if $\{x_\alpha u_\beta \mid \alpha \in \mathbf{R}, \beta \in \mathbf{S}\}$ is a basis for C over k where $\{x_\alpha \mid \alpha \in \mathbf{R}\}$ is a basis for A over k and $\{u_\beta \mid \beta \in \mathbf{S}\}$ is a basis for B over k.

Note that for any two k-algebras A and B, $A \otimes_k B = (A \otimes 1)(1 \otimes B)$ (internal tensor product of k-algebras).

Let A be a k-algebra and K an extension field of k. Then the k-algebra $K \otimes_k A$ can be regarded as a vector space over K. One sees easily that $K \otimes_k A$ is then a K-algebra, called the k-algebra obtained from A by *ascent* from k to K.

Suppose next that K is a field and A_K a K-algebra. If k is a subfield of K and A_k a k-form of A_K as vector space over K, then A_k is an *algebra k-form* of A_K if A_k is also a subring of A_K. Any algebra k-form A_k of A_K can obviously be regarded as a k-algebra such that A_K and $K \otimes_k A_k$ are isomorphic.

A.3 Finite dimensional commutative algebras

A k-algebra A is *commutative* if A is commutative as a ring. We assume for the remainder of this section that A is a finite dimensional commutative k-algebra.

Since A is finite dimensional, A is an integral domain if and only if A is a field. For if A is an integral domain and $x \in A - \{0\}$, then the linear transformation $f: A \to A$ defined by $f(y) = xy$ has kernel $\{0\}$ so that f is injective, hence surjective. Thus, $f(y) = 1_A$ and $xy = 1_A$ for some $y \in A$ and x is a unit.

If M is a maximal ideal of A, then A/M is a field, hence an integral domain, so that M is a prime ideal. In the present context, the converse is also true. For if M is a prime ideal, then A/M is an integral domain, so that A/M is a field by the preceding paragraph.

A subset S of A is *multiplicative* if S is nonempty, S does not contain 0 and $xy \in S$ for all $x, y \in S$. Note that if M is a prime ideal of A, then $S = A\text{-}M$ is a multiplicative subset of A. Now let S be any multiplicative subset of A. An ideal M of A is *S-maximal* if $M \cap S$ is the empty set and the only ideal M' of A containing M such that $M' \cap S$ is empty is $M' = M$. Let M be an S-maximal ideal of A. We claim that M is a prime ideal of A. Suppose that M is not prime and choose $a, b \in A\text{-}M$ such that $ab \in M$. The ideals $M + Aa$ and $M + Ab$ properly contain M, so that there exist $s \in (M + Aa) \cap S$ and $t \in (M + Ab) \cap S$. But then one sees easily that $st \in M \cap S$, a contradiction since $M \cap S$ is empty. Thus, M is a prime ideal of A.

The set $\text{Nil } A = \{a \in A \mid a \text{ is nilpotent } (a^n = 0 \text{ for some } n)\}$ is contained in every prime ideal of A, as one sees immediately from the definition of prime ideal. If $b \notin \text{Nil } A$ and $S = \{b, b^2, \ldots\}$, then S is a multiplicative subset of A and $b \notin M$ where M is any S-maximal (hence prime) ideal of A. It follows that Nil A is the intersection of all prime ideals of A. In particular, Nil A is an ideal of A, called the *nil radical* of A.

We say that A is *local* if A has only one maximal ideal. Since an ideal of A is maximal if and only if it is prime, the maximal ideal M of a local algebra A is $M = \text{Nil } A$. Conversely, if A has a maximal ideal M contained in Nil A, then A is local since every maximal ideal of A is prime, hence contains Nil A, hence coincides with M.

Let I be an ideal of A. Then the k-span I^i of $\{x_1 \cdots x_i \mid x_1, \ldots, x_i \in I\}$ is an ideal of A and $I \supset I^2 \supset \cdots \supset I^{i-1} \supset I^i$ for all i. If $I^i = \{0\}$ for some i, we say that I is *nilpotent*. If I is nilpotent, then certainly $I \subset \text{Nil } A$. Conversely, every ideal I of A contained in Nil A is nilpotent. For if this were not the case, we could take a nonnilpotent ideal I of A contained in Nil A with dimension as small as possible. Since such an I would not be nilpotent, $I^2 \neq 0$. Since I is nilpotent if I^2 is nilpotent, I^2 could not be nilpotent. Since $I^2 \subset I$, it would follow that $I^2 = I$. Thus, there would be ideals J of A contained in I such that the k-span IJ of $\{xy \mid x \in I, y \in J\}$ would be nonzero. Taking a minimal such J and taking $b \in J$ such that the ideal $Ib = \{xb \mid x \in I\}$ is nonzero, we would have $Ib \subset J \subset I$ and $I(Ib) = I^2b = Ib \neq \{0\}$. By the minimality of J, we would have $Ib = J$. Since $b \in J$, we would have $ab = b$ for some $a \in I \subset$ Nil A. But then $(1 - a)b = 0$. Choosing n such that $a^n = 0$, we would have $a'(1 - a) = 1$ where $a' = 1 + \cdots + a^{n-1}$. Thus, $1 - a$ would be a unit, and we would have $b = 0$, a contradiction. Thus, every ideal I of A contained in Nil A is nilpotent.

The ideal Nil A of A is a nilpotent ideal of A which contains every nilpotent ideal of A, by the preceding paragraph. It follows that A is local if and only if A has a nilpotent maximal ideal M.

Ideals I and J of A are *comaximal* if $A = I + J$. If I and J are ideals of A, then $IJ \subset I \cap J$. If, furthermore, I and J are comaximal, then $I \cap J \subset A(I \cap J) = (I + J)(I \cap J) \subset IJ$. Thus, $IJ = I \cap J$ if I and J are comaximal ideals of A.

If I and J_i are comaximal ideals of A for $1 \leq i \leq n$, then I and the k-span $\prod_1^n J_i$ of $\{x_1 \cdots x_n \mid x_1 \in J_1, \cdots, x_n \in J_n\}$ are comaximal ideals of A, as we now show by induction. If $n = 1$, this is trivial. If $n > 1$ and if I and $J = \prod_2^n J_i$ are comaximal, then $A = I + J_1 = I + J$ and

$$A = (I + J_1)(I + J) = I + J_1 J = I + \prod_1^n J_i$$

so that I and $\prod_1^n J_i$ are comaximal.

If I and J are comaximal ideals of A, then I and J^n are comaximal for all $n \geq 1$, by the preceding paragraph, hence I^m and J^n are comaximal for all $m, n \geq 1$.

If ideals $I_1, \ldots, I_n$ are pairwise comaximal, then $\prod_1^n I_i = \bigcap_1^n I_i$. We prove this by induction on n, the case $n = 2$ being already established. Thus, let $n > 2$ and suppose that $\prod_2^n I_i = \bigcap_2^n I_i$. Then I_1 and $J = \prod_2^n I_i$ are comaximal, so that $\prod_1^n I_i = I_1 J = I_1 \cap J = \bigcap_1^n I_i$.

Now suppose that $I_1, \ldots, I_n$ are pairwise comaximal ideals of A such that $\prod_1^n I_j = 0$. Let $A_i = \prod_{j \neq i} I_j$ for $1 \leq i \leq n$. Then $A = A_1 + \cdots + A_n$ (direct) and $A = A_i + I_i$ (direct) for $1 \leq i \leq n$. We prove this by induction on A. If $n = 2$, then it is enough to note that $A_1 = I_2$, $A_2 = I_1$, $A_1 \cap A_2 = I_2 \cap I_1 = I_2 I_1 = \{0\}$ and $A = I_1 + I_2 = A_2 + A_1 = I_1 + A_1 = A_2 + I_2$. Suppose next that $n > 2$ and consider the $n - 1$ pairwise comaximal ideals $I_1 I_2, I_3, \ldots, I_n$. By induction, we have $A = B + A_3 + \cdots + A_n$ (direct) where $B = \prod_3^n I_i$. We may assume that $B \neq \{0\}$. Writing $1_A = e + f$ where $e \in B$ and $f \in A_3 + \cdots + A_n$, one sees easily that $ef = 0$, hence that $e^2 = e$ and $f^2 = f$. And one sees immediately that $B = Ae = \{ae \mid a \in A\}$ can be regarded as a k-algebra with identity $1_B = e$. Now $A_1 = I_2 B$ and $A_2 = I_1 B$, $B = A_1 + A_2$ by the comaximality of I_2 and I_1. Since $A_1 \cap A_2 = A_1 A_2 = \{0\}$, we therefore have $A = A_1 + A_2 + A_3 + \cdots + A_n$ (direct). Since I_i and

A_i are comaximal and $I_i \cap A_i = I_iA_i = \{0\}$, we have $A = A_i + I_i$ (direct) for $1 \leq i \leq n$.

We noted earlier that Nil A is the intersection of all of the prime ideals of A. Since A is finite dimensional, Nil A is the intersection of finitely many distinct prime ideals $M_1, \ldots, M_n$ and these prime ideals are maximal ideals. Choose m such that $(\text{Nil } A)^m = \{0\}$ and let $I_i = M_i{}^m$ for $1 \leq i \leq n$. Since the $M_1, \ldots, M_n$ are distinct maximal ideals of A, they are pairwise comaximal. Thus, the ideals $I_1, \ldots, I_n$ are pairwise comaximal. Moreover,

$$\prod_1^n I_i = \prod_1^n M_i{}^m = \left(\prod_1^n M_i\right)^m = \left(\bigcap_1^n M_i\right)^m = (\text{Nil } A)^m = \{0\}.$$

Thus, $A = A_1 + \cdots + A_n$ (direct) and $A = A_i + I_i$ (direct) where $A_i = \prod_{j \neq i} I_j$ for $1 \leq i \leq n$. One sees easily that A_i may be regarded as a k-algebra isomorphic to $A/I_i = A/M_i{}^m$. The k-algebra $A/M_i{}^m$ is local with nilpotent maximal ideal $M/M_i{}^m$. Thus, A is the direct product k-algebra of the local k-algebras $A_1, \ldots, A_n$. Note that the only maximal ideals of A are $M_1, \ldots, M_n$. For if M were another maximal ideal of A, then M and $\prod_1^n I_i = \{0\}$ would be comaximal.

If M is a maximal ideal of A, then A/M can be regarded as a finite dimensional field extension of k. We say that A is *split* if A/M is of dimension one for every maximal ideal M of A, in which case $A = k1_A + M$ (direct) for every maximal ideal M of A. Note that A is split if and only if the local k-algebras $A_1, \ldots, A_n$ of the preceding paragraph are split.

If Nil $A = \{0\}$, we say that A is *semisimple*. If A is *semisimple*, then we may take $m = 1$ in the proof that $A = A_1 + \cdots + A_n$ (direct), so that A_i is isomorphic to the field A/M_i for $1 \leq i \leq n$. It follows that A is semisimple if and only if A has a decomposition $A = A_1 + \cdots + A_n$ (direct) where the A_i are ideals of A which as k-algebras are finite dimensional field extensions of k. If A is split, then A is semisimple if and only if $A = A_1 + \cdots + A_n$ (direct) where the A_i are ideals which as k-algebras are isomorphic to k.

C Coalgebras

In this appendix, we develop the theory of coalgebras. The preliminary definitions and some introductory observations are given in C.1. Tensor products of coalgebras and the processes of ascent and descent are discussed in C.2. In C.3, a coalgebra C and its dual algebra $A = C^*$ are related. The dual algebra $A = C^*$ is used in studying C. The minimal subcoalgebras of C and the colocal components of C are discussed. Finally, we describe the dual coalgebra of an arbitrary algebra. In C.4, we develop the theory of cocommutative coalgebras.

C.1 Preliminaries

Let k be a field. Then a k-coalgebra is a k-vector space C together with k-linear mappings $\Delta\colon C \to C \otimes_k C$ and $\varepsilon\colon C \to k$, called the *coproduct* and *coidentity* mappings of C respectively, such that the following coassociativity and coidentity diagrams are commutative:

$$\begin{array}{ccc} C \otimes_k C \otimes_k C & \xleftarrow{\Delta \otimes I_c} & C \otimes_k C \\ {\scriptstyle I_c \otimes \Delta}\uparrow & & \uparrow{\scriptstyle \Delta} \\ C \otimes_k C & \xleftarrow[\Delta]{\qquad} & C \end{array}$$

$$\begin{array}{ccccc} C \otimes_k C & \xleftarrow{\Delta} & C & \xrightarrow{\Delta} & C \otimes_k C \\ {\scriptstyle \varepsilon \otimes I_c}\downarrow & & \downarrow & & \downarrow{\scriptstyle I_c \otimes \varepsilon} \\ k \otimes_k C & \longleftarrow & C & \longrightarrow & C \otimes_k k \end{array}$$

Here, I_C denotes the identity mapping from C to C. We denote $\Delta(x)$ by $\Delta(x) = \sum_i {}_ix \otimes_k x_i$ (i ranging over some set and ${}_ix$ or x_i being 0 for all but finitely many i). The ${}_ix$, x_i are certainly not unique. However, at times we shall arbitrarily choose the x_i, say, to be a basis for C, so that the ${}_ix$ for which $\Delta(x) = \sum_i {}_ix \otimes_k x_i$ are then uniquely determined by x. The ${}_ix$ and x_i are referred to informally as the *left* and *right cofactors* of x. It is convenient at times to write Δ_C for Δ and ε_C for ε.

The coproduct Δ and coidentity ε of a k-coalgebra C satisfy the coassociativity and coidentity equations

$$\sum_{i,j} {}_j({}_ix) \otimes ({}_ix)_j \otimes x_i = \sum_{i,j} {}_ix \otimes {}_j(x_i) \otimes (x_i)_j$$

$$x = \sum_i \varepsilon({}_ix)x_i = \sum_i \varepsilon(x_i){}_ix.$$

If A is a finite dimensional algebra, then the product $\pi: A \otimes_k A \to A$ and identity $\iota: k \to A$ induce mappings $\Delta: A^* \to A^* \otimes_k A^*$ and $\varepsilon: A^* \to k$ such that, for $x \in A^*$ and $_1 x, x_1, \ldots, _n x, x_n \in A^*$,

1. $\Delta(x) = \sum_i {}_i x \otimes x_i \Leftrightarrow x(ab) = \sum_i {}_i x(a) x_i(b)$ for all $a, b \in A$;
2. $\varepsilon(x) = x(1_A)$.

These mappings Δ and ε are the adjoints $\Delta = \pi^*$ and $\varepsilon = \iota^*$ of π and ι, if one identifies $A^* \otimes_k A^*$ with $(A \otimes_k A)^*$ and k with k^*. The associativity and identity properties of π and ι lead to the coassociativity and coidentity equations for the mappings Δ and ε, so that $C = A^*$ together with Δ and ε is a k-coalgebra, called the *dual k-coalgebra* of A. If A is infinite dimensional, then $A^* \otimes_k A^*$ and $(A \otimes_k A)^*$ cannot be identified and A^* cannot always be given a k-coalgebra structure in the above prescribed way. It is possible to give the subspace A^0 of "continuous" elements of A^* a k-coalgebra structure in the above prescribed way. The details are given at the end of C.2.

Since we can identify k and its dual space k^*, k may be regarded as k-coalgebra, and we have $\Delta(1_k) = 1_k \otimes 1_k$ and $\varepsilon(1_k) = 1_k$.

If G is any set and if $k[G]$ is a vector space with basis G, then $k[G]$ may be regarded as a k-coalgebra such that $\Delta g = g \otimes g$ and $\varepsilon(g) = 1$ for all $g \in G$. This k-coalgebra is called the *group k-coalgebra* of G. If C is a vector space over k, C^0 a subspace of C of codimension 1 and $e \in C - C^0$, then C may be regarded as a k-coalgebra such that $\Delta e = e \otimes e$, $\varepsilon(e) = 1$ and $\Delta x = e \otimes x + x \otimes e$, $\varepsilon(x) = 0$ for all $x \in C^0$.

If the C_i ($i \in S$) are k-coalgebras, then the vector space direct sum $C = \sum_{i \in S} C_i$ can be regarded as k-coalgebra such that

$$\Delta(x) = \sum_{i \in S} \Delta_i(x_i) \quad \text{and} \quad \varepsilon(x) = \sum_{i \in S} \varepsilon_i(x_i),$$

where $x = \sum_{i \in S} x_i$, $x_i \in C_i$ and the coproduct and coidentity for C_i are Δ_i and ε_i for $i \in S$. Note that $k[G] = \sum_{g \in G} kg$ is an instance of this.

A *homomorphism/isomorphism* from a k-coalgebra C to a k-coalgebra D is a k-linear mapping/bijective k-linear mapping $f: C \to D$ such that for $x \in C$ and $\Delta_C(x) = \sum_i {}_i x \otimes x_i$, we have $\Delta_D(f(x)) = \sum_i f({}_i x) \otimes f(x_i)$; and such that for $x \in C$, $\varepsilon_C(x) = \varepsilon_D(f(x))$. These conditions on f are equivalent to the commutativity of the diagrams

$$\begin{array}{ccc} C & \xrightarrow{\ f\ } & D \\ {\scriptstyle \Delta}\downarrow & & \downarrow{\scriptstyle \Delta} \\ C \otimes_k C & \xrightarrow[f \otimes f]{} & D \otimes_k D \end{array}$$

$$\begin{array}{ccc} C & \xrightarrow{\ f\ } & D \\ {\scriptstyle \varepsilon}\searrow & & \swarrow{\scriptstyle \varepsilon} \\ & k & \end{array}$$

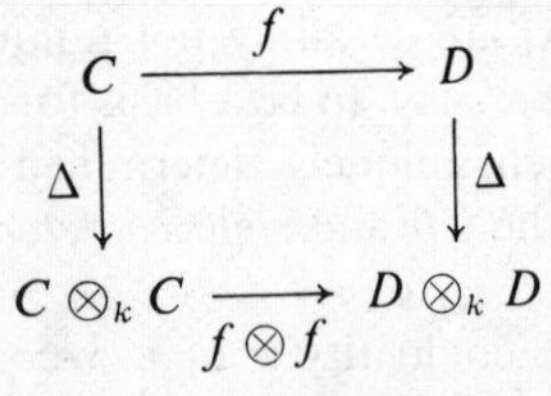

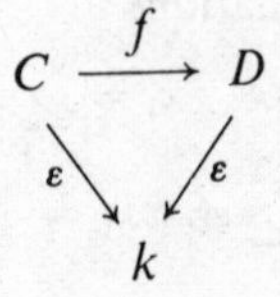

A *subcoalgebra/coideal* of a k-coalgebra C is a subspace D of C such that $\Delta(D) \subset D \otimes_k D$/a subspace P of C such that $\Delta(P) \subset P \otimes C + C \otimes P$ and $\varepsilon(P) = \{0\}$. Any subcoalgebra of a k-coalgebra C can be regarded as k-coalgebra. If P is a coideal of a k-coalgebra C, then C/P can be regarded as a k-coalgebra such that the usual quotient mapping $C \to C/P$ is a homomorphism, since $C/P \otimes_k C/P$ can be identified with $(C \otimes_k C)/(P \otimes_k C + C \otimes_k P)$.

A *measuring representation* of a k-coalgebra C from a k-algebra A_1 to a k-algebra A_2 is k-linear mapping $\rho\colon C \to \mathrm{Hom}_k(A_1, A_2)$ such that $\rho(x)(1_{A_1}) = \varepsilon(x)1_{A_2}$ and $\rho(x)(ab) = \sum_i \rho({}_ix)(a)\rho(x_i)(b)$ for all $x \in C$ and $a, b \in A$. A *measuring representation* of a k-coalgebra C on a k-algebra A is a measuring representation of C from A to A.

Note that a measuring representation ρ of a k-coalgebra C from a k-algebra A to k is a k-linear mapping $\rho\colon C \to A^* = \mathrm{Hom}_k(A, k)$ such that $\rho(x)(1_A) = \varepsilon(x)$ and $\rho(x)(ab) = \sum_i \rho({}_ix)(a)\rho(x_i)(b)$ for all $x \in C$ and $a, b \in A$. In this case, ρ is therefore simply a homomorphism from the k-coalgebra C to the k-coalgebra A^0 described at the end of C.3.

The *kernel* Kernel $\rho = \{x \in C \mid \rho(x) = 0\}$ of a measuring representation ρ of a k-coalgebra C is not always a coideal of C.

C.2 Tensor products of coalgebras

If C and D are k-coalgebras, then there are k-linear mappings $\Delta_{C \otimes D}\colon C \otimes_k D \to (C \otimes_k D) \otimes_k (C \otimes_k D)$ and $\varepsilon_{C \otimes D}\colon C \otimes_k D \to k$ such that $\Delta_{C \otimes D}(x \otimes y) = \sum_{i,j} {}_ix \otimes {}_jy \otimes x_i \otimes y_j$ and $\varepsilon_{C \otimes D}(x \otimes y) = \varepsilon_C(x)\varepsilon_D(y)$ for all $x \in C$ and $y \in D$, where $\Delta_C(x) = \sum_i {}_ix \otimes x_i$ and $\Delta_D(y) = \sum_j {}_jy \otimes y_j$. Together with $\Delta_{C \otimes D}$ and $\varepsilon_{C \otimes D}$, $C \otimes_k D$ is a k-colagebra.

Let C be a k-coalgebra, let K be an extension field of k and regard $K \otimes_k C$ as vector space over K. Let $\Delta_K\colon K \otimes_k C \to (K \otimes_k C) \otimes_K (K \otimes_k C)$ and $\varepsilon_K\colon K \otimes_k C \to K$ be the K-linear mappings such that $\Delta_K(1 \otimes x) = \sum_i (1 \otimes {}_ix) \otimes (1 \otimes x_i)$ and $\varepsilon_K(1 \otimes x) = \varepsilon(x)$ for $x \in C$ and $\Delta x = \sum_i {}_ix \otimes x_i$. Then $K \otimes_k C$ together with Δ_K and ε_K is a K-coalgebra, called the K-coalgebra obtained from C by *ascent* from k to K.

A vector space k-form C_k of a K-coalgebra C_K such that $\Delta(C_k) \subset C_k \otimes_K C_k$ (image of $C_k \times C_k$ in $C_K \otimes_K C_K$ under $\otimes$) and such that $\varepsilon(C_k) \subset k$ is called a *coalgebra k-form* of C_K (see T.4). Any coalgebra k-form C_k of C_K can obviously be regarded as a k-coalgebra such that the K-coalgebras C_K and $K \otimes_k C_k$ are isomorphic.

C.3 Duality

Let C be a k-coalgebra. The field k is isomorphic to its k-dual space k^* by the mapping sending each $c \in k$ to its left translation $c_L \in k^*$. And there is a k-linear injection from $C^* \otimes_k C^*$ into $(C \otimes_k C)^*$ which maps each $a \otimes_k b$ $(a, b \in C^*)$ to the element $a \cdot b$ of $(C \otimes_k C)^*$ such that $(a \cdot b)(x \otimes y) = a(x)b(x)$ for all $x, y \in C$. Thus, the coidentity and coproduct mappings $\varepsilon\colon C \to k$ and $\Delta\colon C \to C \otimes_k C$ induce via their adjoints ε^* and Δ^*, k-linear mappings $\iota: k \to C^*$ and $\pi: C^* \otimes_k C^* \to C^*$ such that $\iota(c) = \varepsilon^*(c_L)$ for $c \in k$ and $\pi(a \otimes b) = \Delta^*(a \cdot b)$ for $a, b \in C^*$. Letting $A = C^*$, $\iota(1_k) = 1_A$ and $\pi(a \otimes b) = ab$ for $a, b \in A$, we have $1_A = \varepsilon$ and $(ab)(x) = \sum_i a({}_ix)b(x_i)$ for all $x \in C$ and $a, b \in A$, where $\Delta x = \sum_i {}_ix \otimes x_i$. It follows easily from the coidentity and coassociativity equations for C that A together with the product ab is an algebra with identity 1_A. Furthermore, one sees easily that for $x \in C$ and ${}_1x, x_1, \ldots, {}_nx, x_n \in C$, $\Delta x = \sum_i {}_ix \otimes x_i$ if and only if $(ab)(x) = \sum_i a({}_ix)b(x_i)$ for all $a, b \in A$. This k-algebra $A = C^*$ is called the *dual k-algebra* of the k-coalgebra C.

Any k-coalgebra C has a measuring representation ρ from its dual k-algebra $A = C^*$ to k. For $x \in C$, $\rho(x)$ is defined by $\rho(x)(a) = a(x)$ for $a \in A$ (so that ρ is the canonical imbedding of C in C^{**}). The equations $\rho(x)(1_A) = \varepsilon(x)$ and $\rho(x)(ab) = \sum_i \rho({}_ix)(a)\rho(x_i)(b)$ are simply translations of the earlier equations $1_A(x) = \varepsilon(x)$ and $(ab)(x) = \sum_i a({}_ix)b(x_i)$ for $x \in C$ and $a, b \in A$. Note that Kernel $\rho = \{0\}$.

If D is a subcoalgebra of C, then $D^\perp = \{a \in A \mid a(x) = 0 \text{ for all } x \in D\}$ is an ideal of A. For if $x \in D$ and $\Delta x = \sum_i {}_ix \otimes x_i$ where the ${}_ix$, x_i are in D, then $a \in D^\perp$ implies that $(ab)(x) = \sum_i a({}_ix)b(x_i) = 0$ and $(ba)(x) = \sum_i b({}_ix)a(x_i) = 0$ for all $b \in A$. Suppose, conversely, that I is an ideal of A and let $D = I^\perp = \{x \in C \mid a(x) = 0 \text{ for all } a \in I\}$. Let $\{x_\alpha \mid \alpha \in \mathbf{R}\}$ be a basis for D and choose a set $\{x_\beta \mid \beta \in \mathbf{S}\}$ such that $\{x_\alpha \mid \alpha \in \mathbf{R}\} \cup \{x_\beta \mid \beta \in \mathbf{S}\}$ (disjoint union) is a basis for C. Let $x \in D$ and $\Delta x = \sum_\alpha {}_\alpha x \otimes x_\alpha + \sum_\beta {}_\beta x \otimes x_\beta$. We claim that ${}_\beta x = 0$ for all $\beta \in \mathbf{S}$ and $x_\alpha \in D$ for all $\alpha \in \mathbf{R}$. For this, note that for $b \in I$,

$$0 = (ab)(x) = \sum_\alpha a({}_\alpha x)b(x_\alpha) + \sum_\beta a({}_\beta x)b(x_\beta) = \sum_\beta a({}_\beta x)b(x_\beta)$$

$$= b\left(\sum_\beta a({}_\beta x)x_\beta\right) \qquad \text{for all } a \in A.$$

Thus, we have $\sum_\beta a({}_\beta x)x_\beta \in I^\perp = D$ for all $a \in A$. Since the set $\{x_\beta \mid \beta \in \mathbf{S}\}$ is a basis for a subspace of C complimentary to D, it follows that $\sum_\beta a({}_\beta x)x_\beta = 0$ for all $a \in A$ and $a({}_\beta x) = 0$ for all $a \in A$ and all $\beta \in \mathbf{S}$. It follows that ${}_\beta x = 0$ for all $\beta \in \mathbf{S}$. Thus, $\Delta x = \sum_\alpha {}_\alpha x \otimes x_\alpha$ where $\{x_\alpha \mid \alpha \in \mathbf{R}\}$ is a basis for D. For $a \in I$, we now have $0 = (ab)(x) = \sum_\alpha a({}_\alpha x)b(x_\alpha) = b(\sum_\alpha a({}_\alpha x)x_\alpha)$ for all $b \in A$. Since $A = C^*$ separates the points of C, we therefore have $\sum_\alpha a({}_\alpha x)x_\alpha = 0$ for all $a \in I$, so that $a({}_\alpha x) = 0$ for all $a \in I$ and all $\alpha \in \mathbf{R}$. Thus, the ${}_\alpha x$ are in $I^\perp = D$ and D is a subcoalgebra of C.

It follows from the preceding paragraph that a subspace D of C is a subcoalgebra of C if and only if $D^\perp$ is an ideal of A, for we always have $D = (D^\perp)^\perp$. Note that the k-algebra $A/D^\perp$ can be regarded as the dual k-algebra of a cosubalgebra D of C.

If P is a coideal of C, then $P^\perp$ is a subalgebra of A. For we have $1_A = \varepsilon \in P^\perp$. And for $a, b \in P^\perp$, we have $ab \in P^\perp$ since $(ab)(x) = \sum_i a({}_ix)b(x_i) + \sum_j a({}_jx)b(x_j) = 0$ for $x \in A$ where the cofactors of x are taken so that ${}_ix \in C$, $x_i \in P$, ${}_jx \in P$, $x_j \in C$ for all i, j. Conversely, let B be a subalgebra of A. Then $P = B^\perp$ is a coideal of C. For $\varepsilon(x) = 0$ for $x \in P$, since $\varepsilon = 1_A \in B$. And we see for $x \in P$, $\{x_\alpha \mid \alpha \in \mathbf{R}\}$ a basis for P and $\{x_\beta \mid \beta \in \mathbf{S}\}$ a basis for a subspace of C

complimentary to P that $\Delta x = \sum_\alpha {}_\alpha x \otimes x_\alpha + \sum_\beta {}_\beta x \otimes x_\beta$ where ${}_\beta x$ is in $P = B^\perp$ for $\beta \in \mathbf{S}$. Namely, we have

$$0 = (ab)(x) = \sum_\beta a({}_\alpha x)b(x_\alpha) + \sum_\beta a({}_\beta x)b(x_\beta) = \sum_\beta a({}_\beta x)b(x_\beta)$$

$$= b\left(\sum_\beta a({}_\beta x)x_\beta\right) \qquad \text{for all } a, b \in B,$$

thus $\sum_\beta a({}_\beta x)x_\beta \in B^\perp = P$ for all $a \in B$, thus $0 = \sum_\beta a({}_\beta x)x_\beta$ for all $a \in B$, thus $a({}_\beta x) = 0$ for all $a \in B$ and all $B \in \mathbf{S}$. Thus, P is a coideal of C.

It follows from the preceding paragraph that a subspace P of C is a coideal of C if and only if $P^\perp$ is a subalgebra of A, for we always have $P = (P^\perp)^\perp$. Note that if P is a coideal of C, then the k-algebra $P^\perp$ can be regarded as the dual k-algebra of the k-coalgebra C/P.

The kernel $C^0 = \{x \in C \mid \varepsilon(x) = 0\}$ is the coideal $1_A{}^\perp$ orthogonal to the subalgebra $k1_A$ of A.

For any collection $\{D_\alpha \mid \alpha \in \mathbf{R}\}$ of subcoalgebras of C, it is obvious that the sum of subspaces $\sum_\alpha D_\alpha$ is a subcoalgebra of C. Furthermore, $\bigcap_\alpha D_\alpha$ is a subcoalgebra of C, since $\bigcap_\alpha D_\alpha = (\sum_\alpha D_\alpha{}^\perp)^\perp$ and $\sum_\alpha D_\alpha{}^\perp$ is an ideal of A. In the same way, $\sum_\beta P_\beta$ is a coideal of C for any collection $\{P_\beta \mid \beta \in \mathbf{S}\}$ of coideals of C, for $(\sum_\beta P_\beta)^\perp = \bigcap P_\beta{}^\perp$ and $\bigcap P_\beta{}^\perp$ is a subalgebra of A.

We now let x be an element of the coalgebra C and let a, b be elements of the dual algebra $A = C^*$. Then x^a, ${}^b x$ are well defined by the equations $x^a = \sum_i a({}_i x)x_i$ and ${}^b x = \sum_i b(x_i){}_i x$ where $\Delta x = \sum_i {}_i x \otimes x_i$. Note that x^a is linear in x and in a and that ${}^b x$ is linear in x and in b. Note also that $b(x^a) = (ab)(x) = a({}^b x)$. It follows that $\Delta x^a = \sum_i {}_i x^a \otimes x_i$, since $(bc)(x^a) = (abc)(x) = \sum_i (ab)({}_i x)c(x_i) = \sum_i b({}_i x^a)c(x_i)$ for all $b, c \in A$. And $\Delta {}^b x = \sum_i {}_i x \otimes {}^b x_i$, since $(ac)({}^b x) = (acb)(x) = \sum_i a({}_i x)(cb)(x_i) = \sum_i a({}_i x)c({}^b x_i)$ for all $a, c \in A$. We have $x^{1_A} = {}^{1_A}x = x$ since $\sum_i \varepsilon({}_i x)x_i = \sum_i \varepsilon(x_i){}_i x = x$. We also have $(x^a)^b = x^{(ab)}$, since $(x^a)^b = \sum_i b({}_i x^a)x_i = \sum_i (ab)({}_i x)x_i = x^{(ab)}$. Similarly, ${}^a({}^b x) = {}^{(ab)}x$, since ${}^a({}^b x) = \sum_i a({}^b x_i){}_i x = \sum_i (ab)(x_i){}_i x = {}^{(ab)}x$. Finally, we have ${}^b(x^a) = ({}^b x)^a$, since $c({}^b(x^a)) = (cb)(x^a) = (acb)(x)$ and $c(({}^b x)^a) = (ac)({}^b x) = (acb)(x)$ for all $c \in A$. We denote ${}^b(x^a) = ({}^b x)^a$ by ${}^b x^a$.

At this point, C is a two-sided *A-module* in the sense that we have defined elements x^a, ${}^b x$ of C for $x \in C$ and $a, b \in A$ such that

1. x^a is linear in x and a, and ${}^b x$ is linear in x and b;
2. $x^{1_A} = {}^{1_A}x = x$ for all $x \in C$;
3. $(x^a)^b = x^{(ab)}$ and ${}^a({}^b x) = {}^{(ab)}x$ for all $x \in C$ and $a, b \in A$;
4. ${}^b(x^a) = ({}^b x)^a$ for all $a, b \in A$, $x \in C$.

An *A-submodule* of C is a subspace D of C such that ${}^b x \in D$ and $x^a \in D$ for all $x \in D$ and $a, b \in A$. Note that for $x \in C$, the k-span ${}^A x^A$ of $\{{}^b x^a \mid a, b \in A\}$ is an A-submodule of C containing x.

For any $x \in C$, the A-submodule ${}^A x^A$ is finite dimensional. To see this, let $\Delta x = \sum_1^n {}_i x \otimes x_i$ and note that the k-span x^A of $\{x^a \mid a \in A\}$ is finite dimensional since $x^a = \sum_1^n a({}_i x)x_i$ is contained in the k-span of $x_1, \ldots, x_n$ for all $b \in A$. Similarly, the k-span ${}^A x$ of $\{{}^b x \mid b \in A\}$ is finite dimensional. Let x^A and

${}^Ax$ be spanned over k by $x^{a_1}, \ldots, x^{a_r}$ and ${}^{b_1}x, \ldots, {}^{b_s}x$ respectively, where we specify that $a_1 = b_1 = 1_A$. Then the k-span D of $\{{}^{b_j}x^{a_i} \mid 1 \leq i \leq r, 1 \leq j \leq s\}$ is an A-submodule of C. For if $a \in A$, we have $({}^{b_j}x^{a_i})^a = {}^{b_j}(x^{(a_i a)}) = {}^{b_j}(c_1 x^{a_1} + \cdots + c_r x^{a_r}) = c_1{}^{b_j}x^{a_1} + \cdots + c_r{}^{b_j}x^{a_r}$, where the $c_1, \ldots, c_r \in A$ are chosen such that the element $x^{(a_i a)}$ of x^A is $c_1 x^{a_1} + \cdots + c_r x^{a_r}$, so that $({}^{b_j}x^{a_i})^a \in D$. And if $b \in A$, we have ${}^b({}^{b_j}x^{a_i}) \in D$ by a similar argument. Since D is an A-submodule and D contains $x = {}^{b_1}x^{a_1}$, it follows that ${}^Ax^A = D$ and therefore that ${}^Ax^A$ is finite dimensional.

A subspace D of C is a subcoalgebra of C if and only if D is an A-submodule of C. For suppose that D is a subcoalgebra of C, and let $x \in D$ and $a, b \in A$. Then $\Delta x \times \sum_i {}_ix \otimes x_i$ where the ${}_ix$, x_i are in D. Thus, D contains $x^a = \sum_i a({}_ix)x_i$ and ${}^bx = \sum b(x_i){}_ix$. Conversely, suppose that D is an A-submodule of C. We claim that $D^\perp$ is an ideal of A. Thus, let $a \in D^\perp$ and $b \in A$. Then we have $(ab)(x) = a({}^bx) = 0$ for all $x \in D$, so that $ab \in D^\perp$. Similarly, we have $(ba)(x) = a(x^b) = 0$ for all $x \in D$, so that $ba \in D^\perp$. Thus, $D^\perp$ is an ideal of A and D is a subcoalgebra of C.

Every element x of a coalgebra C is contained in some finite dimensional subcoalgebra of C. For ${}^Ax^A$ is a finite dimensional subcoalgebra of C containing x, by the preceding two paragraphs.

A subcoalgebra D of C is *minimal* if $D \neq \{0\}$ and the only nonzero subcoalgebra E of C contained in D is $E = D$. If D is a minimal subcoalgebra of C, then $D = {}^Ax^A$ for every nonzero element x of D. Thus, every minimal subcoalgebra D of C is finite dimensional. It follows that a subcoalgebra D of C is minimal if and only if the ideal $D^\perp$ of A is a maximal ideal of A.

Let D be a minimal subcoalgebra of C and suppose that $\{D_\alpha \mid \alpha \in \mathbf{R}\}$ is a collection of subcoalgebras of C such that $D \subset \sum_\alpha D_\alpha$. Then $D \subset D_\alpha$ for some $\alpha \in \mathbf{R}$. To see this, note first that $D \subset \sum_1^n D_{\alpha_i}$ for suitable $\alpha_1, \ldots, \alpha_n$ in $\mathbf{R}$, since D is finite dimensional. Next, let I_i be the ideal $I_i = D_{\alpha_i}{}^\perp$ of A for $1 \leq i \leq n$ and let I be the maximal ideal $I = D^\perp$ of A. Then $I \supset \bigcap_1^n I_i$, since $D^\perp \supset (\sum_1^n D_{\alpha_i})^\perp = \bigcap_1^n D_{\alpha_i}{}^\perp$. It follows that $I \supset I_{i_0}$ for some i_0. For otherwise $A = I + I_i$ for all i, by the maximality of I, so that $A = I + I_1 \cdots I_n$ (for the same reasons as those given in A.3 for commutative algebras) and therefore $A = I + \bigcap_1^n I_i$, a contradiction. But from $I \supset I_{i_0}$, we have $D \subset D_{i_0}$, since $D = D^{\perp\perp} = I^\perp$ and $D_{i_0} = D_{i_0}{}^{\perp\perp} = I_{i_0}{}^\perp$.

Every nonzero coalgebra C has minimal subcoalgebras, since C has nonzero finite dimensional subcoalgebras. If C has precisely one minimal subcoalgebra, then we say that C is *colocal*.

For any minimal subcoalgebra D of C, $C(D)$ denotes the subcoalgebra $\sum_{E \in \mathbf{D}} E$ where $\mathbf{D}$ is the set of colocal subcoalgebras of C containing D. The subcoalgebra $C(D)$ of C is colocal. For if D' is a minimal subcoalgebra of $C(D)$, then $D' \subset E$ for some colocal subcoalgebra E of C containing D, whence $D' = D$. Since the colocal subcoalgebra $C(D)$ of C contains D and every colocal subcoalgebra of C containing D, we call $C(D)$ the *colocal component* of C containing D.

Let M be the set of minimal subcoalgebras of C. The sum $\sum_{D \in \mathbf{M}} C(D)$ is then direct. For if $D \in \mathbf{M}$ and $\mathbf{M}' = \mathbf{M} - \{D\}$, then $C(D) \cap \sum_{D' \in \mathbf{M}'} C(D') =$

$\{0\}$. Otherwise, we would have $D \subset \sum_{D' \in \mathbf{M}'} C(D')$, so that $D \subset C(D')$ and $D = D'$ for some $D' \in M'$, a contradiction. In particular, the sum $\sum_{D \in \mathbf{M}} D$ is direct. We call $\sum_{D \in \mathbf{M}} D$ the *socle* of C and $\sum_{D \in \mathbf{M}} C(D)$ the *hypersocle* of C.

For the sake of completeness, we close this section with a brief description of the *dual k-coalgebra* $C = A^0$ of an arbitrary k-algebra A. For this, let A^* be the k-dual space of a k-algebra A, let $x \in A^*$ and let $a, b \in A$. Let x^a and ${}^bx$ be the functions on A defined by $x^a(c) = x(ac)$ and ${}^bx(c) = x(cb)$ for all c in A. Then x^a and ${}^bx$ are elements of A^* and $x^a(b) = x(ab) = {}^bx(a)$. Note that x^a is linear in x and a, and that ${}^bx$ is linear in x and b. Note also that $x^{1_A} = {}^{1_A}x = x$, $(x^a)^b = x^{(ab)}$, ${}^a({}^bx) = {}^{(ab)}x$ and ${}^b(x^a) = ({}^bx)^a$, as one sees from the equations $x^{1_A}(c) = x(c) = {}^{1_A}x(c)$, $(x^a)^b(c) = x^a(bc) = x(abc) = x^{(ab)}(c)$, ${}^a({}^bx)(c) = {}^bx(ca) = x(cab) = {}^{(ab)}x(c)$, ${}^b(x^a)(c) = x^a(cb) = x(acb)$ and $({}^bx)^a(c) = {}^bx(ac) = x(acb)$ for $c \in A$. Thus, A^* is a two-sided A-module in the sense described earlier in the section. For x in A^*, the k-span ${}^Ax^A$ of $\{{}^bx^a \mid a, b \in A\}$ is an A-submodule of A^*. We let $A^0 = \{x \in A \mid {}^Ax^A \text{ is finite dimensional}\}$ and note that A^0 is an A-submodule of A^*.

It is clear from earlier parts of this section that A^0 is the largest subspace of A^* which possibly could have coidentity and coproduct mappings Δ^0 and ε^0 such that $\varepsilon^0(x) = x(1_A)$ and $\Delta^0(x) = \sum_i {}_ix \otimes x_i$ if and only if $x(ab) = \sum_i {}_ix(a)x_i(b)$ for all $a, b \in A$, where x and the ${}_ix$, x_i are in A^0. We now proceed to produce such mappings Δ^0 and ε^0 and show that $C = A^0$ together with Δ^0 and ε^0 is a k-coalgebra. The mapping ε^0 is, of course, defined by $\varepsilon^0(x) = x(1_A)$ for $x \in C = A^0$. Next, let $x \in C = A^0$, and consider the finite dimensional A-submodule $D = {}^Ax^A$ of $C = A^0$. The subspace $D^\perp = \{a \in A \mid x(a) = 0 \text{ for all } x \in D\}$ is an ideal of A. For if $a \in D^\perp$, $b \in A$ and $y \in D$, we have ${}^by, y^b \in D$, so that $y(ab) = {}^by(a) = 0$ and $y(ba) = y^b(a) = 0$. Since D is finite dimensional, the k-algebra $\bar{A} = A/D^\perp$ is finite dimensional and D may be regarded as the k-dual space $\bar{A}^*$ of $\bar{A}$, an element y of D being regarded as the element of $\bar{A}^*$ which maps the element $\bar{a} = a + D^\perp$ of $\bar{A} = A/D^\perp$ to $y(\bar{a}) = y(a)$ for all $a \in A$. Thus, there is a unique linear mapping Δ_D from D to $D \otimes_k D$ mapping each $y \in D$ into an element $\sum_i {}_iy \otimes y_i$ of $D \otimes_k D$ such that $y(\bar{a}\bar{b}) = \sum_i {}_iy(\bar{a})y_i(\bar{b})$ for all $\bar{a}, \bar{b} \in A$, hence such that $y(ab) = \sum_i {}_iy(a)y_i(b)$ for all $a, b \in A$. Returning to the original element x of $C = A^0$, define $\Delta^0(x) = \sum_i {}_ix \otimes x_i$ (image of x under the mapping Δ_D just described). Thus defined, Δ^0 is a k-linear mapping from C to $C \otimes_k C$ such that for $x \in C$ and ${}_1x, x_1, \ldots, {}_nx, x_n \in C$, $\Delta^0(x) = \sum_i {}_ix \otimes x_i$ if and only if $x(ab) = \sum_i {}_ix(a)x_i(b)$ for all $a, b \in A$. One now easily shows that $C = A^0$ together with Δ^0, ε^0 is a k-coalgebra, the proof being the same as the proof of 5.3.8. The coalgebra $C = A^0$ is called the *dual k-coalgebra* of the k-algebra A.

C.4 Cocommutative coalgebras

A k-coalgebra C is *cocommutative* if $\Delta x = \sum_i {}_ix \otimes x_i$ if and only if $\Delta x = \sum_i x_i \otimes {}_ix$ for all $x \in C$. From the equations $(ab)(x) = \sum_i a({}_ix)b(x_i)$ and $(ba)(x) = \sum_i b({}_ix)a(x_i)$ for $x \in C$, $\Delta x = \sum_i {}_ix \otimes x_i$ and $a, b \in A = C^*$, we see that the k-coalgebra C is cocommutative if and only if its dual k-algebra $A = C^*$ is commutative.

Suppose that C is a finite dimensional cocommutative k-coalgebra. Then the dual k-algebra $A = C^*$ is a finite dimensional commutative algebra. We showed in A.3 that $A = \sum_1^n A_i$ (direct) where the A_i are ideals of A such that A/I_i is a local k-algebra for $I_i = \sum_{j \neq i} A_j$ $(1 \leq i \leq n)$. Letting D_i be the subcoalgebra $D_i = I_i^{\perp}$ of C, we then have $C = \sum_i D_i$ (direct). Since we may regard the local k-algebra $A/D_i^{\perp} = A/I_i$ as the dual k-algebra of D_i, the subcoalgebra D_i is colocal with unique minimal subcoalgebra $M_i^{\perp}$ where M_i is the unique maximal ideal of A containing I_i.

For the remainder of this section, we assume that C is an arbitrary cocommutative k-coalgebra. Then $C = \sum_{D \in \mathbf{M}} C(D)$ (direct) where $\mathbf{M}$ is the set of minimal subcoalgebras of C. To see this, let $x \in C$ and let C_x be the finite dimensional subcoalgebra $C_x = {}^A x^A$ where A is the dual k-algebra of C. Then $C_x = \sum_i D_i$ (direct) where the D_i are colocal subcoalgebras of C_x, by the preceding paragraph. It follows that $C_x \subset \sum_{C \in \mathbf{M}} C(D)$. But then $C = \sum_{D \in \mathbf{M}} C(D)$, since $C = \bigcup_{x \in C} C_x$. That the sum $\sum_{D \in \mathbf{M}} C(D)$ is direct was shown in C.3.

If $C = \sum_{D \in \mathbf{M}} D$ where $\mathbf{M}$ is the set of minimal subcoalgebras of C, we say that C is *cosemisimple*. For C to be cosemisimple, it is necessary and sufficient for the dual k-algebra of every finite dimensional subcoalgebra of C to be semisimple in the sense of A.3.

The subcoalgebras of C of dimension one are the subcoalgebras D of C of the form $D = kg$ (k-span of g) where g is a nonzero element of C such that $\Delta g = g \otimes g$. To see this, let D be a subcoalgebra of dimension one and take $d \in D - \{0\}$. Then $D = kd$, so that $\Delta d = c_1 d \otimes c_2 d = c(d \otimes d)$ for suitable $c_1, c_2, c \in k$. By the coidentity equation, $d = c\varepsilon(d)d$, so that $c \neq 0$. Letting $g = cd$, we therefore have $D = kg$ and $\Delta g = g \otimes g$. The nonzero elements g of C such that $\Delta g = g \otimes g$ are called the *grouplike* elements of C. The set of grouplike elements of C is denoted $G(C)$. Note that for $g \in G(C)$, $\varepsilon(g) = 1$. For the coidentity equation implies that $g = \varepsilon(g)g$ for all $g \in G(C)$. For any grouplike element g of C, we let $C(g)$ denote the colocal component $C(kg)$ containing the minimal subcoalgebra kg of C, and we call $C(g)$ the *colocal component* of C containing g. Note that $G(C)$ is a linearly independent subset of C, since the sum $\sum_{g \in G(C)} C(g)$ is direct.

We say that C is *cosplit* if every minimal subcoalgebra of C is of dimension one. Thus, C is cosplit if and only if $C = \sum_{g \in G(C)} C(g)$ (direct). More generally, C has a unique maximal cosplit subcoalgebra, namely $\sum_{g \in G(C)} C(g)$. Note that C is cosplit if and only if the k-dual algebra of every finite dimensional subcoalgebra of C is split.

Suppose that C is cosplit. Then C is colocal if and only if C has a unique grouplike element e and $C = C(e)$. And C is cosemisimple if and only if $G(C)$ is a basis for C. For, under the assumption that $C = \sum_{g \in G(C)} C(g)$ (direct), $G(C)$ is a basis for C if and only if $C(g) = kg$ for all $g \in G(C)$.

Throughout the remainder of this appendix we assume that C is a colocal cocommutative k-coalgebra with minimal subcoalgebra D. Let $A = C^*$ be the dual k-algebra of C and let M be the maximal ideal $M = D^{\perp}$. Let $M^0 = A$ and let C_i be the subcoalgebra $C_i = (M^{i+1})^{\perp}$ for $i \geq -1$. Note that $C_{-1} =$

$\{0\}$ and $D = C_0 \subset C_1 \subset C_2 \subset \cdots$. Then $C = \bigcup_0^\infty C_i$. To see this, let $x \in C$ and let C_x be the finite dimensional subcoalgebra $C_x = {}^A x^A$. Then we may regard the k-algebra $A/C_x^\perp$ as the dual k-algebra of C_x. Since $A/C_x^\perp$ is finite dimensional, its unique maximal ideal $M/C_x^\perp$ is nilpotent, as we showed in A.3. Thus, some power M^{i+1} of M is contained in $C_x^\perp$, so that $C_x \subset (M^{i+1})^\perp = C_i$ for some $i \geq 0$. It follows that $C = \bigcup_0^\infty C_i$, since $C = \bigcup_{x \in C} C_x$.

Note that for $i \geq 0$, we have $x \in C_i$ if and only if $x^a \in C_{i-1}$ for all a in the maximal ideal M. For if $a \in M$ and $x \in C_i = (M^{i+1})^\perp$, then $0 = x(ab) = x^a(b)$ for all $b \in M^i$, so that $x^a \in (M^i)^\perp = C_{i-1}$. And if x is an element of C such that $x^a \in C_{i-1}$ for all $a \in M$, then $0 = x^a(b) = x(ab)$ for all $a \in M$ and $b \in M^i$, so that $x \in (M^{i+1})^\perp = C_i$.

For $x \in C_n$, we have $\Delta x \in \sum_{i=0}^n C_{n-i} \otimes_k C_i$. We prove this by induction on n. If $n = 0$, we have $\Delta x \in C_0 \otimes_k C_0$ since $C_0 = D$ is a subcoalgebra of C. Next, let $n > 0$ and choose a basis $x_1, \ldots, x_{d_0}, x_{d_0+1}, \ldots, x_{d_1}, \ldots, x_{d_{n-1}+1}, \ldots, x_{d_n}$ for C_n such that $x_{d_{i-1}+1}, \ldots, x_{d_i}$ is a basis for C_i for $1 \leq i \leq n$. Since x is an element of the subcoalgebra C_n of C, we have $\Delta x = \sum_1^{d_n} {}_jx \otimes x_j$ for suitable ${}_jx$ in C_n. We claim that ${}_jx \in C_{n-j}$ and therefore ${}_jx \otimes x_j \in C_{n-i} \otimes_k C_i$ for $d_{i-1} + 1 \leq j \leq d_i$ and $0 \leq i \leq n$. For $a \in M$, we have $x^a \in C_{n-1}$ and $\Delta x^a = \sum_1^{d_n} {}_jx^a \otimes x_j$. Applying induction, we therefore have ${}_jx^a \in C_{n-1-i}$ for $d_{i-1} + 1 \leq j \leq d_i$ and $0 \leq i \leq n$, for all $a \in M$. By the preceding paragraph, it follows that ${}_jx \in C_{n-i}$ for $d_{i-1} + 1 \leq j \leq d_i$ and $0 \leq i \leq n$, so that $\Delta x \in \sum_0^n C_{n-i} \otimes_k C_i$.

Suppose that the colocal cocommutative coalgebra C is cosplit. Let e be the grouplike element of C so that $C_1 = ke$ and $M = \{a \in A \mid a(e) = 0\}$. Let C^0 be the coideal $C^0 = \{x \in C \mid \varepsilon(x) = 0\} = 1_A^\perp$ of C and let C_i^0 be the coideal $C_i^0 = C^0 \cap C_i$ of the cosubalgebra C_i of C for $i \geq 0$. Note that $C = ke + C^0$ (direct) and $C_i = ke + C_i^0$ (direct) for $i \geq 0$. The elements of C_1^0 are those elements x of C such that $\Delta x = x \otimes e + e \otimes x$. For if $x \in C_1^0$, then $\Delta x \in C_1 \otimes_k C_0 + C_0 \otimes_k C_1 = C_1 \otimes e + e \otimes C_1$ so that $\Delta x = y \otimes e + e \otimes z$ where $y, z \in C_1$. Replacing y by $u = y - \varepsilon(y)e$, so that $\varepsilon(u) = 0$, we have $\Delta x = u \otimes e + e \otimes v$ where $v = z + \varepsilon(y)e$. Then, by the coidentity equations, we have $x = \varepsilon(u)e + \varepsilon(e)v = 0 + v$ and $x = u\varepsilon(e) + e\varepsilon(v) = u + \varepsilon(v) = u + \varepsilon(x) = u + 0$. Thus, $u = v = x$ and $\Delta x = x \otimes e + e \otimes x$. Conversely, let x be an element of C such that $\Delta x = x \otimes e + e \otimes x$. Then $\varepsilon(x) = 0$, since $x = \varepsilon(x)e + \varepsilon(e)x = \varepsilon(x)e + x$. And $x \in C_1 = (M^2)^\perp$, since $x(ab) = x(a)e(b) + e(a)x(b) = x(a)0 + 0x(b) = 0$ for all $a, b \in M$. Thus, $x \in C_1^0$. We have now shown that $C_1^0 = P_g(C)$ where $P_g(C)$ is the set $P_g(C) = \{x \in C \mid \Delta x = x \otimes e + e \otimes x\}$ of e-primitive elements of C.

For $n \geq 1$, the elements of C_n^0 are those elements x of C such that $\Delta x \equiv x \otimes e + e \otimes x \pmod{C_{n-1}^0 \otimes_k C_{n-1}^0}$, in fact, such that

$$\Delta x = x \otimes e + e \otimes x + \sum_1^{n-1} {}_ix \otimes x_i$$

where ${}_ix \in C_{n-i}^0$, $x_i \in C_i^0$ for $1 \leq i \leq n - 1$. For suppose that $x \in C_n^0$ and write $\Delta x = y \otimes e + e \otimes z + \sum_1^{n-1} {}_ix \otimes x_i$ where $y, z \in C_n$, ${}_ix \in C_{n-i}$, $x_i \in C_i$

for $1 \leq i \leq n-1$. Since we could replace each x_i by $x_i - \varepsilon(x_i)e(1 \leq i \leq n-1)$ and y by $y + \sum_1^{n-1} \varepsilon(x_i)_i x$ without disturbing the equality of the preceding equation, we may assume without loss of generality that $x_i \in C_i{}^0$ for $1 \leq i \leq n-1$. Similarly, we may assume without loss of generality that ${}_ix \in C_{n-i}^0$ for $1 \leq i \leq n-1$. Replace y by $u = y - \varepsilon(y)e$, as before, so that $\varepsilon(u) = 0$ and $\Delta x = u \otimes e + e \otimes v + \sum_1^{n-1} {}_ix \otimes x_i$ where $v = z + \varepsilon(y)e$. By one coidentity equation, $x = \varepsilon(u)e + \varepsilon(e)v + \sum_1^{n-1} \varepsilon({}_ix)x_i = v$. By the other coidentity equation, $x = \varepsilon(e)u + \varepsilon(v)e + \sum_1^{n-1} \varepsilon(x_i)_ix = u$. Thus, $\Delta x = x \otimes e + e \otimes x + \sum_1^{n-1} {}_ix \otimes x_i$ where ${}_ix \in C_{n-i}^0$, $x_i \in C_i{}^0$ for $1 \leq i \leq n-1$. Suppose, conversely, that $x \in C$ and $\Delta x \equiv x \otimes e + e \otimes x \pmod{C_{n-1}^0 \otimes_k C_{n-1}^0}$. Then $\varepsilon(x) = 0$, since the coidentity equation implies that $x = \varepsilon(x)e + \varepsilon(e)x = \varepsilon(x)e + x$. And for $a_1, \ldots, a_{n+1} \in M$, we have

$$x(a_1 \cdots a_{n+1}) = x(a_1 \cdots a_n)e(a_{n+1}) + e(a_1 \cdots a_n)x(a_{n+1}) = 0,$$

so that $x \in (M^{n+1})^\perp = C_n$. Thus, $x \in C_n{}^0$.

The foregoing material provides a detailed picture of the structure of an arbitrary cosplit cocommutative k-coalgebra C. For we know that $C = \sum_{g \in G(C)} C(g)$. And we know for $g \in G(C)$ that $C(g) = \bigcup_0^\infty C(g)_i$, $C(g)_n = kg + C(g)_n{}^0$ and $\Delta x = x \otimes g + g \otimes x + \sum_1^{n-1} {}_ix \otimes x_i$ with ${}_ix \in C(g)_{n-i}^0$, $x_i \in C(g)_i{}^0$ $(1 \leq i \leq n-1)$ for $n \geq 1$ and $x \in C(g)_n{}^0$. Note that for $g \in G(C)$, $C(g)_1{}^0$ coincides with the set $P_g(C) = \{x \in C \mid \Delta x = x \otimes g + g \otimes x\}$ of *g-primitive* elements of C. For we know that $C(g)_1{}^0 \subset P_g(C)$. And if $x \in P_g(C)$, then $kg + kx$ is a colocal subcoalgebra of C containing g, so that $x \in C(g)$, whence $x \in C(g)_1{}^0$.

B Bialgebras

In this appendix, we develop a theory of bialgebras over an extension field K/k which generalizes the usual theory of bialgebras over a single field k. This theory was developed to study field extensions, and provides the proper framework within which to extensively examine the structure of $H(K/k)$. It also provides a language within which the material of Chapters 5 and 6 can be more effectively understood and applied.

We begin in B.1 with basic definitions and properties of K/k-bialgebras. In B.2, we introduce the conormal K/k-bialgebras, those which arise in studying normal field extensions, and give a rough description of their structure. This description is then made more precise in B.3, where tensor products and semidirect products of bialgebras are discussed. In B.4, we introduce the K-measuring K/k-bialgebras and relate them to $H(K/k)$. We also describe the relationship between the K-measuring k-algebras and the K-measuring K/k-bialgebras. In B.5, we concentrate on the role of K-measuring bialgebras in the study of finite dimensional normal field extensions. The finite dimensional conormal measuring bialgebras are described in terms of the finite dimensional measuring coradical and co Galois bialgebras. The cosplit k-forms of $H(K/k)$ for a finite dimensional normal field extension K/k are described in terms of the k-forms of $H(K_{\text{rad}}/k)$.

B.1 Preliminaries

A *K/k-algebra* is a K-vector space A together with a K-linear product mapping $\pi: A \otimes_k A \to A$ and k-linear identity mapping $\iota: k \to A$ with respect to which A is a k-algebra. Here, $A \otimes_k A$ is regarded as K-vector space with the scalar product such that $c(a \otimes b = (ca) \otimes b$ for $c \in K$ and $a, b \in A$.

A *K/k-bialgebra* is a K/k-algebra H together with K-linear coproduct and coidentity mappings $\Delta: H \to H \otimes_K H$, $\varepsilon: H \to K$ with respect to which H is a K-coalgebra such that

1. $\Delta(1_H) = 1_H \otimes 1_H$;
2. $\Delta(xy) = \sum_{i,j} {}_i x_j y \otimes x_i y_j$ for $x, y \in H$,
   $\Delta x = \sum_i {}_i x \otimes x_i$ and $\Delta y = \sum_j {}_j y \otimes y_j$;
3. $\varepsilon(1_H) = 1_K$;
4. $\varepsilon(xy) = \varepsilon(x)\varepsilon(y)$ for all $x, y \in H$ such that $\varepsilon(y) \in k$.

The K/k-bialgebra $H(K/k)$ introduced in 5.3 will be referred to here as the *K/k-bialgebra* of K/k.

A *k-bialgebra* is a k/k-bialgebra. If H is a k-bialgebra, then $H \otimes_k H$ can be regarded as k-algebra (see A.2) and k-coalgebra (see C.2). It can easily be shown that the mappings $\Delta: H \to H \otimes_k H$, $\varepsilon: H \to k$ are k-algebra homomorphisms and that the mappings $\pi: H \otimes_k H \to H$, $\iota: k \to H$ are k-coalgebra homomorphisms.

A *subbialgebra/biideal* of a K/k-bialgebra H is a K-subspace of H which is a k-subalgebra/k-ideal of H and a K-subcoalgebra/K-coideal of H. A *subbialgebra/biideal* of a k-bialgebra H is a subbialgebra/biideal of H as k/k-bialgebra.

A *homomorphism/isomorphism* from a K/k-bialgebra H to a K/k-bialgebra H' is a mapping $f: H \to H'$ which is a homomorphism/isomorphism of k-algebras and of K-coalgebras from H to H'. A *homomorphism/isomorphism* from a k-bialgebra H to a k-bialgebra H' is a homomorphism/isomorphism of k/k-bialgebras from H to H'.

Any subbialgebra D of a K/k-bialgebra H can obviously be regarded as a K/k-bialgebra such that the inclusion mapping $D \to H$ is a homomorphism. And for any biideal P of a K/k-bialgebra H, H/P can be regarded as a K/k-bialgebra such that the quotient mapping $H \to H/P$ is a homomorphism.

If $f: H \to H'$ is any homomorphism from a K/k-bialgebra H to a K/k-bialgebra H', then Kernel $f = \{x \in H \mid f(x) = 0\}$ is a biideal of H, Image f is a subbialgebra of H' and there is an isomorphism of K/k-bialgebras from $H/$Kernel f to Image f mapping $x +$ Kernel f to $f(x)$ for all $x \in H$.

A *k-form* of a K/k-biaglebra H is a k-subspace H_k of H which is a subalgebra of H as k-algebra and a k-form of H as K-coalgebra. Any k-form H_k of a K/k-bialgebra H can be regarded as a k-bialgebra with the k-algebra structure induced by that of H and the k-coalgebra structure induced by that of H (see C.2).

If H is a k-bialgebra and k_0 is a subfield of k, then a *k_0-form* of H is a k_0-subspace H_{k_0} of H such that H_{k_0} is a k_0-form of H as k-algebra and as k-coalgebra. Any k_0-form H_{k_0} of a k-bialgebra H can be regarded as a k_0-bialgebra.

We now give some examples of k-bialgebras and K/k-bialgebras. As a first example, let G be a group and let $k[G]$ be a vector space over k with basis G. Then $k[G]$ with its structure as group k-algebra of G (see A.1) and its structure as group k-coalgebra of G (see C.1) is a cocommutative k-bialgebra, called the group *k-bialgebra* of G.

As a second example, let G be a finite commutative group. Let T be the k-dual space $k[G]^*$ of $k[G]$, together with the dual k-algebra structure of $k[G]$ as group k-coalgebra of G and the dual k-coalgebra structure of $k[G]$ as group k-algebra of G. Then T is a commutative and cocommutative k-bialgebra called the *toral k-bialgebra* of G. Note that $\varepsilon(t) = t(1_G)$ for $t \in T$, $\Delta(t) = \sum_i {}_it \otimes t_i$ if and only if $t(gh) = \sum_i {}_it(g)t_i(h)$ for all $g, h \in G$ for $t \in T$, $(st)(g) = s(g)t(g)$ for $s, t \in T$ and $g \in G$ and $1_T(g) = 1$ for $g \in G$. Note also that the toral k-bialgebra T has a π-form $T_\pi = \{t \in T \mid t(G) \subset \pi\}$, and that $T_\pi = \{t \in T \mid t^p = t\}$.

As a third example, let G be a group and $\rho: G \to \mathrm{Aut}_k K$ a homomorphism of groups. Let $K[G]$ be the group K-coalgebra of G. Then $K[G]$ can be given the structure of k-algebra such that $(ag)(bh) = (a\rho(g)(b))(gh)$ for all

$a, b \in K$ and $g, h \in G$ (see B.3). Then $K[G]$ is a K/k bialgebra, called the *group K/k-bialgebra* of G via ρ. The k-span $k[G]$ of G in $K[G]$ is a k-form of $K[G]$. The k-bialgebra structure of $k[G]$ induced by the K/k-bialgebra structure of $K[G]$ coincides with the group k-bialgebra structure of $k[G]$. The K/k-bialgebra structure of $K[G]$ can be recovered from the k-bialgebra structure of $k[G]$ and the action ρ of G on K.

As a final example, let T be the toral k-bialgebra of a finite commutative group G. Suppose that K has a graded decomposition $K = \sum_{g \in G} K_g$ (direct sum of k-subspaces of K) where $ab \in K_{gh}$ for $a \in K_g$, $b \in K_h$ for all $g, h \in G$ (compare with 6.2.5). We can regard $K \otimes_k T$ as K-coalgebra (see C.2) and as k-algebra with identity $1_K \otimes 1_T$ such that

$$(a \otimes s)(b \otimes t) = \sum_i a_i s(g) b \otimes s_i t$$

for all $g \in G$, $a \in K$, $b \in K_g$, $s, t \in T$ where $\Delta s = \sum_i {}_i s \otimes s_i$ (see 6.2.7 and B.3). Then $K \otimes_k T$ is a K/k-bialgebra, called the *K/k-bialgebra* determined by T and the graded decomposition $K = \sum_{g \in G} K_g$. Note that $1 \otimes_k T$ is a k-form of the K/k-bialgebra $K \otimes_k T$.

Let H be an arbitrary K/k-bialgebra. Then the colocal component $H(1_H)$ of H containing 1_H is a subbialgebra of H. To see this, we must show that $H(1_H)$ is closed under multiplication. Thus, let $C = H(1_H)$, let $A = H^*$ be the dual K-algebra of H as K-coalgebra and let M be the maximal ideal $M = 1_H{}^\perp$. Then $C_i = (M^{i+1})^\perp$ and $C = \bigcup_0^\infty C_i = K1_H + \bigcup_1^\infty C_i{}^0$ (see C.4). It suffices to show that for $x \in C_m$ and $y \in C_n$, we have $xy \in C_{m+n}$ for $m, n \geq 0$. If $m + n \leq 1$, this is certainly true. Next, suppose that $m + n > 1$. It obviously suffices to consider the case $x \in C_m{}^0$, $y \in C_n{}^0$. Then $\Delta x = \sum_0^m {}_{m-i}x \otimes x_i$ and $\Delta y = \sum_0^n {}_{n-j}y \otimes y_j$ where ${}_i x, x_i \in C_i{}^0$ for $1 \leq i \leq m-1$, ${}_j y, y_j \in C_j{}^0$ for $1 \leq j \leq n-1$, ${}_0 x = x_0 = {}_0 y = y_0 = 1_H$, ${}_m x = x_m = x$ and ${}_n y = y_n = y$. Let $a_0, a_1, \ldots, a_{m+n}$ be $m+n+1$ arbitrary elements of M. Then

$$(a_0 \cdots a_{m+n})(xy) = \sum_{i=0}^{m} \sum_{j=0}^{n} a_0({}_{m-i}x_{n-j}y)(a_1 \cdots a_{m+n})(x_i y_j) = 0$$

since each term in the sum is 0 by induction. (Either $i + j \leq m + n - 1$ and $x_i y_j \in C_{m+n-1}$, so that $(a_1 \cdots a_{m+n})(x_i y_j) = 0$, or $i = m$ and $j = n$, so that $a_0({}_0x_0 y) = a_0(1_H) = 0$.) It follows that $xy \in C_{m+n}$.

In an arbitrary K/k-bialgebra H, the set $G(H)$ of grouplike elements of H is closed under products and is therefore a monoid.

Let H be an arbitrary K/k-bialgebra and let $x \in H$. If $\Delta x = \sum_i^n {}_i x \otimes x_i$ where the elements of $\{{}_i x \mid 1 \leq i \leq n\} \cup \{x_i \mid 1 \leq i \leq n\}$ commute pairwise, then $\Delta x^p = \sum_i {}_i x^p \otimes x_i{}^p$. To see this, note that

$$\Delta x^p = \sum {}_{i_1}x \cdots {}_{i_p}x \otimes x_{i_1} \cdots x_{i_p} = \sum (e_1, \ldots, e_n)\, {}_1x^{e_1} \cdots {}_n x^{e_n} \otimes x_1{}^{e_1} \cdots x_n{}^{e_n}$$

where the coefficients $(e_1, \ldots, e_n)$ are those such that

$$\begin{aligned}\left(\sum_i^n {}_i X X_i\right)^p &= \sum {}_{i_1}X \cdots {}_{i_p}X X_{i_1} \cdots X_{i_p} \\ &= \sum (e_1, \ldots, e_n)\, {}_1X^{e_1} \cdots {}_n X^{e_n} X_1{}^{e_1} \cdots X_n{}^{e_n}\end{aligned}$$

in the polynomial ring $k[X]$ in $2n$ algebraically independent elements $X = \{{}_iX \mid 1 \leq i \leq n\} \cup \{X_i \mid 1 \leq i \leq n\}$ over k. Since

$$\left(\sum_i^n {}_iXX_i\right)^p = \sum_i {}_iX^p X_i{}^p \text{ in } k[X],$$

it follows, by comparison of coefficients, that

$$\Delta x^p = \sum_i {}_ix^p \otimes x_i{}^p \text{ in } H \otimes_K H.$$

B.2 Conormal bialgebras

A K/k-bialgebra H is *conormal* if H is cosplit and cocommutative (as K-coalgebra) and $G(H)$ is a group. A K/k-bialgebra H is *co Galois/coradical* if H is conormal and H is cosemisimple/colocal.

If H is a cocommutative K/k-bialgebra and $G(H)$ is a group, then H is conormal/co Galois/coradical if and only if

$$H = \sum_{g \in G(H)} H(g)/H = \sum Kg/H = H(1_H).$$

It is clear that the K/k-bialgebra $H(K/k)$ introduced in 5.3 is conormal/co Galois/coradical in the present sense if and only if $H(K/k)$ is conormal/co Galois/coradical in the sense of 5.3.

Let H be a K/k-bialgebra. We let $Dg = \{ug \mid u \in D\}$ for $D \subset H$ and $g \in G(H)$. Let g be an invertible element of the semigroup $G(H)$. A subset D of H is a subcoalgebra of H if and only if Dg is a subcoalgebra of H, as one sees from the equations $\Delta(ug) = \sum_i ({}_iug) \otimes u_ig$ $(u \in D)$ and the invertibility of g. (Note that g is put on the right hand side so that $u \mapsto ug$ is K-linear.) It follows that $H(h)g \subset H(hg)$ for $h \in G(H)$. Similarly, $H(hg)g^{-1} \subset H(h)$ for $h \in G(H)$, so that $H(h)g = H(hg)$ for $h \in G(H)$. In particular, we have $H(g) = H(1_H)g$.

Let H be a conormal K/k-bialgebra. Then every element $g \in G(H)$ is invertible and $H = \sum_{g \in G(H)} H(g)$ (direct), so that $H = \sum_{g \in G(H)} H(1_H)g$ (direct), by the preceding paragraph. Letting $k[G(H)]$ be the k-span of $G(H)$, it follows that $H = H(1_H)k[G(H)]$ (k-span of $\{xy \mid x \in H(1_H), y \in k[G(H)]\}$]. In fact, the k-linear mapping $f\colon H(1_H \otimes_k k[G(H)] \rightarrow H$ such that $f(x \otimes y) = xy$ for $x \in H(1_H)$, $y \in k[G(H)]$ is an isomorphism of vector spaces over k, as one sees from the direction of the decomposition $H = \sum_{g \in G(H)} H(1_H)g$ (direct). This generalizes to K/k-bialgebras part of a theorem of Bertram Kostant on k-bialgebras (see [18] and B.3).

It is instructive at this point to look more closely at the conormal k-bialgebras H. Then $k[G(H)]$ is a subbialgebra. Moreover, the proof that $H(1_H)g = H(g)$ $(g \in G(H))$ has, for k-bialgebras, a left sided analogue which shows that $gH(1_H) = H(g)$ $(g \in G(H))$. Thus, we have $gH(1_H)g^{-1} = H(1_H)$ for $g \in G(H)$. Letting $\rho(g)\colon H(1_H) \rightarrow H(1_H)$ be defined by $\rho(g)(x) = gxg^{-1}$ for $x \in H(1_H)$ and $g \in G$, ρ is a homomorphism from $G(H)$ into the group of bijective linear transformations of $H(1_H)$. The product in the k-algebra

$H = H(1_H)k[G(H)]$ is determined by ρ and the products in $H(1_H)$, $k[G(H)]$, since $(xg)(yh) = x(gyg^{-1})(gh) = (x\rho(g)(y))(gh)$ for $x, y \in H(1_H)$, $g, h \in G(H)$. In the sense of the next section, this means that H as k-algebra is the internal semidirect product of $H(1_H)$ and $k[G(H)]$.

B.3 Tensor products and semidirect products

Let H_K be a K/k-bialgebra and H_k a k-bialgebra. Then there are K-linear mappings $\Delta_{H_K \otimes H_k}\colon H_K \otimes_k H_k \to (H_K \otimes_k H_k) \otimes_K (H_K \otimes_k H_k)$ and $\varepsilon_{H_K \otimes H_k}\colon H_K \otimes_k H_k \to K$ such that $\Delta_{H_K \otimes H_k}(x \otimes y) = \sum_{i,j} ({}_ix \otimes_k {}_jy) \otimes_K (x_i \otimes_k y_j)$ and $\varepsilon_{H_K \otimes H_k}(x \otimes y) = \varepsilon_{H_K}(x)\varepsilon_{H_k}(y)$ for $x \in H_K$, $y \in H_k$. Together with $\Delta_{H_K \otimes H_k}$ and $\varepsilon_{H_K \otimes H_k}$, $H_K \otimes_k H_k$ is a K-coalgebra. Moreover, $H_K \otimes_k H_k$ together with this K-coalgebra structure and the tensor product k-algebra structure is a K/k-bialgebra, called the *tensor product K/k-bialgebra* of H_K and H_k. If $K = k$, so that $H = H_K$ and H_k are both k-bialgebras, then $H \otimes_k H_k$ is called the *tensor product k-bialgebra* of H and H_k.

A *representation* of a k-algebra H on a k-vector space A is a k-linear homomorphism $\rho\colon H \to \operatorname{End}_k A$ from H to the ring $\operatorname{End}_k A$ of k-linear endomorphisms of A. A *measuring representation* of a K-coalgebra H on a K/k-algebra A is a K-linear mapping $\rho\colon H \to \operatorname{End}_k A$ such that $\rho(x)(1_A) = \varepsilon(x)1_A$ and $\rho(x)(ab) = \sum_i \rho({}_ix)(a)\rho(x_i)(b)$ for $x \in H$ and $a, b \in A$ (compare with C.1). A *measuring representation* of a K/k-bialgebra H on a K/k-algebra A is a mapping $\rho\colon H \to \operatorname{End}_k A$ which is a representation of H as k-algebra and a measuring representation of H as K-coalgebra. A *measuring representation* of a k-bialgebra H on a k-algebra A is a measuring representation of H as k/k-bialgebra on A as k/k-algebra.

Let H_k be a k-bialgebra. Then an *H_k-module algebra* is a k-algebra A together with a measuring representation ρ_A of H_k on A. We let $x(a)$ denote $\rho_A(x)(a)$ for $x \in H_k$, $a \in A$ and note that $x(1_A) = \varepsilon(x)1_A$ for $x \in H_k$ and $x(ab) = \sum_i {}_ix(a)x_i(b)$ for $x \in H_k$ and $a, b \in A$. An *H_k-module K/k-bialgebra/H_k-module k-bialgebra* is a K/k-bialgebra/k-bialgebra H together with a measuring representation ρ_H of H_k on H as k-algebra. Any H_k-module bialgebra H is, of course, also regarded as H_k-module algebra and the notation $x(a)$ for $\rho_H(x)(a)(x \in H_k, a \in H)$ is used as for H_k-module algebras.

Let H be a k-bialgebra and A an H-module algebra. Let $H_A = A \otimes_k H$ and let $\pi\colon H_A \otimes_k H_A \to H_A$ be the k-linear mapping such that

$$\pi((a \otimes x) \otimes (b \otimes y)) = \sum_i a_ix(b) \otimes x_iy \qquad \text{for } a, b \in A,\ x, y \in H.$$

Then H_A is a k-algebra with identity $1_A \otimes 1_H$, called the *semidirect product* algebra of A and H. To see that H_A is a k-algebra, note first that $(a \otimes x) \times (1_A \otimes y) = a \otimes xy$ and $(a \otimes 1_H)(b \otimes y) = ab \otimes y$ for $a, b \in A$, $x, y \in H$. In particular, $1_A \otimes 1_H$ is an identity with respect to the product in H_A. To show that $((a \otimes x)(b \otimes y))(c \otimes z) = (a \otimes x)((b \otimes y)(c \otimes z))$ for $a, b, c \in A$ and $x, y, z \in H$, we show that

$$\sum_{i,r,s} a_ix(b)_r(x_i)({}_sy)(c) \otimes_k (x_i)_r y_s z = \sum_{i,r,s} a_r({}_ix)(b)({}_ix)_{r\,s}y(c) \otimes_k x_iy_sz.$$

Letting $d = {}_sy(c)$ and $w = y_sz$, it suffices to show that

$$\sum_{i,r,s} {}_ix(b)_r(x_i)(d) \otimes_K (x_i)_r = \sum_{i,r,s} {}_r({}_ix)(b)({}_ix)_r(d) \otimes_K x_i,$$

for we could first multiply by $a \otimes_K 1_H$, then by $1_A \otimes_K w$. Since H is a k-bialgebra, the above equation follows from the coassociativity equation

$$\sum_{i,r} {}_ix \otimes_K {}_r(x_i) \otimes_K (x_i)_r = \sum_{i,r} {}_r({}_ix) \otimes_K ({}_ix)_r \otimes_K x_i.$$

Thus, H_A is a k-algebra.

Let H_k be a k-bialgebra and let H be an H_k-module K/k-bialgebra/k-bialgebra. Whenever $H \otimes_k H_k$ as K-coalgebra/k-coalgebra with respect to the mappings $\Delta_{H\otimes H_k}$, $\varepsilon_{H\otimes H_k}$ described earlier in this section and as semidirect product algebra of H and H_k is a K/k-bialgebra/k-bialgebra, it is called the *semidirect product bialgebra* of H and H_k.

For any conormal k-bialgebra H, $H(1_H)$ is a coradical subbialgebra of H and the k-span $k[G(H)]$ of $G(H)$ is a co Galois subbialgebra of H (see B.1 and B.2). We let H_{rad} and H_{Gal} denote $H(1_H)$ and $k[G(H)]$ respectively. We call H_{rad} the *coradical* component of H and H_{Gal} the *co Galois* component of H. The k-linear mapping $\rho\colon \mathrm{H}_{\mathrm{Gal}} \to \mathrm{End}_k\, H_{\mathrm{rad}}$ such that $\rho(g)(x) = gxg^{-1}$ for $g \in G(H)$ and $x \in H_{\mathrm{rad}}$ is a measuring representation of the k-bialgebra H_{Gal} on H_{rad} as k-algebra. For ρ is a representation of H_{Gal} as k-algebra on H_{rad} (see B.2) and ρ is a measuring representation of H_{Gal} as k-coalgebra on H_{rad} as k-algebra since $\rho(g)(xy) = gxyg^{-1} = gxg^{-1}gyg^{-1} = \rho(g)(x)\rho(g)(y) = \sum_i \rho({}_ig)(x)\rho(g_i)(y)$ for $x, y \in H_{\mathrm{rad}}$ and $g \in G(H)$, and $\rho(g)(1_H) = g1_Hg^{-1} = 1_H = \varepsilon(g)1_H$ for $g \in G(H)$. Furthermore, the k-linear mapping $f\colon H_{\mathrm{rad}} \otimes_k H_{\mathrm{Gal}} \to H$ from the semidirect product k-bialgebra $H_{\mathrm{rad}} \otimes_k H_{\mathrm{Gal}}$ to H such that $f(a \otimes x) = ax$ for $a \in H_{\mathrm{rad}}$, $x \in H_{\mathrm{Gal}}$ is an isomorphism of k-bialgebras. For f is an isomorphism of vector spaces over k, as was shown in B.2. And we have $(a \otimes g)(b \otimes h) = \sum_i a\rho({}_ig)(b) \otimes g_ih = a\rho(g)(b) \otimes gh$ in the semidirect product $H_{\mathrm{rad}} \otimes_k H_{\mathrm{Gal}}$ whereas $(ag)(bh) = agbg^{-1}gh = a\rho(g)(b)gh$ in H for $a, b \in H(1_H)$ and $g, h \in G(H)$. In this sense, $H = H_{\mathrm{rad}}H_{\mathrm{Gal}}$ (*internal semidirect product bialgebra*). This proves Bertram Kostant's splitting theorem for k-bialgebras [18].

B.4 Measuring bialgebras

A *K-measuring K/k-bialgebra/k-bialgebra* is a *K/k-bialgebra/k-bialgebra* H together with a measuring representation ρ_H of H on K. We denote $\rho_H(x)(a)$ by $x(a)$ for any element x of a K-measuring K/k-bialgebra/k-bialgebra H of any element a of K. Note than $H(K/k)$ may be regarded as K-measuring K/k-bialgebra where $\rho_{H(K/k)}$ is the inclusion mapping.

A *homomorphism/isomorphism* from a K-measuring K/k-bialgebra/k-bialgebra H to a K-measuring K/k-bialgebra/k-bialgebra H' is a homomorphism/isomorphism f of K/k-bialgebras/k-bialgebras from H to H' such that the diagram

$$\begin{array}{ccc} H & \xrightarrow{f} & H' \\ & {\scriptstyle \rho_H}\searrow \quad \swarrow{\scriptstyle \rho_{H'}} & \\ & \operatorname{End}_k K & \end{array}$$

is commuative.

The *commutant* of subset C of a K-measuring bialgebra H in K is the subfield $K^C = \{a \in K \mid x(ab) = ax(b) \text{ for all } x \in C, \text{b} \in K\}$ of K. Note that if C is closed under Δ, that is, if for $x \in C$ there exist ${}_ix, x_i \in C$ such that $\Delta x = \sum_i {}_ix \otimes x_i$, then the commutant K^C of C in K is $K^C = \{a \in K \mid x(a) = \varepsilon(x)a \text{ for all } x \in C\}$. For if $\Delta x = \sum_i {}_ix \otimes x_i$, then

$$x(ab) = \sum_i {}_ix(a)x_i(b) = a \sum_i \varepsilon({}_ix)x_i(b) = ax(b)$$

for any $a \in K$ such that ${}_ix(a) = \varepsilon({}_ix)a$ for all i and for any $b \in K$. And if $a \in K$ and $x(ab) = ax(b)$ for all $x \in C$ and $b \in K$, then $x(a) = x(a1) = ax(1) = \varepsilon(x)a$ for all $x \in C$.

Let H be a K-measuring K/k-bialgebra and let Kernel $H = \{x \in H \mid x(a) = 0 \text{ for all } a \in K\}$. Then ρ_H is a homomorphism of K-measuring K/k-bialgebras from H to $H(K/k)$ with kernel Kernel H. For we have $\varepsilon(\rho_H(x)) = \rho_H(x)(1_K) = \varepsilon_H(x)1_K = \varepsilon_H(x)$. And for $\Delta_H(x) = \sum_i {}_ix \otimes x_i$, we have

$$\rho_H(x)(ab) = \sum_i \rho_H({}_ix)(a)\rho_H(x_i)(b) \qquad \text{for all } a, b \in K,$$

so that $\Delta(\rho_H(x)) = \sum_i \rho_H({}_ix) \otimes \rho_H(x_i)$. Finally, the diagram

$$\begin{array}{ccc} H & \xrightarrow{\rho_H} & H(K/k) \\ & {\scriptstyle \rho_H}\searrow \quad \swarrow{\scriptstyle \rho_{H(K/k)}} & \\ & \operatorname{End}_k K & \end{array}$$

is commutative since $\rho_{H(K/k)}$ is the inclusion mapping.

It follows that for any K-measuring K/k-bialgebra H, Kernel H is a bi-ideal of H and $H/\text{Kernel } H$ can be regarded as a K-measuring K/k-bialgebra isomorphic to a subbialgebra of $H(K/k)$. If K/k is finite dimensional, then $H/\text{Kernel } H$ is isomorphic to $H(K/K^H)$, by 5.3.12.

The above discussion has no counterpart for K-measuring k-bialgebras H.

A K-measuring K/k-bialgebra H is *semilinear* if $x(by) = \sum_i {}_ix(b)x_iy$ for all $k \in K$ and $x, y \in H$. Note that $H(K/k)$ is semilinear.

Suppose that H is a semilinear K-measuring K/k bialgebra and let g be an invertible element of $G(H)$. If D is a K-subspace of H, then $gD = \{gu \mid u \in D\}$ is a K-subspace of H, since $a(gu) = g(g^{-1}(a))(gu) = g(g^{-1}(a)u) \in gD$ for $a \in K$, $u \in D$, $g \in G(H)$. It follows that a subset D of H is a subcoalgebra of H if and only if gD is a subcoalgebra of H, hence that $gH(h) \subset H(gh)$ and $g^{-1}H(gh) \subset H(h)$ for all $h \in G(H)$ as in B.2. Thus, $gH(h) = H(gh)$ for $h \in G(H)$. In particular, $H(g) = gH(1_H)$. Since $H(g^{-1}) = H(1_H)g^{-1}$ (see B.2), we therefore have $gH(1_H)g^{-1} = H(1_H)$.

Suppose that H is a conormal semilinear K-measuring K/k-bialgebra, so that $H = H(1)k[G]$ where $H(1) = H(1_H)$ and $G = G(H)$ (see B.2). Let $\rho: k[G] \to \operatorname{End}_k H(1)$ be the k-linear mapping such that $\rho(g)(x) = gxg^{-1}$ for $x \in H(1)$ and $g \in G$. (Here we use the observation of the preceding paragraph that $gH(1)g^{-1} = H(1)$ for $g \in G$ in the presence of the semilinearity of H.) Then ρ is a representation of $k[G]$ on $H(1)$ as k-algebra. In fact, ρ is a measuring representation of $k[G]$ on $H(1)$, since $\rho(xy) = g(xy)g^{-1} = gxg^{-1}gyg^{-1} = \rho(g)(x)\rho(g)(y)$ for $x, y \in H(1)$ and $g \in G$, and $\rho(g)(1_H) = g1_Hg^{-1} = 1_H = \varepsilon(g)1_H$ for $g \in G$. Now it is clear that the vector space isomorphism $f: H(1) \otimes_k k[G] \to H$ of B.2 such that $f(x \otimes y) = xy$ for $x \in H(1)$, $y \in k[G]$ is an isomorphism of K/k-bialgebras from $H(1) \otimes_k k[G]$ (semidirect product K/k-bialgebra) to H. For we have $(x \otimes g)(y \otimes h) = x\rho(g)(y) \otimes gh$ in $H(1) \otimes_k k[G]$, whereas $(xg)(yh) = xgyg^{-1}gh = x\rho(g)(y)(gh)$ in H for $x, y \in H(1)$ and $g, h \in G$. In this sense, we have $H = H(1)k[G]$ (*internal semidirect product bialgebra*). This generalizes to K/k-bialgebras Bertram Kostant's splitting theorem for k-bialgebras (see B.3).

We next relate the K-measuring k-bialgebras and the semilinear K-measuring K/k-bialgebras. If H_k is a K-measuring k-bialgebra, and if $H_K = K \otimes_k H_k$ as semidirect product k-algebra (see B.3) and K-coalgebra by ascent (see C.2), then H_K together with the K-linear mapping $\rho_{H_K}: H_K \to \operatorname{End}_k K$ such that $\rho_{H_K}(a \otimes x) = a\rho_{H_k}(x)$ for $a \in K$, $x \in H_k$ is a semilinear K-measuring K/k-bialgebra. Moreover, $1 \otimes H_k$ is a k-form of H_K as K/k-bialgebra. We defer the details until the next paragraph. It is convenient to be able to pass from the k-bialgebra H_k to the K/k-bialgebra $H_K = K \otimes_k H_k$, since then Kernel H_K is a biideal of H_K, and H_K/Kernel H_K can be imbedded in $H(K/k)$ and in $\operatorname{End}_k K$. Conversely, suppose that H is a semilinear K-measuring K/k-bialgebra. Suppose further that H_k is a k-form of H as K/k-bialgebra. Then H_k can be regarded as K-measuring k-bialgebra. Let $H_K = K \otimes_k H_k$ be the k-measuring K/k-bialgebra described above. Then the k-linear mapping f from $H_K = K \otimes_k H_k$ to H such that $f(a \otimes x) = ax$ for $a \in K$ and $x \in H_k$ is an isomorphism of K-measuring K/k-bialgebras, as one easily sees from the equations

$$(a \otimes x)(b \otimes y) = \sum_i a_i x(b) \otimes x_i y \text{ (product in } H_K)$$

and

$$(ax)(by) = \sum_i a_i x(b) x_i y \text{ (product in } H) \text{ for } a, b \in K,\ x, y \in H_k.$$

In this sense, $H = KH_k$ (*internal semidirect product*).

We now give the details which establish that the H_K constructed in the preceding paragraph from the K-measuring k-bialgebra H_k is a semilinear K-measuring K/k-bialgebra. Let the structure mappings for H_K be denoted $\pi_K, \iota_K, \Delta_K, \varepsilon_K, \rho_{H_K}$. Then ρ_{H_K} is a representation of H_K as k-algebra, since

$$((a \otimes x)(b \otimes y))(c) = \left(\sum_i a_i x(b) \otimes x_i y\right)(c) = \sum_i a_i x(b) x_i y(c)$$

and

$$(a \otimes x)((b \otimes y)(c)) = (a \otimes x)(by(c)) = a(x(by(c))) = \sum_i a_i x(b) x_i y(c)$$

for all $c \in K$ and for all $a, b \in K$, $x, y \in H_k$. And $\varepsilon_K(y) = y(1_K)$ for $y \in H_K$, since $\varepsilon_K(a \otimes x) = a\varepsilon(x) = ax(1_K) = (a \otimes x)(1_K)$ for $a \in K, x \in H_k$. For $y \in H_K$ and $b, c \in K$, $y(bc) = \sum_j {}_jy(b)y_j(c)$, since

$$\Delta_K(a \otimes x) = \sum_i a_ix \otimes x_i$$

and

$$(a \otimes x)(bc) = a(x(bc)) = a \sum_i {}_ix(b)x_i(c) = \sum_i (a_ix)(b)x_i(c)$$

for $x \in H_k$ and $a \in K$. Thus, ρ_{H_K} is a measuring representation of H_K as K-coalgebra. We next show that H_K is a K/k-bialgebra. For $x, y \in H_K$ and $\varepsilon_K(y) \in k$, we have $\varepsilon_K(xy) = \varepsilon_K(x)\varepsilon_K(y)$, since $\varepsilon_K(xy) = (xy)(1_K) = x(y(1_K)) = x(y(1_K)1_K) = y(1_K)x(1_K) = \varepsilon_K(y)\varepsilon_K(x)$. And $\Delta_K(xy) = \sum_{i,j} {}_ix_jy \otimes_K x_iy_j$ for $x, y \in H_K$. To prove this, it suffices to show that

$$\Delta((a \otimes x)(b \otimes y)) = \sum_{i,s} {}_i(a \otimes_k x)_s(b \otimes_k y) \otimes_K (a \otimes_k x)_i(b \otimes_k y)_s$$

for $a, b \in K$ and $x, y \in H_k$.

This amounts to showing that

$$\sum_{i,r,s} (a_ix(b) \otimes_k {}_r(x_i)_sy) \otimes_K (1 \otimes_k (x_i)_ry_s)$$
$$= \sum_{i,r,s} (a_r({}_ix)(b) \otimes_k ({}_ix)_r\, {}_sy) \otimes_K (1 \otimes_k x_iy_s),$$

which follows from the equation

$$\sum_{i,r} {}_ix(b) \otimes_k {}_r(x_i) \otimes_k (x_i)_r = \sum_{i,r} {}_r({}_ix)(b) \otimes_k ({}_ix)_r \otimes_k x_i$$

which in turn follows from the coassociativity equation

$$\sum_{i,r} {}_ix \otimes_k {}_r(x_i) \otimes_k (x_i)_r = \sum_{i,r} {}_r({}_ix) \otimes_k ({}_ix)_r \otimes_k x_i.$$

Thus, H_K is a K/k-bialgebra. It is clear that $1 \otimes_k H_k$ is a k-form of H_K. Finally, we show that $x(by) = \sum_i {}_ix(b)x_iy$ for $x, y \in H_K$ and $b \in K$. What we must show is that

$$(a \otimes x)(b(c \otimes y)) = \sum_i {}_i(a \otimes x)(b)(a \otimes x)_i(c \otimes y)$$

for $a, c \in K$, $x, y \in H_k$ and $b \in K$. This amounts to showing that

$$\sum_{i,r} a_r({}_ix)(b)({}_ix)_r(c) \otimes x_iy = \sum_{i,r} a_ix(b)_r(x_i)(c) \otimes (x_i)_ry,$$

which follows from

$$\sum_{i,r} {}_r({}_ix)(b)({}_ix)_r(c) \otimes x_i = \sum_{i,r} {}_ix(b)_r(x_i)(c) \otimes (x_i)_r,$$

which in turn follows from the coassociativity equation

$$\sum_{i,r} {}_r({}_ix) \otimes ({}_ix)_r \otimes x_i = \sum_{i,r} {}_ix \otimes {}_r(x_i) \otimes (x_i)_r.$$

We conclude this section with some examples of K-measuring bialgebras. If G is a group and $\rho: G \to \mathrm{Aut}_k K$ a homomorphism of groups, then the group k-bialgebra $k[G]$ together with the k-linear mapping $\rho_k: k[G] \to \mathrm{End}_k K$ such that $\rho_k(g) = \rho(g)$ for $g \in C$ is a K-measuring k-bialgebra. The semidirect product K-measuring K/k-bialgebra $K \otimes_k k[G]$ is canonically isomorphic to the group K/k-bialgebra $K[G]$ of G via ρ described in B.1. Thus, $K[G]$ may be regarded as a K-measuring K/k-bialgebra with k-form $k[G]$ and $K[G] = Kk[G]$ (internal semidirect product).

As a second example, let G be a finite commutative group and $K = \sum_{g \in G} K_g$ (direct) a graded decomposition of K into k-subspaces K_g such that $K_g K_h \subset K_{gh}$ for $g, h \in G$. Let T be the k-bialgebra $k[G]^*$ with k-algebra and k-coalgebra structures dual to the k-coalgebra and k-algebra structures of the k-bialgebra $k[G]$. For $t \in T$, let $\rho(t)$ be the element of $\mathrm{End}_k K$ such $\rho(t)(a) = t(g)a$ for $a \in K_g$ and $g \in G$. Then T together with ρ is a K-measuring k-bialgebra, called the *toral K-measuring k-bialgebra* determined by G and the grading $K = \sum_{g \in G} K_g$. The semidirect product K-measuring K/k-bialgebra $K \otimes_k T$ is canonically isomorphic to the K/k-bialgebra described in B.1.

B.5 Bialgebras and the structure of finite dimensional field extensions

We have introduced K-measuring bialgebras for the purpose of studying the structure of field extensions K/k. We know from B.4 that if H_k is a K-measuring k-bialgebra, then H_k gives rise to a semilinear K-measuring K/k-bialgebra $H_K = K \otimes_k H_k$. And if H is any K-measuring K/k-bialgebra, then Kernel H is a biideal and $H/\text{Kernel } H$ may be regarded as a K-measuring K/k-bialgebra isomorphic to a subbialgebra of $H(K/k)$. In particular, $H/\text{Kernel } H$ is isomorphic to $H(K/K^H)$ if $H/\text{Kernel } H$ is finite dimensional over K or K/k is finite dimensional. Thus, the K-measuring K/k-bialgebra $H(K/k)$ is, ultimately, what should be used in studying the extension K/k.

It is convenient, nevertheless, to know the behavior of K-measuring K/k-bialgebras more general than $H(K/k)$, since K-measuring K/k-bialgebras other than $H(K/k)$ do occur naturally. One instance of this is the $H(K/k)^T$ of 6.3, which can be regarded naturally as a K^T-measuring K^T/k-bialgebra if K/k is a finite dimensional normal extension, T is a toral k-subbiring of $H(K/k)$ and the extension K/K^T is radical.

The K-measuring K/k-bialgebras of most interest in studying finite dimensional normal extensions K/k are the finite dimensional conormal semilinear K-measuring bialgebras such that $G(H)$ is *faithfully represented* on K in the sense that the only element g in $G(H)$ such that $g(a) = a$ for all $a \in K$ is $g = 1_H$. The reason for this is that the present emphasis in studying K/k is on the radical component K_{rad} of K over k. For instance, let T be a nonzero toral k-subbiring of $H(K/k)$ such that K/K^T is radical. Then, in passing from the K-measuring K/k-bialgebra $H(K/k)$ to the K^T-measuring K^T/k-bialgebra $H(K/k)^T$, $G(H(K/k)^T)$ is faithfully represented on K^T, as was $G(H(K/k))$ on K, even though Kernel $H(K/k)^T$ is nonzero.

We now consider a finite dimensional conormal semilinear K-measuring K/k-bialgebra H such that $G(H)$ is faithfully represented. Note that H may be regarded as a K/K^H-bialgebra. For if $b \in K^H$ and $x, y \in H$, then we have $x(by) = b(xy)$, since

$$x(by) = \sum_i {}_ix(b)x_iy = \sum_i \varepsilon({}_ix)bx_iy = b\left(\left(\sum_i \varepsilon({}_ix)x_i\right)y\right) = b(xy).$$

Recall that $H(1)$ is a coradical subbialgebra of H (see B.1 and B.2). And the K-span $K[G]$ of the group $G = G(H)$ of grouplike elements of H is a co Galois subbialgebra of H, since $(ag)(bh) = \sum_i a_ig(b)g_ih = ag(b)gh$ for $a, b \in K$ and $g, h \in G$. Recall that $\rho_H: H \to H(K/k)$ is a homomorphism of K/k-bialgebras and that $\rho_H(H) = H(K/K^H)$. Similarly, $\rho_H(H(1)) = H(K/K^{H(1)})$ and $\rho_H(K[G]) = H(K/K^G)$. Since $H/H(1)/K[G]$ are conormal/coradical/co Galois, it follows from 5.3 that K/K^H is normal, $K/K^{H(1)}$ is radical and K/K^G is Galois. Letting $K_{\text{rad}} = K^G$ and $K_{\text{Gal}} = K^{H(1)}$, it follows that $K = K_{\text{rad}}K_{\text{Gal}}$, since $K/K_{\text{rad}}K_{\text{Gal}}$ is radical and Galois. We observe that the group G acts as a group of bijective k-linear transformations on H, an element $g \in G$ sending h to ghg^{-1} for all $h \in H$. Moreover, we observed in B.4 that $gH(1_H)g^{-1} \subset H(1_H)$ for $g \in G(H)$, so that the subbialgebra $H(1)$ of H is G-stable. Since G is faithfully represented on K, we may regard G as a subgroup of $\text{Aut}_k K$. Then the mapping $G \times H \to H$ sending (g, h) to ghg^{-1} is a G-product on H as vector space over K, since $g(ah)g^{-1} = (g(a)gh)g^{-1} = g(a)(ghg^{-1})$ for all $a \in K$, $h \in H$ and $g \in G$ (see 3.2). It follows that $H^G = \{h \in H \mid ghg^{-1} = h \text{ for all } g \in G\}$ is a K^G-form of H as vector space over K. It is clear that H^G is a k-subalgebra of H. Let $x_1, \ldots, x_n$ be a basis for H^G over K^G, hence a basis for H over K. Let $x \in H^G$ and $\Delta x = \sum_i {}_ix \otimes x_i$. Note that for $g \in G$, $\Delta g(x) = \sum_i g({}_ix) \otimes g(x_i)$ by the product preservation $\Delta gxg^{-1} = \sum_i g_ixg^{-1} \otimes gx_ig^{-1}$. Thus, we have

$$\sum_i {}_ix \otimes x_i = \Delta x = \Delta g(x) = \sum_i g({}_ix) \otimes g(x_i) = \sum_i g({}_ix) \otimes x_i$$

so that ${}_ix = g({}_ix)$ for all $g \in G$. Thus, the ${}_1x, \ldots, {}_nx$ are elements of H^G. It follows that there is a K^G-linear mapping $\Delta^G: H^G \to H^G \otimes_{K^G} H^G$ such that for $x \in H^G$ and ${}_1x, x_1, \ldots, {}_nx, x_n \in H^G$, $\Delta^G(x) = \sum_1^n {}_ix \otimes_{K^G} x_i$ if and only if $\Delta(x) = \sum_1^n {}_ix \otimes_K x_i$. The mapping $\varepsilon^G = \varepsilon|_{H^G}$ maps H^G to K^G, since $g(\varepsilon^G(x)) = g(\varepsilon(x)) = g(x(1_K)) = g(x)(g(1_K)) = x(1_K) = \varepsilon(x)$ for $x \in H^G$, $g \in G$. Now it is easy to see that H^G as K^H-algebra together with Δ^G and ε^G is a K^G/K^H-bialgebra. Let $H_{\text{rad}} = H(1)^G$ and recall that $K_{\text{rad}} = K^G$. Since $H(1)$ is a G-stable subbialgebra of H, H_{rad} is a K_{rad}-form of $H(1)$ as K-vector space. There exists a K_{rad}-linear mapping $\Delta_{\text{rad}}: H_{\text{rad}} \to H_{\text{rad}} \otimes_{K_{\text{rad}}} H_{\text{rad}}$ such that for $x \in H_{\text{rad}}$ and ${}_1x, x_1, \ldots, {}_nx, x_n \in H_{\text{rad}}$, $\Delta_{\text{rad}}(x) = \sum_1^n {}_ix \otimes_{K_{\text{rad}}} x_i$ if and only if $\Delta(x) = \sum_1^n {}_ix \otimes x_i$, which is proved as above in the case of H^G. The mapping $\varepsilon_{\text{rad}} = \varepsilon|_{H_{\text{rad}}} = \varepsilon^G|_{H_{\text{rad}}}$ maps H_{rad} into $K_{\text{rad}} = K^G$. It is now easy to see that H_{rad} as K^H-algebra together with Δ_{rad} and ε_{rad} is a coradical K_{rad}/K^H-bialgebra. We next let H_{Gal} be the K_{Gal}-span $K_{\text{Gal}}[G]$ of G. Note that H_{Gal} is a K_{Gal}-form of $K[G]$ as vector space. Let $\varepsilon_{\text{Gal}} = \varepsilon|_{H_{\text{Gal}}}$ and let Δ_{Gal} be the K_{Gal}-linear mapping from H_{Gal} to $H_{\text{Gal}} \otimes_{K_{\text{Gal}}} H_{\text{Gal}}$ such that $\Delta_{\text{Gal}}(g) = g \otimes g$ for $g \in G$, so that H_{Gal} together with Δ_{Gal}, ε_{Gal} is a K_{Gal}-coalgebra. We have $K_{\text{rad}} = K^{H_{\text{Gal}}}$ and $K_{\text{Gal}} = K^{H_{\text{rad}}}$. Furthermore, the elements of $H_{\text{rad}} = H(1)^G$

commute with the element of $H_{\text{Gal}} = K_{\text{Gal}}[G]$. It follows that $H = H_{\text{rad}}H_{\text{Gal}}$ (k-span of $\{xy \mid x \in H_{\text{rad}}, y \in H_{\text{Gal}}\}$), since $H = H(1)G = KH_{\text{rad}}G = K_{\text{rad}}K_{\text{Gal}}H_{\text{rad}}G = K_{\text{rad}}H_{\text{rad}}K_{\text{Gal}}G = H_{\text{rad}}H_{\text{Gal}}$. The subfield $K_{\text{rad}} = K^{H_{\text{Gal}}}$ is H_{rad}-stable. For if $a \in K_{\text{rad}}$ and $x \in H_{\text{rad}}$ then

$$y(x(a)) = (yx)(a) = (xy)(a) = x(y(a)) = x(\varepsilon_{\text{Gal}}(y)a) = \varepsilon_{\text{Gal}}(y)x(a)$$

for all $y \in K_{\text{Gal}}$. Similarly, the subfield $K_{\text{Gal}} = K^{H_{\text{rad}}}$ is H_{Gal}-stable. It follows that H_{Gal} is a K^H-subalgebra of H, hence that H_{Gal} can be regarded as a K_{Gal}/K^H-bialgebra. For $(ag)(bh) = ag(b)gh \in H_{\text{Gal}}$ for $a, b \in K_{\text{Gal}}$, $g, h \in G$. Let $\rho_{\text{rad}}: H_{\text{rad}} \to \text{End}_k K_{\text{rad}}$ and $\rho_{\text{Gal}}: H_{\text{Gal}} \to \text{End}_k K_{\text{Gal}}$ be defined by $\rho_{\text{rad}}(x)(a) = x(a)$ for $a \in K_{\text{rad}}$ and $x \in H_{\text{rad}}$ and $\rho_{\text{Gal}}(y)(b) = y(b)$ for $b \in K_{\text{Gal}}$ and $y \in H_{\text{Gal}}$. Then H_{rad} together with ρ_{rad} is a coradical semilinear K_{rad}-measuring K_{rad}/K^H-bialgebra and H_{Gal} together with ρ_{Gal} is a co Galois semilinear K_{Gal}-measuring K_{Gal}/K^H-bialgebra such that $G(H_{\text{Gal}})$ is faithfully represented. The extension K_{rad}/K^H is radical, since $\rho_{\text{rad}}(H_{\text{rad}})$ is coradical and is isomorphic to $H(K_{\text{rad}}/K^H)$ (see 5.3). And the extension K_{Gal}/K^H is Galois, since $\rho_{\text{Gal}}(H_{\text{Gal}})$ is co Galois and is isomorphic to $H(K_{\text{Gal}}/K^H)$. Since $K = K_{\text{rad}}K_{\text{Gal}}$, it follows that K_{rad} and K_{Gal} are the radical and Galois components of K over K^H, so that $K = K_{\text{rad}}K_{\text{Gal}}$ (internal tensor product of K^H-algebras). In particular, we have $K: K^H = (K_{\text{rad}}:K^H)(K_{\text{Gal}}:K^H)$. Since $H = H_{\text{rad}}H_{\text{Gal}}$, we therefore have $H = H_{\text{rad}}H_{\text{Gal}}$ (internal tensor product of K^H-algebras). For we know that the elements of H_{rad} commute with the elements of H_{Gal}, so that it remains only to show that

$$H:K^H = (H_{\text{rad}}:K^H)(H_{\text{Gal}}:K^H).$$

But $H = H(1)G$, so that

$$H:K = (H(1):K)(G:1) = (H_{\text{rad}}:K_{\text{rad}})(H_{\text{Gal}}:K_{\text{Gal}})$$

and

$$\begin{aligned} H:K^H &= (H:K)(K:K^H) = (H_{\text{rad}}:K_{\text{rad}})(H_{\text{Gal}}:K_{\text{Gal}})(K_{\text{rad}}:K^H)(K_{\text{Gal}}:K^H) \\ &= (H_{\text{rad}}:K^H)(H_{\text{Gal}}:K^H). \end{aligned}$$

The observations of the preceding paragraph enable us to describe the finite dimensional conormal semilinear measuring bialgebras H such that $G(H)$ is faithfully represented in terms of the finite dimensional coradical and co Galois semilinear measuring bialgebras. For suppose that K_{rad}/k and K_{Gal}/k are finite dimensional radical and Galois extensions respectively, and that H_{rad} and H_{Gal} are finite dimensional coradical and co Galois semilinear K_{rad}- and K_{Gal}-measuring K_{rad}/k- and K_{Gal}/k-bialgebras respectively. Let K be the field $K = K_{\text{rad}} \otimes_k K_{\text{Gal}}$ (tensor product of k-algebras) and let H be the k-algebra $H = H_{\text{rad}} \otimes_k H_{\text{Gal}}$ (tensor product of k-algebras). Then there are K-linear mappings $\Delta_H: H \to H \otimes_K H$ and $\varepsilon_H: H \to K$ such that

$$\Delta_H(x \otimes_k y) = \sum_{i,j} ({}_ix \otimes_{K_{\text{rad}}} {}_jy) \otimes_K (x_i \otimes_{K_{\text{Gal}}} y_j)$$

and

$$\varepsilon_H(x \otimes_k y) = \varepsilon_{H_{\text{rad}}}(x) \otimes_k \varepsilon_{H_{\text{Gal}}}(y)$$

for $x \in H_{\mathrm{rad}}$ and $y \in H_{\mathrm{Gal}}$, and H together with Δ_H, ε_H is a conormal semilinear K/k-bialgebra called the *tensor product bialgebra* of H_{rad} and H_{Gal}. Letting $\rho_H: H \to \mathrm{End}_k K$ be the K-linear mapping such that $\rho_H(x \otimes_k y) = \rho_{H_{\mathrm{rad}}}(x) \otimes_k \rho_{H_{\mathrm{Gal}}}(y)$, H together with ρ_H is a finite dimensional conormal semilinear K-measuring K/k-bialgebra such that $K^H = k$. Moreover, $G(H)$ is faithfully represented if $G(H_{\mathrm{Gal}})$ is faithfully represented. Conversely, let H be any finite dimensional conormal semilinear K-measuring K/k-bialgebra such that $G(H)$ is faithfully represented and such that $K^H = k$. Let K_{rad}, K_{Gal}, H_{rad}, H_{Gal} be as obtained from K and H in the preceding paragraph. Thus, K_{rad}/k and K_{Gal}/k are finite dimensional radical and Galois extensions respectively, H_{rad} and H_{Gal} are finite dimensional coradical and co Galois semilinear K_{rad}- and K_{Gal}-measuring K_{rad}/k- and K_{Gal}/k-bialgebras respectively. Moreover, $G(H_{\mathrm{Gal}})$ is faithfully represented. Then the k-linear mapping $\alpha: K_{\mathrm{rad}} \otimes_k K_{\mathrm{Gal}} \to K$ such that $\alpha(a \otimes b) = ab$ for $a \in K_{\mathrm{rad}}$, $b \in K_{\mathrm{Gal}}$ is an isomorphism of fields. Upon identifying $K_{\mathrm{rad}} \otimes_k K_{\mathrm{Gal}}$ and K by way of α, the k-linear mapping $\beta: H_{\mathrm{rad}} \otimes_k H_{\mathrm{Gal}} \to H$ such that $\alpha(x \otimes y) = xy$ for $x \in H_{\mathrm{rad}}$, $y \in H_{\mathrm{Gal}}$ is an isomorphism from the K-measuring K/k-bialgebra $H_{\mathrm{rad}} \otimes_k H_{\mathrm{Gal}}$ (tensor product bialgebra) to the K-measuring K/k-bialgebra H.

We conclude with some comments on k-forms of K-measuring K/k-bialgebras. The k-forms of $H(K/k)$ are of particular importance, since their structure in some important cases is simple enough that they can be used to study the structure of K/k in great detail. This is the case, for instance, when $H(K/k)$ has a k-form of the form $Tk[G]$ (internal tensor product of k-bialgebras) where T is a diagonalizable toral k-subbiring of $H(K/k)$ and $G = \mathrm{Aut}_k K$ (see 6.2).

We begin with a finite dimensional conormal semilinear K-measuring K/k-bialgebra H with k-form H_k. We know that H_k is a K-measuring k-bialgebra and that $H = KH_k$ (semidirect product bialgebra). If H_k contains $G(H)$, then H_k is cosplit (as k-coalgebra). For if D_k is a subcoalgebra of H_k, then $D = KD_k$ is a subcoalgebra of $H = KH_k$ and $D_k \cap G(H_k) = D \cap G(H)$ is nonempty. Suppose, conversely, that H_k is cosplit. Then H_k is conormal. For H_k is cosplit and cocommutative, so that $H_k = \sum_{g \in G(H_k)} H_k(g)$. And $G(H_k)$ is a subsemigroup of the finite group $G(H)$, hence a group. It follows that $H_k = H_k(1)k[G(H_k)]$ (semidirect product bialgebra). Since $H = KH_k$ and $H = H(1)k[G(H)]$ (semidirect product bialgebra) (see B.4), it follows that $H(1) = KH_k(1)$ and $G(H) = G(H_k)$.

It follows from the above paragraph that the cosplit k-forms of a finite dimensional conormal semilinear K-measuring K/k-bialgebra H are those k-subspaces H_k of H of the form $H_k = H(1)_k k[G(H)]$ where $H(1)_k$ is a k-form of $H(1)$ such that $gxg^{-1} \in H(1)_k$ for all $x \in H(1)_k$ and $g \in G(H)$.

Let K/k be a finite dimensional formal field extension. Then a cosplit k-form H_k of the K/k-bialgebra $H(K/k)$ is of the form $H_k = H_k(1)k[G]$ (internal semidirect product of k-bialgebras) where $G = G(H(K/k)) = \mathrm{Aut}_k K$ and $H_k(1)$ is a k-form of $H(K/k)(1) = H(K/K_{\mathrm{Gal}})$. For H_k to stabilize K_{rad} and K_{Gal}, it is therefore necessary and sufficient that $H_k(1)$ stabilize K_{rad}. And for $H_k(1)$ to stabilize K_{rad}, it is necessary and sufficient that $H_k(1)$ be a

k-form of $H(K/k)_{\text{rad}}$, in which case $H_k(1)$ may be regarded as a k-form of $H(K_{\text{rad}}/k)$ (see 5.3.21) and $H_k = H_k(1)k[G]$ (internal tensor product of k-bialgebras).

The problem of finding the cosplit k-forms H_k of $H(K/k)$ which stabilize the radical and Galois components K_{rad} and K_{Gal} of a finite dimensional normal extension K over k is reduced by the preceding paragraph to the problem of finding the k-forms of $H(K_{\text{rad}}/k)$. For the latter are automatically cosplit and give rise, upon tensoring over k with the group k-bialgebra $k[\text{Aut}_k K_{\text{Gal}}]$, to cosplit k-forms of $H(K/k)$ stabilizing K_{rad} and K_{Gal}.

It is not known whether $H(K_{\text{rad}}/k)$ has a k-form for every finite dimensional radical extension K_{rad}/k. If T is a toral k-subbiring of $H(K_{\text{rad}}/k)$ such that $K_{\text{rad}}^T = k$, then T is a k-form of $H(K_{\text{rad}}/k)$ and the structure of the extension K_{rad}/k can be studied in great detail using T as in Chapter 6. More generally, for any toral k-subbiring T of $H(K_{\text{rad}}/k)$, the centralizer $H(K_{\text{rad}}/k)^T$ of T is a K^T-form of $H(K_{\text{rad}}/k)$, which may be regarded as a K_{rad}^T-measuring K_{rad}^T/k-bialgebra. Since for any nontrivial finite dimensional radical extension K_{rad}/k, $H(K_{\text{rad}}/k)$ contains nontrivial toral k-subbirings, the extension K/k_{rad} can be studied inductively by taking sequences of toral forms.

REFERENCES

[1] *Allan, Harry and Sweedler, Moss.* A theory of linear descent based upon Hopf algebraic techniques J. of Alg. 12 (1969), 242-294.

[2] *Cartier, Pierre.* Questions de rationalité des diviseurs en geometrie algébrique. Bull. Soc. Math. France 86 (1958), 177-251.

[3] *Dieudonné, Jean.* Les semi- derivations dans les extensions radicielles, C. R. Acad. Sci. Paris, 227 (1948), 1319-1320.

[4] *Fraenkel, Abraham.* Set theory and logic, Addison-Wesley, Reading, 1966.

[5] *Gerstenhaber, Murray.* On the Galois theory of inseparable extensions, Bull. Amer. Math. Soc. 70 (1964), 561-566.

[6] *Gerstenhaber, Murray and Zaromp, Avigdor.* On the Galois theory of purely inseparable field extensions. Bull. Amer. Math. Soc. 76 (1970), 1011-1014.

[7] *Halmos, Paul.* Naive Set Theory, Van Nostrand, Princeton, 1960.

[8] *Hasse, Helmut and Schmidt, Friedrich.* Noch eine Begründung der Theorie der höheren Differential-quotienten in einem algebraischen Funktionkörper einer Unbestimmten, J. Reine u. Angew. Math. 177 (1937), 215-237.

[9] *Hochschild, Gerhardt.* Simple algebras with purely inseparable splitting fields of exponent 1, Trans. Amer. Math. Soc. 79 (1955), 477-489.

[10] *Jacobson, Nathan.* Forms of Algebras, Yeshiva Sci. Confs. 7 (1966), 41-71.

[11] *Jacobson, Nathan.* Galois theory of purely inseparable fields of exponent 1, Amer. J. Math. vol. 66 (1944), 645-648.

[12] *Jacobson, Nathan.* Lectures in Abstract Algebra Vol. III Van Nostrand, Princeton (1964).

[13] *Jacobson, Nathan.* Lie Algebras, 1962.

[14] *Kaplanski, Irving.* Fields and Rings, University of Chicago Press, Chicago, 1969.

[15] *Mordeson, John and Vinograd, Bernard.* Structure of Arbitrary Purely Inseparable Extension Fields, Lecture Notes in Mathematics, Springer-Verlag, New York, 1970.

[16] *Pickert, Günther.* Neue Methoden in der Structurtheorie der Kommutativen Assoziativen Algebren, Math. Ann. 116 (1939), 217-280.

[17] *Seligman, George*. Modular Lie Algebras, Springer, New York, 1967.

[18] *Sweedler, Moss*. Hopf Algebras, Benjamin, New York, 1969.

[19] *Sweedler, Moss*. Structure of inseparable extensions, Amer. Math. (21) 87 (1967-8), 401-410.

[20] *Sweedler, Moss*. The Hopf algebra of an algebra applied to field Theory J. of Alg. 8 (1968), 262-276.

[21] *Weisfeld, Morris*. Purely inseparable extensions and higher derivations, Trans. Amer. Math. Soc. 116 (1965), 435-449.

[22] *Winter, David J*. Abstract Lie Algebras, MIT Press, Cambridge, 1972.

[23] *Winter, David J*. A generalization of the normal basis theorem, Mathematische Nachrichten, 54 (1972), 75-77.

[24] *Winter, David J*. On the toral structure of Lie p-algebras, Acta Math. 123 (1969), 69-81.

[25] *Winter, David J*. Separability in an algebra with semi-linear homomorphism, Canadian Journal of Math, 24 No. 4 (1972), 668-671.

INDEX OF SYMBOLS

INDEX OF TERMINOLOGY